AQUACULTURE AND GENETIC IMPROVEMENT

AQUACULTURE AND GENETIC IMPROVEMENT

Edited by:

Bruno Augusto Amato Borges

Delve
PUBLISHING

www.delvepublishing.com

Aquaculture and Genetic Improvement

Bruno Augusto Amato Borges

Delve Publishing
2010 Winston Park Drive,
2nd Floor
Oakville, ON L6H 5R7
Canada
www.delvepublishing.com
Tel: 001-289-291-7705
001-905-616-2116
Fax: 001-289-291-7601
Email: orders@arclereducation.com

ISBN: 978-1-77361-041-2 (Hardcover)

DECLARATION

Some content or chapters in this book are open access copyright free published research work, which is published under Creative Commons License and are indicated with the citation. We are thankful to the publishers and authors of the content and chapters as without them this book wouldn't have been possible.

ABOUT THE EDITOR

Bruno Augusto Amato Borges

Bruno Augusto is Aquaculture Engineer from Federal University of Santa Catarina. His focus is on writing projects and development of new technologies in aquaculture. He is currently project service provider, consultant and develops research in the field of Biofloc Technology (BFT) for freshwater fish farming.

TABLE OF CONTENT

SECTION 3 – SELECTION STRATEGIES

LIST OF CONTRIBUTORS

Bruno Augusto A. Borges
Aquaculture Engineer Florianopolis, Brazil

Agim Rexhepi
University of Prishtina, Faculty of Agriculture and Veterinary, Kosova

Kurtesh Sherifi
University of Prishtina, Faculty of Agriculture and Veterinary, Kosova

Hysen Bytyqi
University of Prishtina, Faculty of Agriculture and Veterinary, Kosova

Behlul Behluli
University of Prishtina, Faculty of Agriculture and Veterinary, Kosova

João Costa Filho
Zootecnista; Mestrando no Programa de Pós-Graduação em Ciência Animal Departamento de Produção Animal e Alimentos do Centro de Ciências Agroveterinárias da Universidade do Estado de Santa Catarina - CAV/UDESC; Endereço: Av. Luiz de Camões, 2090 - Conta dinheiro, CEP: 88.520-000, Lages, Santa Catarina, Brasil

Leila de Genova Gaya
Dra.; Médica Veterinária; Professora do Departamento de Engenharia de Biossistemas, Universidade Federal de São João del-Rei, UFSJ; Endereço: Praça Dom Helvécio, 74, Campus Dom Bosco, CEP: 36.301-160, São João del-Rei, Minas Gerais, Brazil

Zaijie Dong
Freshwater Fisheries Research Centre of Chinese Academy of Fishery Sciences, Key Laboratory of Freshwater Fisheries and Germplasm Resources Utilization, Ministry of Agriculture, Wuxi 214081, China.

Nguyen Hong Nguyen
University of the Sunshine Coast, Locked Bag 4, Maroochydore, DC QLD 4558, Australia.

Wenbin Zhu
Freshwater Fisheries Research Centre of Chinese Academy of Fishery Sciences, Key Laboratory of Freshwater Fisheries and Germplasm Resources Utilization, Ministry of Agriculture, Wuxi 214081, China.

Hisham Abdelrahman
School of Fisheries, Aquaculture and Aquatic Sciences, Auburn University, Auburn, AL 36849, USA.

Mohamed ElHady
Department of Biological Sciences, Auburn University, Auburn, AL 36849, USA.

Acacia Alcivar-Warren
Environmental Genomics Inc., Southborough, MA 01772-1801, USA.

Standish Allen
Aquaculture Genetics & Breeding Technology Center, Virginia Institute of Marine Science, Gloucester Point, VA 23062, USA.

Rafet Al-Tobasei
Department of Biology, Middle Tennessee State University, Murfreesboro, TN 37132, USA.

Lisui Bao
School of Fisheries, Aquaculture and Aquatic Sciences, Auburn University, Auburn, AL 36849, USA.

Ben Beck
Aquatic Animal Health Research Unit, USDA-ARS, 990 Wire Road, Auburn, AL 36832, USA.

Harvey Blackburn
USDA-ARS-NL Wheat & Corn Collections at a Glance GRP, National Animal Germplasm Program, 1111 S. Mason St., Fort Collins, CO 80521-4500, USA.

Brian Bosworth
USDA-ARS/CGRU, 141 Experimental Station Road, Stoneville, MS 38701, USA.

John Buchanan
Center for Aquaculture Technologies, 8395 Camino Santa Fe, Suite E, San Diego, CA 92121, USA.

Jesse Chappell
School of Fisheries, Aquaculture and Aquatic Sciences, Auburn University, Auburn, AL 36849, USA.

William Daniels
School of Fisheries, Aquaculture and Aquatic Sciences, Auburn University, Auburn, AL 36849, USA.

Sheng Dong
School of Fisheries, Aquaculture and Aquatic Sciences, Auburn University, Auburn, AL 36849, USA.

Rex Dunham
School of Fisheries, Aquaculture and Aquatic Sciences, Auburn University, Auburn, AL 36849, USA.

Evan Durland
Department of Fisheries and Wildlife, Oregon State University, Corvallis, OR 97331, USA.

Ahmed Elaswad
School of Fisheries, Aquaculture and Aquatic Sciences, Auburn University, Auburn, AL 36849, USA.

Marta Gomez-Chiarri
Department of Fisheries, Animal & Veterinary Science, 134 Woodward Hall, 9 East Alumni Avenue, Kingston, RI 02881, USA.

Kamal Gosh
School of Fisheries, Aquaculture and Aquatic Sciences, Auburn University, Auburn, AL 36849, USA.

Ximing Guo
Haskin Shellfish Research Laboratory, Department of Marine and Coastal Sciences, Rutgers University, 6959 Miller Avenue, Port Norris, NJ 08349, USA.

Perry Hackett
Department of Genetics, Cell Biology and Development, 5-108 MCB, 420 Washington Avenue SE, Minneapolis, MN 55455, USA.

Terry Hanson
School of Fisheries, Aquaculture and Aquatic Sciences, Auburn University, Auburn, AL 36849, USA.

Dennis Hedgecock
Department of Biological Sciences, University of Southern California, Los Angeles, CA 90089-0371, USA.

Tiffany Howard
School of Fisheries, Aquaculture and Aquatic Sciences, Auburn University, Auburn, AL 36849, USA.

Leigh Holland
School of Fisheries, Aquaculture and Aquatic Sciences, Auburn University, Auburn, AL 36849, USA.

Molly Jackson
Taylor Shellfish Farms, 130 SE Lynch RD, Shelton, WA 98584, USA.

Yulin Jin
School of Fisheries, Aquaculture and Aquatic Sciences, Auburn University, Auburn, AL 36849, USA.

Karim Khalil
School of Fisheries, Aquaculture and Aquatic Sciences, Auburn University, Auburn, AL 36849, USA.

Thomas Kocher
Department of Biology, University of Maryland, 2132 Biosciences Research Building, College Park, MD 20742, USA.

Tim Leeds
National Center for Cool and Cold Water Aquaculture, Agricultural Research Service, United States Department of Agriculture, Kearneysville, WV 25430, USA.

Ning Li
School of Fisheries, Aquaculture and Aquatic Sciences, Auburn University, Auburn, AL 36849, USA.

Lauren Lindsey
School of Fisheries, Aquaculture and Aquatic Sciences, Auburn University, Auburn, AL 36849, USA.

Shikai Liu
School of Fisheries, Aquaculture and Aquatic Sciences, Auburn University, Auburn, AL 36849, USA.

Zhanjiang Liu
School of Fisheries, Aquaculture and Aquatic Sciences, Auburn University, Auburn, AL 36849, USA.

Kyle Martin
Troutlodge, 27090 Us Highway 12, Naches, WA 98937USA.

Romi Novriadi
School of Fisheries, Aquaculture and Aquatic Sciences, Auburn University, Auburn, AL 36849, USA.

Ramjie Odin
School of Fisheries, Aquaculture and Aquatic Sciences, Auburn University, Auburn, AL 36849, USA.

Yniv Palti
National Center for Cool and Cold Water Aquaculture, Agricultural Research Service, United States Department of Agriculture, Kearneysville, WV 25430, USA.

Eric Peatman
School of Fisheries, Aquaculture and Aquatic Sciences, Auburn University, Auburn, AL 36849, USA.

Dina Proestou
USDA ARS NEA NCWMAC Shellfish Genetics at the University Rhode Island, 469 CBLS, 120 Flagg Road, Kingston, RI 02881, USA.

Guyu Qin
School of Fisheries, Aquaculture and Aquatic Sciences, Auburn University, Auburn, AL 36849, USA.

Benjamin Reading
Department of Applied Ecology, North Carolina State University, Raleigh, NC 27695-7617, USA.

Caird Rexroad
USDA ARS Office of National Programs, George Washington Carver Center Room 4-2106, 5601 Sunnyside Avenue, Beltsville, MD 20705, USA.

Steven Roberts
School of Aquatic and Fishery Sciences, University of Washington, Seattle, WA 98105, USA.

Mohamed Salem
Department of Biology, Middle Tennessee State University, Murfreesboro, TN 37132, USA.

Andrew Severin
Genome Informatics Facility, Office of Biotechnology, Iowa State University, Ames, IA 50011, USA.

Huitong Shi
School of Fisheries, Aquaculture and Aquatic Sciences, Auburn University, Auburn, AL 36849, USA.

Craig Shoemaker
Aquatic Animal Health Research Unit, USDA-ARS, 990 Wire Road, Auburn, AL 36832, USA.

Sheila Stiles
USDOC/NOAA, National Marine Fisheries Service, NEFSC, Milford Laboratory, Milford, Connectcut 06460, USA.

Suxu Tan
School of Fisheries, Aquaculture and Aquatic Sciences, Auburn University, Auburn, AL 36849, USA.

Kathy F. J. Tang
School of Animal and Comparative Biomedical Sciences, University of Arizona, Tucson, AZ 85721, USA.

Wilawan Thongda
School of Fisheries, Aquaculture and Aquatic Sciences, Auburn University, Auburn, AL 36849, USA.

Terrence Tiersch
Aquatic Germplasm and Genetic Resources Center, School of Renewable Natural Resources, Louisiana State University Agricultural Center, Baton Rouge, LA 70820, USA.

Joseph Tomasso
School of Fisheries, Aquaculture and Aquatic Sciences, Auburn University, Auburn, AL 36849, USA.

Wendy Tri Prabowo
School of Fisheries, Aquaculture and Aquatic Sciences, Auburn University, Auburn, AL 36849, USA.

Roger Vallejo
[17]National Center for Cool and Cold Water Aquaculture, Agricultural Research Service, United States Department of Agriculture, Kearneysville, WV 25430, USA.

Hein van der Steen
Stonebridge breeding Ltd, Gate House, Abbotswood, Evesham WR11 4NS, UK.

Khoi Vo
School of Fisheries, Aquaculture and Aquatic Sciences, Auburn University, Auburn, AL 36849, USA.

Geoff Waldbieser
USDA-ARS/CGRU, 141 Experimental Station Road, Stoneville, MS 38701, USA.

Hanping Wang
Aquaculture Genetics and Breeding Laboratory, The Ohio State University South Centers, Piketon, OH 45661, USA.

Xiaozhu Wang
School of Fisheries, Aquaculture and Aquatic Sciences, Auburn University, Auburn, AL 36849, USA.

Jianhai Xiang
Key Laboratory of Experimental Marine Biology, Institute of Oceanology, Chinese Academy of Sciences, Qingdao 266071, China.

Yujia Yang
School of Fisheries, Aquaculture and Aquatic Sciences, Auburn University, Auburn, AL 36849, USA.

Roger Yant
Hybrid Catfish Company, 1233 Montgomery Drive, Inverness, MS 38753, USA.

Zihao Yuan
School of Fisheries, Aquaculture and Aquatic Sciences, Auburn University, Auburn, AL 36849, USA.

Qifan Zeng
School of Fisheries, Aquaculture and Aquatic Sciences, Auburn University, Auburn, AL 36849, USA.

Tao Zhou
School of Fisheries, Aquaculture and Aquatic Sciences, Auburn University, Auburn, AL 36849, USA.

Jesús Fernández
Departamento de Mejora Genética Animal, INIA, Madrid, Spain

Miguel Á. Toro
Departamento de Producción Animal, ETSI Agrónomos, Universidad Politécnica de Madrid, Madrid, Spain

Anna K. Sonesson
Nofima, Ås, Norway

Beatriz Villanueva
Departamento de Mejora Genética Animal, INIA, Madrid, Spain

Yaw B. Ansah
Department of Fish and Wildlife Conservation, Virginia Polytechnic Institute and State University, 100 Cheatham Hall, Blacksburg, VA 24061, USA
Department of Agricultural and Applied Economics, Virginia Polytechnic Institute and State University, 208 Hutcheson Hall, Blacksburg, VA 24061, USA

Emmanuel A. Frimpong
Department of Fish and Wildlife Conservation, Virginia Polytechnic Institute and State University, 100 Cheatham Hall, Blacksburg, VA 24061, USA

Eric M. Hallerman
Department of Fish and Wildlife Conservation, Virginia Polytechnic Institute and State University, 100 Cheatham Hall, Blacksburg, VA 24061, USA

Kang Xu
Key Laboratory of Protein Chemistry and Fish Developmental Biology of Ministry of Education of China, College of Life Sciences, Hunan Normal University, Changsha 410081, China

Wei Duan
Key Laboratory of Protein Chemistry and Fish Developmental Biology of Ministry of Education of China, College of Life Sciences, Hunan Normal University, Changsha 410081, China

Jun Xiao
Key Laboratory of Protein Chemistry and Fish Developmental Biology of Ministry of Education of China, College of Life Sciences, Hunan Normal University, Changsha 410081, China

Min Tao
Key Laboratory of Protein Chemistry and Fish Developmental Biology

of Ministry of Education of China, College of Life Sciences, Hunan Normal University, Changsha 410081, China

Chun Zhang
Key Laboratory of Protein Chemistry and Fish Developmental Biology of Ministry of Education of China, College of Life Sciences, Hunan Normal University, Changsha 410081, China

Yun Liu
Key Laboratory of Protein Chemistry and Fish Developmental Biology of Ministry of Education of China, College of Life Sciences, Hunan Normal University, Changsha 410081, China

ShaoJun Liu
Key Laboratory of Protein Chemistry and Fish Developmental Biology of Ministry of Education of China, College of Life Sciences, Hunan Normal University, Changsha 410081, China

JinGou Tong
State Key Laboratory of Freshwater Ecology and Biotechnology, Institute of Hydrobiology, Chinese Academy of Sciences, Wuhan 430072, China;

XiaoWen Sun
Heilongjiang Fisheries Research Institute, Chinese Academy of Fishery Sciences, Harbin 150070, China

JianFang Gui
State Key Laboratory of Freshwater Ecology and Biotechnology, Institute of Hydrobiology, Chinese Academy of Sciences, Wuhan 430072, China

ZuoYan Zhu
State Key Laboratory of Freshwater Ecology and Biotechnology, Institute of Hydrobiology, Chinese Academy of Sciences, Wuhan 430072, China

Jovana Marjanovic
Animal Breeding and Genomics Centre, Wageningen University and Research, 6700 AH Wageningen, The Netherlands.

Department of Animal Breeding and Genetics, Swedish University of Agricultural Sciences, 75007 Uppsala, Sweden.

Han A. Mulder
Animal Breeding and Genomics Centre, Wageningen University and Research, 6700 AH Wageningen, The Netherlands.

Hooi L. Khaw
WorldFish, Jalan Batu Maung, 11960 Bayan Lepas, Penang, Malaysia.

Piter Bijma
Animal Breeding and Genomics Centre, Wageningen University and Research, 6700 AH Wageningen, The Netherlands.

Cheryl D. Quinton
MTT Agrifood Research Finland, Biotechnology and Food Research, Biometrical Genetics, 31600 Jokioinen, Finland

Antti Kause
MTT Agrifood Research Finland, Biotechnology and Food Research, Biometrical Genetics, 31600 Jokioinen, Finland

Juha Koskela
Finnish Game and Fisheries Research Institute Jyväskylä, Survontie 9, 40500 Jyväskylä, Finland

Ossi Ritola
Finnish Game and Fisheries Research Institute, Tervo Fisheries Research and Aquaculture, 72210 Tervo, Finland

Cecilia J. Jaspe
Brackishwater Aquaculture Center, Institute of Aquaculture, College of Fisheries and Ocean Sciences, University of the Philippines Visayas, Miag-ao 5023, Iloilo, Philippines

Christopher M. A. Caipang
Faculty of Biosciences and Aquaculture, University of Nordland, Bodø 8049, Norway

Kasper Janssen
Animal Breeding and Genomics, Wageningen University and Research,

Droevendaalsesteeg 1, 6708 PB Wageningen, The Netherlands.

Paul Berentsen
Business Economics Group, Wageningen University and Research, Hollandseweg 1, 6706 KN Wageningen, The Netherlands.

Mathieu Besson
Animal Breeding and Genomics, Wageningen University and Research, Droevendaalsesteeg 1, 6708 PB Wageningen, The Netherlands.
Génétique Animale Biologie Intégrative, INRA, AgroParisTech, Université Paris-Saclay, 78350 Jouy-en-Josas, France

Hans Komen
Animal Breeding and Genomics, Wageningen University and Research, Droevendaalsesteeg 1, 6708 PB Wageningen, The Netherlands.

José M. Yáñez
Faculty of Veterinary and Animal Sciences, University of Chile, Santiago, Chile,
Aquainnovo, Puerto Montt, Chile,

Scott Newman
Genus plc, Hendersonville, TN, USA,

Ross D. Houston
The Roslin Institute and Royal (Dick) School of Veterinary Studies, University of Edinburgh, Midlothian, UK

Anna K Sonesson
Nofima Marine AS, 1432 Ås, Norway and

Theo HE Meuwissen
University of Life Sciences, 1432 Ås, Norway

Elisa Boscari
Department of Biology, University of Padova, Padova, Italy,

Jose Martin Pujolar
Department of Bioscience, Aarhus University, Aarhus C, Denmark,

Isabelle Dupanloup
Institute of Ecology and Evolution, University of Berne, Berne, Switzerland,
Swiss Institute of Bioinformatics, Lausanne, Switzerland,

Riccardo Corradin
Department of Statistics, University of Padova, Padova, Italy

Leonardo Congiu
Department of Biology, University of Padova, Padova, Italy,

Bambang Iswanto
Research Institute for Fish Breeding. Jl. Raya 2 Sukamandi-Patokbeusi, Subang, West Java, Indonesia

Imron
Research Institute for Fish Breeding. Jl. Raya 2 Sukamandi-Patokbeusi, Subang, West Java, Indonesia

Huria Marnis
Research Institute for Fish Breeding. Jl. Raya 2 Sukamandi-Patokbeusi, Subang, West Java, Indonesia

Rommy Suprapto
Research Institute for Fish Breeding. Jl. Raya 2 Sukamandi-Patokbeusi, Subang, West Java, Indonesia

Imron
Research Institute for Fish Breeding

Harry Wuwungan
Department of Aquaculture, Faculty of Fisheries and Marine Sciences, Bogar Agricultural University

Dinar Tri Soelistyawati
Department of Aquaculture, Faculty of Fisheries and Marine Sciences, Bogar Agricultural University

PREFACE

With the world population growth, food demand grows. Aquatic foods provide high quality protein, essential for human health. It is clearly visible that the production of aquatic organisms using aquaculture will have to increase, to the extent that the fishery productivity holds steady over the past few years, even increasing fishing effort.

The total production of aquatic organisms by-catch (including fish, molluscs, crustaceans, etc.) represents 83.5 million tonnes in the year 2014, with a stable production for some years. On the other hand, aquaculture has grown enough in recent years. In the year 2006 the total output of aquatic organisms through cultivation was approximately 45 million tonnes in 2014 the production over 73 million tones.

Since the establishment and technical domain about reproduction and hatchery of marine fish, has begun a process of genetic improvement, as the best individuals were selected for mating and reproduction in search of the best possible offspring.

According to Dunham et al. (2000), breeding programs began in the 1960s. Molecular-based knowledge emerged in the 1980s and has continued to gain momentum. Efforts are now well established in traditional selective breeding, biotechnology and molecular genetics of finfish, and are rapidly developing for aquatic invertebrate domestication.

Genetic improvement is the theory that aims to find the animals that produce the best offspring generation after generation, increasing the population average to some trait. The main objectives of the breeding programs are: increase animal health and increase productivity and product quality. However, on aquaculture, breeding programs are not common, and in many species the production is based only on the capture of wild animals. This can be explained, primarily because of the complexity of the reproductive cycle; the rapid increase of inbreeding in breeding stocks; and the few knowledge of the theory of genetic improvement for researchers, educators, etc. In addition to all this, the breeding programs of aquatic organisms face severe problems compared to terrestrial animals.

The genetic improvement for aquatic organisms really is a challenging path ahead of us, for that we shall seek to carry out research to help this area to develop, benefiting the entire aquaculture chain.

Editor

SECTION I

AQUACULTURE BREEDING PROGRAMS: REVIEW AND CURRENT STATE

CHAPTER

1

BREEDING PROGRAMS ON AQUACULTURE

Bruno Augusto A. Borges

Aquaculture Engineer Florianopolis, Brazil

DEVELOPMENT OF GENETIC IMPROVEMENT ON AQUACULTURE

With the world population growth, food demand grows. Aquatic foods provide high quality protein, essential for human health. It is clearly visible that the production of aquatic organisms using aquaculture will have to increase, to the extent that the fishery productivity holds steady over the past few years, even increasing fishing effort. With the increased demand for aquaculture products, must be developed ways to improve the existing farming systems and the interesting features of the productive point of view.

Historical records show that the emergence of aquaculture occurred about 4000 years ago in China, with the cultivation of marine seaweeds and subsequently the cultivation of carps. Since that time until today, many new production and management techniques were created in search to facilitate and improve processes; however, many methods created at that time are used until today.

The total production of aquatic organisms by-catch (including fish, molluscs, crustaceans, etc.) represents 83.5 million tonnes in the year 2014, with a stable production for some years. On the other hand, aquaculture has grown enough in recent years. In the year 2006 the total output of aquatic organisms through cultivation was approximately 45 million tonnes in 2014 the production over 73 million tones. Currently, Chinese production of aquatic organisms almost reaches 40% of the world total production from aquaculture and capture (FAO, 2016). Also according to FAO (2016), world production from aquaculture has already surpassed 50% of total production of aquatic organisms, including fish, crustaceans, molluscs and aquatic plants. The most produced species are carps, tilapias, salmons, red and brown seaweeds, clams, oysters and shrimps.

Since the establishment and technical domain about reproduction and hatchery of marine fish, has begun a process of genetic improvement, as the best individuals were selected for mating and reproduction in search of the best possible offspring. Currently, this method is known as selective breeding, and is widely used by several companies. The basic principle of this method of selection is to use individuals that express the desired trait and formed couples who will keep such trait to future generations. However, it's not as simple as it seems. Should be avoided crossing of individuals with similar genetics, which favors the inbreeding. It is important to have individuals from different populations in search of greater genetic variability, which makes the method more complex because they must be formed several populations.

There are basically two approaches to improve productive performance: one is through a breeding program to increase genetic gain in fish being grown and the other is to improve how these fish are being managed (environmental improvement). The environmental improvements promoted on the property, such as a better food, proper management and control of water quality, among other, more immediate positive effects on productivity, however, this does not mean that the genetic potential of the property being improved, incorporating performance phenotypes that will be inherited throughout generations definitively.

According to Dunham et al. (2000), breeding programs began in the 1960s. Molecular-based knowledge emerged in the 1980s and has continued to gain momentum. Efforts are now well established in traditional selective breeding, biotechnology and molecular genetics of finfish, and are rapidly developing for aquatic invertebrate domestication.

One of the most common ways to establish a breeding program is through the selection. According to Fjalestad (2005), the aim of selection is to identify and select as parents for the next generation the individuals whose progeny, as a group, have the highest possible additive genetic merit for the trait or traits in question. With advances in studies in this area, there are currently several selection methods, including individual selection, pedigree selection, family selection, within family selection, progeny test, among others. All of them with the same objective: increase the genetic value of the population for a determinate trait.

The method of individual selection is relatively simple and quick response by selecting individuals who stand out in the population for the generation of offspring, but this method does not control the inbreeding. Selection of pedigree is based more remote ancestors assessment to select the individual. Selection of family is one of the most used methods because it takes into account the genetic variability of the population to avoid inbreeding; the choice of individuals is based on the average family and the crossing is done only among families. The selection within the family is based on the deviation of the individual in relation to the average family, selecting fewer individuals and saving space in relation to selection for families. Ever the progeny test is the selection based on the performance of the offspring of individuals.

When you want to start a breeding program, you must have some basic knowledge related to population: it is essential to know the reproduction in order to close the cycle; the goal of the breeding program; there must be genetic variation of characteristics with economic purpose; and the magnitude of this phenotypic variation must be known (Gjedrem, 2010). A detail that cannot be forgotten in the choice of phenotypes for a selection program is have knowledge about how the market or the processing units are willing to pay for a product with added value. For example: If the producer selects a tilapia with most of steak or with less fat, which is the price to be paid for this improved product, which compensate for the investments of the breeding program? A second aspect of the evaluation is to decide if the production of fingerlings genetically enhanced will be exclusively for own use in property, or if the fish farmer will produce fingerlings for marketing. This decision determines how much the fish farmer should invest in the program. If the farmer produces fingerlings for your own use, it uses only a part of your property to the establishment of the program. On the other hand, if it intends to commercialize their fry genetically enhanced, the farmer should invest in a larger infrastructure and technology for productivity differentials of these fingerlings are achieved.

According to Gjedrem et al. (2012) the genetic improvement is very important to supply the fish production chain, mainly due to the increase of genetic gain increases productivity, benefiting the sector's growth.

In this way, it is clear that should be invested in search efforts to develop effective genetic improvement programs in aquatic organisms for aquaculture industry continues to grow. However, it is essential that such searches are conducted in a responsible manner without compromising the natural populations still exist. It can be observed with the agricultural production of pigs and cattle as selective breeding improved productivity over the years. As the genetic improvement in aquatic organisms is relatively new, there is still a long way to go and a vast field to be explored.

BREEDING PROGRAMS TECHNIQUES

Genetic improvement is the theory that aims to find the animals that produce the best offspring generation after generation, increasing the population average to some trait. The main objectives of the breeding programs are: increase animal health and increase productivity and product quality. However, on aquaculture, breeding programs are not common, and in many species the production is based only on the capture of wild animals. This can be explained, primarily because of the complexity of the reproductive cycle; the rapid increase of inbreeding in breeding stocks; and the few knowledge of the theory of genetic improvement for researchers, educators, etc.

Generally, the genetic improvement in a productive sector involves two sectors: the first are the farms with the genetic material (usually a few, core farms), and the second the producers. It is vital to the spread of that genetic material from the nucleus to the producers for the entire industry benefits with the improvement of the species.

The main steps involved when you want to start a breeding program are: analyze the production system and set the objectives; choose the selection criteria; arrange the services of records (pedigree and phenotypic); evaluate candidates for selection (to estimate the genetic values); select the best individuals as parents of the next generation.

There are several agents that can promote changes in gene frequency. They can be systematic or dispersive. The systematic, can be predicted in quantity and direction: migration, mutation, selection. The dispersive already can be predicted in quantity but not in direction: sampling. The migratory processes are events that collaborate with the genetic variability, since the entry of an individual in a population group enables the introduction of new genes and therefore differentiated from existing features. There are two types of mutation. The non-recurring mutation: it gives rise to only one representative of the gene or mutant chromosome in the population. It is of minor importance, because the product of a mutation only has an infinitely small chance of survival, in a large population, unless you have a selective advantage. Already the recurring mutation: is the mutational event that repeats frequently feature regularly, and in a large population, the mutant frequency is never so low that may occur by sampling complete loss. The selection causes change in the gene frequency and genotype frequency therefore. To describe the effects of selection must take into consideration the degree of dominance with respect to adaptation.

The main features that breeding programs seek to improve are: size or growth rate, age of sexual maturity, resistance to disease, tolerance to temperature, salinity or other attributes of the water quality of steak and feed conversion. Of these, the rate of growth (or size in a certain age) has been the most popular because of your importance in the production system.

To evaluate the effective size of a population, the allele frequencies must be observed. The allele frequency in a group of individuals is the ratio of this allele in relation to other alleles of that gene. Thus, the allele frequency can be calculated from the observed number of different genotypes at locus or from genotype frequencies. From the number of genotypes are the numbers of alleles of a given kind in a given locus and divided by the total number of alleles from the population. Homozygous individual shave two copies of a same allele, while heterozygous individuals possess a copy of different alleles.

In addition to the allele frequency, must also be evaluated the response to selection for a certain characteristic, just so you can be verified the success or not of the selection method used. Several factors may influence the response to selection, among them: selective intensity; genetic prediction accuracy; heritability of the trait-indicates how much of the phenotypic variance is due to the additive gene action; range of generations-is the speed with which young animals replace parents; number of observations in the feature.

In general, animals genetically superior feature similarities productive and probably carry genes in common. As a result, the selected animals can be related and the high degree of kinship, called inbreeding, can compromise the quality of the squad, from the point of view of diversity and genetic variability. The more inbreeding, less are the chances to acquire genetic gain satisfactory, due to similarity of individuals.

A problem that affects a lot of breeding programs, in order to make it difficult to quantify genetic gain and improved features is the genotype-environment interaction. In General, the genotype-environment interaction is the result of the differential response of a genotype in different environments, mainly due to quality, food availability, temperature, and other factors that may affect or improve the performance of individuals. The main causes of the interaction are some genotypes adapt well to an environment and harm to another (the best genotype for an environment is not always the best for all); this results in difficulty in the establishment of the genetic value, reducing the efficiency in the selection. According to Falconer and Mackay (1996), to estimate the interaction genotype-environment must be established the genetic correlation between the performances of genotypes in different environments.

In addition to all this, the breeding programs of aquatic organisms face severe problems compared to terrestrial animals. Starting from the most basic, the land animals can be viewed and tracked much more easily when compared to aquatic, collaborating with the choice of breeders and animal health monitoring. Another factor is that aquatic organisms have high fecundity; few breeders generate large proles, hindering with relation to the space to packaging and possibly reducing the genetic diversity of stocks.

Several fish species are in focus for transgenic experiments and can be divided into two main groups: animals used in aquaculture and fish used in basic research. Among the major food fish species are carp (*Cyprinus* sp.), tilapia (*Oreochromis* sp.), salmon (*Salmo* sp., *Oncorhynchus* sp.) and channel catfish (*Ictalurus punctatus*) while zebrafish (*Danio rerio*), medaka (*Oryzias latipes*) and goldfish (*Carassius auratus*) are used in basic research. Transgenic has been showing better gross food conversion, increase in fish weight per unit of food fed than their unmodified relatives (Importance…, 2017).

Thus, we can point out that there are several ways to establish a breeding program, depend on what is the purpose. In addition, it is essential to understand the basic concepts of genetics to formulate the program model and what particulars shall be the target trait to improve

GENETIC IMPROVEMENT ON MARINE FISH

In Europe, the main marine fish (including anadromous salmon) that have selective breeding programs are the Atlantic salmon, European seabass, gilthead seabream and turbot (Janssen et al., 2016). Currently, 70% of the salmon consumed in the world comes from aquaculture, being the main producers are Norway, Chile and Canada and represents the third largest economy in aquaculture, behind shrimp and cyprinids. His production in 2014 exceeded 3,000,000 tons (FAO, 2014). Is a high-value fish and very well accepted by the market.

The genetic improvement is fundamental to the development of the production of marine fish. A large number of enterprises use genetically enhanced individuals in your production. The trend now is to improve the techniques and developing programs for species that do not have. The first generations are the ones obtained genetic gain, greatly improving production rates. For Atlantic salmon, in ten generations was achieve a total of 196% genetic gain in ten generations (Solberg et al., 2013). For turbot (Danacher, & Garcia-Vazquez, 2007) and seabass (Thorland et al., 2015a) the genetic gain achieved was 45% and 65% in three and two generations, respectively.To the seabream was achieved a 30% genetic gain in three generations (Knibb, 2000). There are encouraging results that in large volume bring a significant increase in production.

Several problems related to the productive chain of the salmon are pressing producers and the scientific community to propose changes in search of better results using resources in a responsible manner. The search for an effective diet that does not use fishmeal is a big problem, as other species are fished to feed farmed salmon. Many ingredients of plant origin are being tested to replace fishmeal, making increasingly sustainable salmon farming. In addition, the salmon productive chain has established a large amount of breeding programs in last years that bring several benefits and improvements to production.

Several studies have been and are being carried out to assess the genetic gain per generation in marine fish (including salmon). For coho salmon, there was achieved a 10% genetic gain per generation in four generations (Hershberger et al., 1990). For rainbow trout, in two generations was acquired 13% genetic gain per generation (Gjerde, 1986). To the Atlantic salmon, in a generation was achieved a gain of 14.4% (Gjerde, 1986). Klause et al. (2005) reported a genetic gain for 7.9% per generation trout in 3 generations. Chatain & Chavanne (2009) presented in a work 30% genetic gain per generation for seabass. To the seabream, Thorland et al. (2015b) reported a 13% genetic gain per generation.

After observing how important has been the evolution of knowledge of genetic improvement and its various methods to increase the productivity of marine fish, we can state clearly that the breeding programs are essential to maintain the growth of aquaculture. Choose which features will be improved is also extremely important, since some features have higher heritability, genetic gain and response to selection than others.

GENETIC IMPROVEMENT ON FRESHWATER FISH

According to FAO (2016), the cultivation of freshwater fish globally is almost reaching a production of 70 million tons, doubling from the year of 2006. The major species cultured worldwide are tilapia, catfish and carp.

Tilapia fish farming is one of the most practiced worldwide, and is recognized as one of the main species in various cropping systems (Paulin, 1995). In tilapias, vast majority of breeding programs have a focus on growth rate feature (Ponzoni et al., 2005). Due to the high growth in production, and seeking improvements in the productive performance of Nile tilapia, there is need to establish selection criteria that are associated with the growth rate, feed conversion, meat quality, disease resistance and age at sexual maturity (Gjedrem, 2000).

Some work with tilapias has been developed. Eknath et al. (2007) assessed the composition and the genetic parameters of the third generation of genetically Improved Farmed Tilapias (GIFT) to form a population base for selective breeding. This strain of tilapia (GIFT), according to Dey et al. (2000) has proven growth potential. Results of Ponzoni et al. (2005) research shows that even being subjected to several generations of selection, the population of tilapias still possessed genetic additive gain allows improvements. Yoshida et al. (2013) showed that in Nile tilapia for performance and morphometry, the responses to indirect selection were 87% higher than those obtained by direct response, indicating that the selection can be performed early, using as a criterion the total daily weight gain measured in biometrics, without intermediate expressive losses in genetic gain.

In work done by Thodesen et al. (2013), discussed the need to adapt to selective breeding for different species, and that the program that was occurring with tilapia had genetic gain considerable resultdado in growth and survival after four generations of multi-trait selection. In work performed by Zac et al. (2014), the response to selection for final weight at harvest was 17.7% over three generations.

Oliveira et al. (2015) related that heritability estimates obtained were of average magnitude: 0.273 to daily weight gain; and 0.282, to the harvest. The participation of the common environment effects on total variation was from 15% to 5% and to hatchery rearing. The phenotypic and genetic correlations were 95%, which indicates a strong association between the features. The genetic gain was 3.8% for generation, with cumulative 15% in four generations. There was genetic progress and changes, caused by Nile tilapia selection, cultivation conditions evaluated. The selection about the weight gain allows indirect gains weight-related.

Similar to what was shown on the breeding programs in marine fish, we can point out that for freshwater fish such programs have the same importance, in order to increase productivity making more profitable activity.

GENETIC IMPROVEMENT ON SHRIMP

Since the emergence of the shrimp farming worldwide, there is a growing concern with the use of resources and the search for optimization of the cultivation and commercialization processes. The production of shrimp by aquaculture is already greater than the total from fisheries; aquaculture production exceeds 55% (FAO, 2016). Also according to FAO (2016), in 2014 the total world production of shrimp is reaching 4,200,000 tons, with *Peneaus vannamei* (85%) and *Peneaus monodon* (15%). Currently the main producing countries are China, Thailand, Indonesia, Ecuador, Vietnam, India, Mexico and Brazil as well as other tropical countries. Of the food grown by man, the shrimp are among the largest market values. Among the penaeids, *Litopeneaus vannamei* is the most cultivated species and of extreme importance in Latin America and Asia (Rojas & Alfaro, 2007).

In work done by Goyard et al. (2008), was shown that in populations of shrimp that was inserted a new strain, due to the increase in heterosis and the performance of pure lineage was low due to inbreeding. In addition, the population there was insertion of the new line growth rates were 37% larger than pure population. In other work done by Argue et al. (2002), was studied selective breeding for the white shrimp growth and resistance to Taura Syndrome Virus, a disease that heavily impacts the cultivation of shrimp in the world.

As well as for other aquatic organisms covered in the chapter, the shrimp has immense potential for development in breeding programs.

GENETIC IMPROVEMENT AND ENVIRONMENT

Currently, we see a growing demand for aquatic foods and as a consequence the growth of aquaculture production worldwide. It is imperative that this growth is properly managed in a sustainable way, supporting the environmental, economic and social aspects. For this, several works have been made to improve management and create new techniques that allow for the sustainable development of the activity.

One of the main discussions regarding the possible problem of genetic improvement programs involving aquatic species are, in addition to the possible escape of individuals improved to the natural environment in many breeding programs are captured wild organisms to form the squad of players. In both situations cited serious problems may occur. In the case of an individual from a program for the natural environment, there may be imbalances in the allele frequency of the wild population, and can compromise it in any way. In the case of withdrawal of the wild to compose the breeding in batch programs, certainly the population composition is modified to reduce the genetic diversity of the population. Therefore, control measures should be laid down to ensure that such problems do not develop.

Like any business, there will always be challenges and methodologies that can improve or modify processes. Currently, it is remarkable the lack of work and projects that take environmental issues seriously reconciled production. The main focus is productivity; the environment is in the background.

In this way, strengthening the importance of seeking sustainable solutions to the problems currently encountered in the sector. The genetic improvement is seen by as much as a bad thing. In General, genetic manipulation is something that is not yet accepted by the majority of the population, and for that, we must take further studies aimed at the dissemination of knowledge so that this matter be understood and increasingly accepted. When comparing with other terrestrial animal production systems, it is clear that this is the way for development.

REFERENCES

1. Argue, B.J., Arce, S.M., Lotz, J.M., Moss, S.M. (2002). Selective breeding of Pacific white shrimp (*Litopenaeus vannamei*) for growth and resistance to Taura Syndrome Virus. Aquaculture, v. 204, issues 3-4, 447-460.
2. Chatain, B., Chavanne, H., 2009. Genetics of European seabass (Dicentrarchus labrax L.). Cah. Agric. 18, 249–255.
3. Danacher, D. & Garcia-Vazquez, E. (2007). Turbot - Scophthalmus maximus. In: Svåsand, T., Crosetti, D., García-Vázquez, E., Verspoor, E. (Eds.), Genetic Impact of Aquaculture Activities on Native Populations. Genimpact Final Scientific Report. 55–61.
4. Dunham, R.A., Majumdar, K., Hallerman, E., Bartley, D., Mair, G., Hulata, G., Liu, Z., Pongthana, N., Bakos, J., Penman, D., Gupta, M., Rothlisberg, P. & Hoerstgen-Schwark, G. 2001. Review of the

status of aquaculture genetics. In R.P. Subasinghe, P. Bueno, M.J. Phillips, C. Hough, S.E. McGladdery & J.R. Arthur, eds. Aquaculture in the Third Millennium. Technical Proceedings of the Conference on Aquaculture in the Third Millennium, Bangkok, Thailand, 20-25 February 2000. pp. 137-166. NACA, Bangkok and FAO, Rome.

5. Eknath, A.E., Tayamen, M.M., Palada-de Vera, M.S., Danting, J.C., Reyes, R.A., Dionisio, E.E., Capili, J.B., Bolivar, H.L., Abella, T.A., Circa, A.V., Bentsen, H.B., Gjerde, B., Gjedrem, T. and Pullin, R.S.V., 1993. Genetic improvement of farmed tilapias: the growth performance of eight strains of *Oreochromis niloticus* tested in different farm environments. *Aquaculture,* 111: 17 1- 188.
6. Eknath, A.E, Bentsen, H.B., Ponzoni, R.W., Rye, M., Nguyen, N.H., Thodesen, J., Gjerde, B. (2007). Genetic improvement of farmed tilapias: Composition and gentic parameters of a synthetic base population of *Oreochromis niloticus* for selective breeding. Aquaculture, 273, 1-14.
7. FAO (2016). FAO yearbook. Fishery and Aquaculture Statistics. 2014.
8. Falconer, D.S.; Mackay, T.F.C. (1996). Introduction to quantitative genetics. 4.ed. Edinburgh : Longman Group Limited. 464p.
9. Fjalestad, K.T. (2005). T. Gjedrem (ed.), Selection and Breeding Programs in Aquaculture, 159–171.
10. Goyard, E., Goarant, C., Ansquer, D., Brun, P., Decker, S., Dufour, R., Galinié, C., Peignon, J., Pham, D., Vourey, E., Harache, Y., Patrois, J. Cross breeding of different domesticated lines as a simple way for genetic improvement in small aquaculture industries: Heterosis and inbreeding effects on growth and survival rates of the Pacific blue shrimp *Penaeus* (*Litopenaeus*) *stylirostris.* Aquaculture, v. 278, issues 1-4, 43-50.
11. Gjedrem, T. (2000). Genetic improvement of cold-water fish species. Aquaculture research, 31, 25-33.
12. Gjedrem, T., Robinson, N., Rye,M., (2012). The importance of selective breeding in aquaculture to meet future demands for animal protein: a review. Aquaculture 350-353, 117–129.
13. Gjerde, B. (1986). Growth and reproduction in fish and shellfish. Aquaculture, 57, 37-55.

14. Hershberger, W.K., Meyers, J.M., McAuley, W.C., Saxton, A.M. (1990). Genetic changes in growth of coho salmon (*Oncorhynchus kisutch*) in marine netpens, produced by ten years of selection. Aquaculture, 85, 187-197.

15. Importance of Transgenic Fish to Global Aquaculture: A Review. The Fish Site. N.p., n.d. Web. 11 May 2017.

16. Janssen, K., et al., Impact of selective breeding on European aquaculture, Aquaculture (2016), http://dx.doi.org/10.1016/j.aquaculture.2016.03.012.

17. Kause, A., Ritola, O., Paananen, T., Wahlroos, H., Mantysaari, E., 2005. Genetic trends in growth, sexual maturity and skeletal deformations, and rate of inbreeding in a breeding programme for rainbow trout (Oncorhynchus mykiss). Aquaculture 247, 177–187.

18. Knibb, W. (2000). Genetic improvement of marine fish — which method for industry? Aquaculture Research. 31, 11–23.

19. Ponzoni, R.W., Hamzah, A., Tan, S., Kamaruzzaman, N. (2005). Genetic parameters and response to selection for live weigth in the GIFT strain of Nile tilapia (*Oreochromis niloticus*). Aquaculture, 247, 203-210.

20. Pullin, R.S.V. (1985). Tilapias: 'everyman's fish'. Biologist, 32(2): 84-88.

21. Rojas, E.; Alfaro, J. In vitro manipulation of egg activation in the open thelycum shrimp Litopenaeus. Aquaculture, v.264, p.469-474, 2007.

22. Solberg, M.F., Skaala, O., Nilsen, F., Glover, K.A. (2013). Does domestication cause changes in growth reaction norms? A study of farmed, wild and hybrid Atlantic salmon families exposed to environmental stress. PLoS ONE 8, 1–11.

23. Thodesen, J., Rye, M., Wang, Y., Li, S., Bentsen, H.B., Gjedrem, T. (2013). Genetic improvement of tilapias in China: Genetic parameters and selection responses in growth, pond survival and cold-water tolerance of blue tilapia (*Oreochromis aureus*) after four generations of multi-trait selection. Aquaculture, v. 396-399, p. 32-42.

24. Thorland, I., Kottaras, L., Refstie, S., Papanna, K., Papaharisis, L., Rye, M. (2015a). Response to Selection for Harvest Weight in

European Sea Bass (Dicentrarchus labrax) in Greece. ISGA XII, Santiago de Compostella, Spain.

25. Thorland, I., Kottaras, L., Refstie, S., Dimitroglou, A., Papaharisis, L., Rye, M., (2015b). Response to Selection for Harvest Weight in a Family Based Selection Program of Gilthead Seabream (Sparus aurata). ISGA XII, Santiago de Compostella, Spain.
26. Yioshida, G.M., Oliveira, C.A.L., Kunita, N.M., Oliveira, S.N., Alexandre Filho, L., Resende, E.K., Lopera-Barrero, N.M.,Ribeiro, R.P. (2013). Resposta a seleção de características dedesempenho e morfométricas de tilápia-do-nilo ao longo do período de cultivo. Arq. Bras. Vet. Zootec., v.65, n.6, p. 1815-1822.
27. Zac, T., Deshev, R., Benet-Perlberg, A., Naor, A., Magen, I., Shapira, Y., Ponzoni, R.W., Hulata, G. (2014). Genetic improvement of Israeli blue (Jordan) tilapia, *Oreochromis aureus*(Steindachner), through selective breeding for harvest weight. Aquaculture research, v.45, issue 3, 546-557.

CHAPTER

2

REVIEW: IMPORTANCE OF ESTABLISHING AND RUNNING A BREEDING PROGRAM IN THE DEVELOPING FISH FARMING INDUSTRY

Agim Rexhepi, Kurtesh Sherifi, Hysen Bytyqi, Behlul Behluli

University of Prishtina, Faculty of Agriculture and Veterinary, Kosova

ABSTRACT

Fish farming in developing countries including the Republic of Kosova is largely based on unimproved fish strains. In aquaculture research the main focus has been on increasing productivity through improvements in management, technology, disease control etc. Anyhow, is accepted worldwide that the full benefits can be obtained thorough genetically improved fish. In many countries are given evidence indicating the potential of genetic improvement programs and a range of selection methods may be used. In a survey during year 2012 in fish farms in Kosova, we found that 93.33% (28 of 30) fish farms use own brood fish for future generation, and most of the farms have started fishing farming with buying fish from the same fish farm without any precaution to avoid inbreeding. The main objective of this review is to highlight the importance of running a breeding program in fish farming, strategies for improvements and by controlling inbreeding accumulation.

Key words: Breeding program; inbreeding; fish farming; Kosova.

INTRODUCTION

The science of applied selective breeding and genetics has contributed greatly and steadily in increasing productivity in fish, domestic animals and plants. The profitable and sustainable genetic improvement is documented in several fish species. This potential is exploited through the effective and sustainable genetic improved strains with better performance, resource efficiency and product quality. The productivity may increase through improvement in management, feed and feeding practices, control of disease, which must completed with genetically improved animals in other side in order to increase benefits [12, 18]. The commercial breeding program may be partitioned into 10 elements: industry assessment, establish goals, measure goals, initial stocks, estimate parameters, mating system, selection criteria, ancillary studies, monitoring procedures, and periodic reconsideration [29].

The high reproductive capacity in most aquaculture species is a great advantage compared to farm animals. These, together with external fertilization, give the possibility to design more efficient breeding programs, than for farm animals. The high selection intensities can be practiced and possibilities for genetic improvement for most examined traits are very good, for traits which show genetic variation [16, 1].

Improvement in selective breeding need a good understanding of genetics, because improper breeding practices used have an adverse effect. There are several strategies that lead to improvements of breeding goals. And we have to choose the best one to achieve that, meantime we have to consider constrains like inbreeding accumulation and negative genetic correlation of the traits, to be sure that the breeding program is progressing in desired direction. The survey in all current 30 fish farms in Kosova it is found that none of fish farms is running any planned breeding program and none of them is aware about harmful inbreeding effects. The main source of broodfish was from one oldest fish farm in Kosova and they have used for many years own broodfish. The objective of this review is briefly to present the main breeding goals on fish farming, selection methods that may be used, inbreeding accumulation and its harmful effects in breeding programs which is often constrained in inadequate hatchery management practices and selection.

OBJECTIVE OF BREEDING PROGRAM

The main objectives of a breeding program are to move the mean value of the trait in the desired direction for a normally distributed trait, or to increase the frequency of the desired class (es) for a trait with discrete classes. Hence, changes in the population mean or in class frequencies from one generation to the next are termed genetic gain [25, 16].

BREEDING STRATEGY

Breeding strategy which is also known from traditional farm animals is that additive genetic improvement is the best strategy for long term progress. The basic requirement of this strategy is that we want to utilize an additive genetic variation. The important is to have a continuous response to selection which depends on the maintenance of genetic variation which has to maintain through properly designed programs [12].

BREEDING GOALS

Definition of the breeding goal is an important part of a breeding program and is set by following steps as: definition of trait to be improved, evaluation a breeding capacity (genetic capacity) of potential breeding animal and selecting highest ranking individuals as parents for the next generation. For a trait to be included in the breeding goal the following prerequisites must hold: a) The trait must be of economic and ethical importance; b) The trait must show genetic variance; c) It must be possible to measure or judge (score) the trait, to a reasonable price [11, 16].

Growth rate

Growth is a heritable trait and of the most important trait in farmed fish. So, selective breeding program may be taken up for higher growth rate, which will decrease the duration of the production cycle, reduce the cost of production and increase the profit, reported from many authors. Reported heritability for body weight in juveniles is rather low around 0.10 [14], while medium heritabilities 0.2-0.4 for body weight in adults [13].

Feed conversion

One of the most expensive constituents in total fish production is fed. It is finding a high positive genetic correlation in young rainbow trout between growth rate and gross feed conversion (growth/feed consumed) and a negative genetic correlation between growth rate and net feed conversion (growth/feed consumed less the maintenance ration) [20].

Disease resistance

The mortality and morbidity in fish production may cause serious problems. In general the heritability is low for this trait. The application of the challenge test has shown the large genetic variation in disease resistance [11, 22].

Age at sexual maturity

Sexual maturation in salmonids result in reduced growth rate, reduced meat quality and increased mortality [13, 11]. On average for rainbow trout the heritability for early maturity is 0.05 and 0.20 for late maturity respectively, with considerable additive genetic variation [14]. Thus, the potential for genetic improvement for this trait is promising.

Fecundity

The high reproductive rate in salmonids makes the expenses low for keeping broodfish [13]. However selection for egg size and egg quality could be used which have correlation with early survival and growth rate [14]. The data give a strong evidence of significant additive genetic variance for egg size, egg number and egg volume.

Quality traits

As the industry is developing and peoples awareness of product quality is increasing. The different regions and different culture for fish consumption have an effect on what should be produced. Fish quality is usually a question of size, fat percentage, fat distribution, flesh color, taste, structure, firmness, shape of body and filet yield, dressing percentage and so on [15, 11].

INBREEDING

Inbreeding means the mating of individuals that are related to each other by ancestry. The degree of relationship between the individuals in a population depends partly on the size of population [7, 4]. Using the small number of broodfish can lead on increasing the risk for inbreeding accumulation. The basic impact of inbreeding on the genetic constitution of a population is a shift in the distribution of genotypes toward fewer heterozygotes and more homozygotes [31, 3, 27, 6, 8].

The reproductive characteristics of fish allow high selection intensities, which may result in a rapid accumulation of inbreeding. One prominent consequence of increased inbreeding is the reduction in mean phenotypic value of one or more traits with respect to fitness. This phenomenon is known as inbreeding depression [21]. Inbreeding depression causes reduced growth, viability and survival and increased numbers of abnormalities [30, 22, 23].

Effect on Growth Rate

The reported estimates imply a reduction in the expected genetic gain per generation of about 1% for the highest inbreeding rate ($\Delta F = 2\%$) [13, 17].

Effect on Fitness

It is reported that inbreeding on the male had no deleterious influence on fertility and hatchability this, at least in low levels of inbreeding, in contrast in female had a strong negative effect [30]. As reported in [16] the average inbreeding depression was found to be 10.0% for survival of eyed-eggs, 5.3% for alevins and 11.1 for fry.

BASE POPULATION

The first step to start breeding program is to collect the genetic material that form base population. When developing breeding program in aquaculture it is very important to start with a broad genetic base, because there are examples of failure as a narrow genetic variation in the base. Selecting from different farms, in this case we reduce risk of inbreeding. Other possible contributors for base population are wild stock with a moderate genetic difference from cultured animals [26, 2, 12, 18, 11].

THE IMPORTANCE OF VARIATION

Genetic variation is important for genetic change. If there is a little genetic variation in a trait-more specifically, if there is little genetic variation in a trait response will be low (selection will be difficult) because no individual is much better than any other as genetic parent [4]. If there exists tremendous genetic variation in a population for a particular trait, and if breeders select only the best individuals based on accurate measures of the trait, then the selected individuals and their offspring will be far better than average, and the rate of genetic change will be faster. For this purpose it is recommended (10) that taking the broodfish from different genetic origin can differ great in average performance, therefore the variation within broodstock is generally large and provides the highest potential for within population selection.

SELECTION AND MATING

The method chosen for selection depend on several factors, among which the heritability of the trait(s) and the reproductive capacity of species are most important [16]. So far the additive genetic variation has been found for all traits of economic importance in all species. The selection of breeders in a breeding program is based on its breeding value, how much superior the selected individual is from population average, because the breeding value of an individual cannot measure directly. It can only be estimate from the phenotypic values of the trait(s). They are several alternatives available to fish breeders: individual (mass) selection, family selection and combined family and within-family selection, pedigree selection, progeny selection [10, 22]. By controlling mating is necessary to avoid mating of close relatives and inbreeding accumulation in population. The choice of the mating system is a primary tool of the breeder. Inbreeding, outcrossing, population sizes, and choice of mates are but a few areas of decision for the geneticist.

Individual Selection

When breeding candidates are selected according their own performance, this is called individual or mass selection [12, 22]. The best performing individuals compared to the population mean are selected. This method is most widely used method of selection in aquaculture [16, 11].

Within Family Selection

Whole families are selected or rejected as units according to the mean phenotypic value of the family. The family selection is relatively efficient compared to individual selection for traits with low heritability. Each family must be reared in separate tanks before they can be tagged, since it is impossible to tag newly hatched larvae or fry [16, 11]. Another important factor affecting the efficiency of family selection is the number of individuals in the families, or the family size. This accuracy depends on the heritability of the trait, family size, family type (full and/ or half sibs) and the variation due to the environment [16, 22, 12].

Combined Selections

In this method of selection, the criterion for selection is the deviation of each individual from the mean value of the family to which belongs. Selection within families would eliminate large non-genetic component of variation operated on by selection [16, 22], and high selection intensities may be applied without increasing the rate of inbreeding.

Progeny Testing and Pedigree Selection

These methods of selection on breeding animals are chosen based on the performance or breeding values on their parents, grandparents or more remote ancestors. The progeny testing as reported [11] is not so valuable since the generation interval will be increased and usually doubled. This method can be practiced by using cryopreserved spermatozoa in selective breeding [22], which show good results in egg fertilization [28].

Testing Strategies

In the breeding program the standardization of environmental conditions and facilities during the testing period for the genetic groups under comparison should be given priority. They are many factors that should be standardized like water temperature and quality, amount, content and quality of feed, light intensity and photoperiod, density of fish and so on.

Registration of Records

Incorrect records will give false value of breeding candidates. Thus, records, measures and scores of trait should be taken according the breeding goal. This is done to ensure that important traits don't change in an undesirable way due to unfavorable genetic correlation with traits selected for. Very often it is also necessary to record traits that are not selected for. This for example growth rate can be recorded as body weight or carcass weight. Appropriate experimental designs and statistical methods [19] are the basis for estimation of genetic parameter from breeding populations.

Correlated effects

The correlated effects among traits must be well defined before the trait of the selection is included in breeding goal. There are cases when we have to select indirectly for interest trait. A good example is feed conversion in order to increase weight gain. Therefore, accuracy of selection may greater for the correlated trait that for the trait of interest [4]. Such the correlated response may be advantageous to the productivity and disadvantageous to the productivity in antagonist correlation [11].

CONCLUDING REMARKS

In order to utilize the maximum farm production potential, the breeding plan must start as soon as possible in order to achieve highest yield. Investments in this field need to be multidirectional such as establish breeding station, well-educated personnel in quantitative genetics, and training of fish farmers in order to maintain what is achieved. Individual selection is the most common method of selection used in fish breeding, because is simple to apply and efficient in improving growth rate. Reports of very good genetic gain per generation in average should encourage farmers and government support promising genetic improvement. Taking into account the heritability of economically important traits, breeding program must be established in fish farming which improves feed conversion, improve growth rate and disease resistance, including improvements of quality traits.

The initiatives could come from several directions including the fish farmers, genetic researchers, or government in order to stimulate aquaculture production. The dissemination of improved fish could be directly with eggs, fry or broodfish from breeding station. This might have a direct impact on developing industry and direct increase of fish farmer's income. In countries where fish farming is in developing phase and without any established fish breeding program, inbreeding can be a serious constraint. The improper use of selection methods may cause negative and unfavorable inbreeding depression and consequently the negative effects reducing genetic variation, fitness and growth rate. The rate of inbreeding using mass selection is determined from the number used as parents. Thus, recommended in different authors by selecting 50 pairs for each sex the rate of inbreeding might be kept at 1% per generation, is the recommended limit to minimize loss of genetic variation or to avoid loss of fitness [2] and minimize an immediate loss of heterozygosity or genetic variation [5, 8, 31].

REFERENCES

1. Bentsen HB, Gjerde B: Design of fish breeding programs. Proc. 5th world congress on genetics applied to livestock prod. Guelph. XIX: 1994, 353-359.
2. Bentsen HB, Olesen I: Designing aquaculture mass selection programs to avoid high inbreeding rates. Aquaculture 2002, 204: 349- 359
3. Bijlsma R, Bundgaard J, Boerema AC: Does inbreeding affect the extinction risk of small populations?: predictions from Drosophila. J. Evol. Biol. 2000, 13: 502–514.
4. Bourdon RM: Understanding animal breeding. Charles E. Steweard, Jr, USA; 1997.
5. Caballero A: Developments in the prediction of effective populations size. Heredity 1994, 73:657–679.
6. Campton DE: Genetic effects of hatchery fish on wild populations of Pacific salmon and steelhead: What do we really know?. Am. Fish. Soc. Symp. 1995, 15, 337–353.
7. Falconer DS, Mackay TFC: Introduction to quantitative genetics. Longman group Ltd, England; 1996.
8. Frankham R: Conservation genetics. Annu. Rev. Genet., 1995, 29, 305–327.
9. Franklin IR, Frankham R: How large must populations be to retain evolutionary potential?. Anim. Conserv., 1998. 1, 69–73.
10. Gall GAE, Huang N: Heritability and selection schemes for rainbow

trout: body weight. Aquaculture 1988, 74: 43-56.

11. Gjedrem T: Selection and Breeding Programs in Aquaculture. Springer, Netherland; 2005.
12. Gjerde B, Rye M: Design of breeding programs in aquaculture species- possibilities and constrains. Genetics and breeding of Mediterranean aquaculture species. Zaragoza, Spain. April, 1997, 28-30.
13. Gjerde B, Gunnes K., Gjedrem T: Effect of inbreeding on survival and growth in rainbow trout. Aquaculture 1983, 34: 327-332.
14. Gjerde B: Growth and reproduction in fish and shellfish. Aquaculture1986, 57: 37-55.
15. Gjerde B: Body traits in rainbow trout I. Phenotypic means and standard deviation and sex effect. Aquaculture 1989, 80: 7-24.
16. Gjerde B: Breeding and selection. In: Salmon aquaculture. Fishing news books, USA; 1993: 187-208.
17. Gjerde B, Gjoen HM, Villanueva B: Optimum designs for fish breeding progammes with constrained inbreeding, Mass selection for normally distributed trait. Livestock production science 1996, 47: 59-72.
18. Gjerde B, Villanueva B, Bentsen HB: Opportunities and challenge in designing sustainable fish breeding programs. 7th World congress on genetic applied to livestock production. Montpellier, France, August 2002: 19-23.
19. Huang N, Gall GAE: Correlation of body weight and reproductive characteristics in rainbow trout. Aquaculture1990, 86: 191-200.
20. Kinghorn BP: A review of quantitative genetics in fish breeding. Aquaculture 1981, 31: 283-304.
21. Loeschcke V, Tomiuk J, Jain SK: Introductory remarks: Genetics and conservation biology. In: Conservation Genetics. Basel/ Switzerland: Birkhäuser Verlag, 1994: 3–8.
22. Padhi BK, Mandal RK:. Applied fish genetics. Fishing chimes, India; 2000.
23. Pante MJR, Gjerde B, Mcmillan I: Inbreeding levels in selected populations of rainbow trout, Oncorhynchus mykiss. Aquaculture 2001, 192: 213-224.

24. Ponzoni RW, Nguyen NH, Khaw HL: Genetic improvements for aquaculture species in developing countries: Prospect and challenges. Proc. Assoc. Advmt. Anim. Breed. Genet. 2009, 18:342-349.

25. Refstie T: Application of breeding schemes. Aquaculture 1990, 85: 163-169.

26. Ryman N, Laikre L: Effects of supportive breeding on the genetically effective population size. Conserv. Biol. 1991, 5, 325–329.

27. Saccheri I, Kuussaari M, Kankare M, Vikman P, Fortelius W, Hanski I: Inbreeding and extinction in a butterfly metapopulation. Nature 1998, 392, 491–494.

28. Salte R, Galli A, Falaschi U, Fjalestad KT, Aleandri R: A protocol for the on-site use of frozen milt from rainbow trout (Oncorhynchus mykiss Walbaum) applied to the production of progeny groups: comparing males from different populations. Aquaculture 2004, 231: 337-345.

29. Shultz FT: Developing a Commercial Breeding Program. Aquaculture 1986, 57, 65-76

30. Su GS, Liljedahl LE, Gall, GAE: Effect of inbreeding on growth and reproductive traits in rainbow trout (Oncorhynchus mykiss). Aquaculture 1996, 142: 139-148.

31. Wang Sh, Jeffrey JH, Fred U: Salmonid inbreeding: a review. Reviews in Fish Biology and Fisheries, 2001, 11: 301–319.

CHAPTER

3

RECENT APPROACHES OF FISH BREEDING

João Costa Filho[1] and Leila de Genova Gaya[2]

[1]Zootecnista; Mestrando no Programa de Pós-Graduação em Ciência Animal Departamento de Produção Animal e Alimentos do Centro de Ciências Agroveterinárias da Universidade do Estado de Santa Catarina - CAV/UDESC; Endereço: Av. Luiz de Camões, 2090 - Conta dinheiro, CEP: 88.520-000, Lages, Santa Catarina, Brasil

[2]Dra.; Médica Veterinária; Professora do Departamento de Engenharia de Biossistemas, Universidade Federal de São João del-Rei, UFSJ; Endereço: Praça Dom Helvécio, 74, Campus Dom Bosco, CEP: 36.301-160, São João del-Rei, Minas Gerais, Brazil

ABSTRACT

This review aims at addressing the main issues related to genetic improvement applied to fish species with zootechnical potential, such as selection, genetic parameters and use of biotechnology as well as address the importance of this area of study for the expansion of aquaculture and its development trend. Fish breeding is one of the most important areas of study for fish production development. Important techniques are applied in order to advance genetic progress in fish breeding (e.g., individual selection, genetic parameters estimation and the use of biotechnology). Traditional breeding in fish farming has been conducted mainly through biometric analysis, which helps in choosing selection criteria and in the assessment of genetic parameters such as heritability and correlations between interesting traits. The use of biotechnology in fish breeding has been successful; it is an important strategy that allows breeding results to be achieved faster, although at a greater cost than traditional breeding. Fish breeding in Brazil is currently in its initial stages; however, it is a field with great growth expectations, and there is still significant research to be conducted in this area.

Key words: Biometry; biotechnology; carcass; genetic parameters; performance.

INTRODUCTION

Total continental aquaculture production was about 31.6 million tons in 2006, and it is quickly expanding as the demand for food of aquatic origin has been increasing considerably (FAO, 2008). In aquaculture worldwide, fish represent 50.1% of total production (CAMARGO; POUEY, 2005). The farming of aquatic species is a quite recent phenomenon. In recent years, several companies have invested in intensive farming of these organisms. Studies in this field have been conducted mainly with the purpose of increasing environmental control of aquatic organisms' habitats and researching fundamental aspects of biology such as feeding and reproduction (GONZAGA; ZANIBONI FILHO, 2006). However, domestication of the species used in aquaculture will not be complete until all biological aspects of these animals are controlled; this includes genetics that, together with animal breeding, are among the priority areas of research and development in fish farming, especially for fish, freshwater prawns and marine shrimps (QUEIROZ et al., 2002).

The tools and study techniques used in fish breeding, such as selection and genetic parameters estimation, are similar to those used for other species, like mammals and poultry. Moreover, some techniques, such as chromosome manipulation and sexual reversion, are easily performed in several species of fish, which might contribute to their breeding (GALL, 1991). Thus, this review aims at addressing the main issues related to genetic improvement applied to fish species with zootechnical potential, such as selection, genetic parameters and use of biotechnology as well as address the importance of this area of study for the expansion of aquaculture and its development trend.

SELECTION AND FISH BREEDING

According to Resende et al. (2008), fish breeding might increase productive gain to every generation, which is already considered high compared to other animals. The main benefit, according to these authors, is the increase in the fish growth, which directly influences production cost decrease and fish farmers income improvement.

Thus, one of the tools for increasing production in order to meet the increasing demand for aquatic species is the development of improved fish lineages. One breeding method for doing so is the use of selection. Every improved generation of fish is the basis for the next generation; therefore, the seventh generation of selected fish is estimated to achieve double the growth rate of the original population. Improved lineages are distributed to producers who are then able to produce high quality fry and juveniles to be sold for fattening. In this way, animal breeding is spread to future generations, which results in gains for the productive system (Resende et al., 2008). According to Hinrichs et al. (2006), the genetic gain of fish populations might be maximized if genetic evaluation is used, thus obtaining a value that can be used to represent the true genetic merit of individuals; animals are then selected on the basis of this merit.

The genetic selection for native fish such as tilapia (Tilapia rendalli) resulted in the rapid spread of improved lines that replaced non-improved lines in Brazil until the end of 1990s (KUBITZA et al., 2007). On the other hand, zootechnical gains in fish are still in their initial phases, and there is much yet to be studied in this area. Another study using nile tilapia conducted by Leonhardt al et. (2006) found that the existence of genetic variability in a population is an initial requirement for applying for selection. Once this genetic variability is confirmed, the families and desired traits of the target population are identified. According to Resende et al. (2008), identification of genetic variability makes it possible to select reproducers by choosing the ones that present the highest growth rate, then using them for consecutive breeding.

Tilapias are disease-resistant individuals that exhibit good adaptability in captivity and reproduce easily. All tilapia producers searching for good reproducers should consider growth speed, resistance to disease and body shape in addition to filet yield due to market demands when aiming for selection (GONÇALVES et al., 2003). As for carnivorous fish that present a strong tendency for cannibalism, such as pintado (Pseudoplatystoma corruscans) and dourado (Salminus maxillosus), selection of individuals conditioned on feed intake in some generations might result in lines with less cannibalistic behavior and better feed acceptance. This has been demonstrated in other species of carnivorous fish around the world (KUBITZA et al., 2007). According to the same authors, some producers of juvenile pintado and dourado in Brazil have already noticed differences, such as a decrease in cannibalism among juveniles produced from flocks of domesticated matrix in comparison to flocks of wild matrix.

In Houston et al. (2008) study, progress in easily measurable traits, such as growth of sexual maturity in salmon (Salmo salar), was rapid and significant. Studies related to molecular genetics currently aim to increase resistance to diseases like infectious pancreatic necrosis, which is the main disease in this specie. With regard to catfish (Ictalurus punctatus), Liu et al. (2003) noted that the existence of numerous families for these fish leads to a great variability of the specie, making it possible to apply phenotypic selection. However, according to the same authors, genotypic selection with molecular markers would be very efficient for identifying those animals with the best traits in terms of performance.

SELECTION FOR CARCASS YIELD IN FISH

Evaluation methods for animal productivity that are effective and allow for fast evolution in breeding programs have been researched in an effort to achieve greater efficiency in farm fishing as a zootechnical activity (CREPALDI et al., 2008). Carcass yield has been one of the main goals of research on increasing efficiency in animal production systems, especially for fishes. Selection programs have made it possible to improve traits of economic interest, not only by searching for animals with better carcass yield, but also by increasing yield in noble cuts, which are more valued in the consumer market. Therefore, several studies have been conducted with the aim of improving this trait (CREPALDI et al., 2008).

Studies related to selection for carcass yield increase in tilapia, according to Santos et al. (2007), can lead to greater interest in breeding these species or lines and serve as a reference for other possible studies on fish breeding. In addition, information obtained from this kind of research might be of importance in determining the ideal weight for slaughtering and the final characterization of the product as well as establishing fish breeding programs.

Different methods of assessing carcass yield are used in animal production, including assessment by weighting, biometry of specific areas such as the loin eye, and techniques that keep the animals alive, like carcass evaluation by ultrasonography (CREPALDI et al., 2008). Moreover, if these techniques are used in the initial selection of a reproducer's flock, it can advance animal breeding programs, according to the same authors, and carcass evaluation of traits is a reliable technique that makes it possible to keep fish with superior zootechnical traits inside the flock. Over many years of breeding systems, catfish have undergone several selection processes in the search for greater carcass yield; ultrasonography has also been used with this specie (BOSWORTH et al., 2001).

According to Reidel et al. (2004), analysis of carcass traits and morphometric features in fish species allow for comparisons to differentiate individual capacity of production and, consequently, the selection process. According to Bosworth et al. (2001), catfish with smaller head, deeper body and thick in the visceral cavity have greater yield based on correlation estimates. The selection for this body shape could, therefore, increase filet yield. Carcass yield does not depend on the length or height of Nile tilapias at the time of slaughter (GONÇALVES et al., 2003). Only weight and size of filet would be related to carcass yield. According to the same authors, meat production is related more to fish height than to fish length; therefore, taller fish would tend to present a better slaughter weight as well as a higher filet weight. During a study conducted by Reidel et al. (2004), male curimbatás (Prochilodus lineatus) were noted to be smaller in body size than females and to present a lower filet yield. Head length in comparison to body length was bigger for male curimbatás than for females, indicating that males have a relatively larger head, which resulted in the reduced yield. Among piavuçus (Leporinus macrocephalus), another native species analyzed by Reidel et al. (2004), the males were also found to be smaller than females; however, the filet yield for males was higher than that of females. Thus, female curimbatás and male piavuçus tend to be more likely to obtain better carcass yield when breeding these species.

Although the selection process for fish species is efficient, its productive performance might be affected by environmental factors. Therefore, a favorable fish breeding environment is essential so that animals can express their real genetic potential (MAREGONI et al., 2008). Working with juvenile Nile tilapia, Margoni et al. (2008) showed that increased stock density causes alterations in feed consumption and, consequently, production costs. Type of feeding can also significantly influence fish production, and performance is frequently affected by the food supplied to animals in production system. Piedras and Pouey (2004) reported that the growth and survival of peixe-rei fry (Odontesthes bonariensis) tended to be affected by the kind of feed they were given. Temperature is another important environmental factor in fish development. In a study conducted by Piedras et al. (2004), the performance traits of jundiá fry (Rhamdia quelen) were influenced by oscillations in water temperature. Therefore, the environmental conditions to which fish are subjected must be controlled so that the selection process of reproducers and, consequently, their breeding, become more efficient.

GENETIC PARAMETERS

Heritability corresponds to the portion of total variation in a trait that is of additive genetic nature, is considered an important genetic parameter for defining more adequate breeding methods (PEREIRA, 1999). The response to selection of individuals is a linear function of the heritability of the trait to be improved (FUNK et al., 2005). Traits with low heritability, such as survival and fertility in fish, have little additive genetic action, and are therefore more influenced by the environment in which animals are raised (VEHVILÄINEN et al., 2008). These same authors subjected rainbow trout (Oncorhynchus mykiss) to two different environments in an attempt to estimate the heritability of survival traits; in both cases, heritability was low, suggesting that selection response to these traits tend to be unfavorable.

Silva (2007), working with Nile tilapias, observed high heritability for performance and carcass traits such as weight and length at 60 and 150 days old, daily weight gain, daily length gain, eviscerated carcass weight and carcass yield. According to this study, these traits tend to respond to selection. Genetic correlation measures the probability of two different traits being affected by the same genes, i.e., the correlation between the genetic values of two traits (PEREIRA, 1999). Genetic correlations often make it possible for two traits to be improved at the same time (FUNK et al., 2005), most often when they are evaluated in fish of different ages (ROCHA et al., 2002). Therefore, it is essential to understand correlations between traits of interest in a population when studying animal breeding; it is also important to establish zootechnical rates that can support other studies on selection and breeding (ROMAGOSA et al., 2003).

Estimates of genetic correlations between weight, body length, and body height and head length ranged from moderate to high for most associations at different developmental stages of common carp (Cyprinus carpio). Therefore, when selecting for one of these traits, the others will also benefit, according to Rocha et al. (2002). Genetic correlations in Nile tilapia regarding the live weight and length of fish of different ages as well as carcass weight were also high, according to Silva (2007), making it clear that indirect selection for performance and carcass traits in these species is possible.

BIOTECHNOLOGIES USED IN FISH BREEDING

Research in the field of biotechnology in fish farming is of great importance, according to Igarashi and Magalhães Neto (2001). Progress in fish breeding has been the subject of many discussions and reviews throughout the last decade (LEONHARDT et al., 2006). The main goal of this research has been to increase the number of new lines through the development of modern technologies.

With advances in molecular genetics and genetically modified organism techniques, it has become possible to obtain genetically modified individuals, i.e., organisms that have had genes from another organism inserted in their genome and then transferred to their descendents (ARAGÃO, 2003). Genetically modified organisms often have advantages over other non-improved individuals. For example, it is possible to increase the survival time in salmon through genetic modification. However, it is necessary to take great care to avoid releasing these genetically modified fish into the environment, as it could cause an ecological unbalance in the local populations (HOWARD et al., 2004). According to Rexroad et al. (2008), rainbow trout have achieved great importance in aquaculture as well as in fishing sports. Genome mapping in this specie has made it possible to identify important genes that affect traits of economic interest.

Quinn et al. (2008) report that sequencing the salmon genome makes it possible to better understand the genetics of this fish, which is a species of great economic interest that has an important role in aquatic ecology when in its native habitat. Resistance to disease has also been studied in fish breeding. Several species, like salmon, have their production damaged because they are vulnerable to some disease. Therefore, alternatives have been researched through genetic mapping to identify individuals that might present this trait (MOEN et al., 2004).

Moreira et al. (2007) conducted a study that evaluated the genetic variability of two varieties of Nile tilapia (Oreochromis niloticus), Chitralada and Red Stirling, and their progeny, which were subjected to breeding on intensive raising systems using microsatellite markers. This study used DNA extraction and amplification by PCR (Polimerase Chain Reaction) to determine that the variability between these two Nile tilapia strains was high and that heterosis was able to improve zootechnical performance of progenies. Knowledge about DNA markers has evolved beyond the experimental phase and been incorporated into aquaculture in a practical and efficient way, according to Leonhardt et al. (2006).

Silva et al. (2011) used F1 generations of silver catfish of the genus Rhamdia, for a study to identify individuals with triploid cytogenetic tools, showing that in the analyzed species occurred a high number of triploid individuals, 46.8%. Taking into account that the triploid species have partial or total sterility, this study highlights the importance of cytogenetic monitoring in repopulating programs aimed at genetic conservation.

Catfish production is subject to a constant breeding process for trait selection as much as for molecular genetics techniques (QUINIOU et al., 2007) and is an excellent candidate for genomic studies (LIU et al., 2003). According to Cereali et al. (2008), the biggest families of catfish are of great interest in cytogenetic studies due to significant variations in chromosome number among the species in these families, as well as the rearrangement of chromosomes, which suggests an intense caryotypic evolution. The catfish genome has regions that control important production traits (QUINIOU et al., 2007). Among these are growth rate, food conversion, resistance to disease, body shape and filet yield, and tolerance for low oxygen and low quality water (LIU et al., 2003). Thus, with complete construction of the genome, it will be possible to locate these regions and search for new ways to improve this specie. Thus, Odegard et al. (2011) reported that breeding, using tools of genomic analysis, is essential for achieving genetic progress in fish species. In order to improve these individuals productivity is necessary to understand their genomic structures, so that it becomes possible to extend the methods of genetic selection and identification of molecular markers associated with production traits (KANG et al., 2008).

According to Hemmer-Hansen et al. (2011), the new sequencing technologies have revolutionized the discovery of genes and polymorphisms throughout the genome of the species. Quinn et al. (2008) report that by sequencing the genome of fish is possible to obtain better results in the genetic understanding of these species. An important tool for genomic studies is the use of candidate genes, which consist of genes from known biological function and are involved in the development or physiology a particular characteristic (PEIXOTO et al., 2009). Permits the link between genotype and biological function, extending the observation of genetic variation (HEMMER-HANSEN et al., 2011). A major effort has been applied on the isolation and analysis of genes of interest in several species of fish, including salmon, trout, halibut (KANG et al., 2002) and tilapia (BLANCK et al., 2009).

As a result of the increase of technology in the biological area, especially regarding to genomic sequencing, a large individual genetic variation can be seen established. The variations that occur with greater frequency in the genome are highlighted as single nucleotide polymorphisms or SNPs, which consist of mutations point created in different alleles (KONGCHUM et al., 2010). For its part, these regions throughout these individual's genome are abundant molecular markers and have been receiving great attention in recent studies (KITAOKA et al., 2010). Many of these polymorphisms are associated with productive traits of economic importance. According to Blanck et al. (2009) study of polymorphisms in candidate genes has contributed information that can be applied as selection criteria in breeding programs of species, especially aquaculture.

This practice aims to identify molecular markers that can be used in selection programs for improvement of traits difficult to measure (NINOV et al., 2007). Similarly, studies based on the identification of markers can be applied to the detection of variability between strains and association with the phenotype (MARKLIUND; CARLBORG, 2010), and such studies potentiate more genomic research (GROENEN et al. 2008). The identification of polymorphisms can serve as a starting point for marker assisted selection (MAS), a very important tool in today's scientific scene (KITAOKA et al., 2010). In modern animal breeding programs it is used the association of a marker gene or a set of genes, so the selection for this marker is more efficient than for the very characteristic (JANGARELLI et al., 2011).

Several studies have shown the importance of identifying polymorphisms in genes of fishes. Blanck et al. (2009), by working with Nile tilapia, have identified a significant correlation between growth hormones' gene polymorphisms and body characteristics. Likewise, Kocour and Kohlmann (2011), by studying Tinca tinca L., have identified polymorphisms in genes of this species. Kang et al. (2002) pointed out that there are variations in genes and their association with body weight in halibuts. In this same segment, further investigations have been conducted with some genes that affect growth characteristics, such as the genes of prolactin, growth factor similar to insulin and myostatin.

Other studies aimed at broadening the understanding of the function of some important aquatic organisms' genes have used the post-transcriptional silencing of genes, known as RNA interference or RNAi, which consists of a dsRNA that links to a target sequence promoting their degradation and blocking the expression (ESTRADA et al. 2007). Thus, some researchers have shown the use of RNAi as an important tool for studies. Estrada et al. (2007) evaluated the biological function of the myostatin gene in aquatic organisms, noting that the gene silencing, promoted by RNAi, results in an increase in muscle, since myostatin is involved in muscle deposition. Lee et al. (2009) also used this methodology to study the suppression of the myostatin gene in transgenic zebrafish, where the fish had a double muscling phenotype. These results may contribute to a breakthrough for aquaculture if they achieve commercially important species.

Triploidization can also be used to improve fish production. This technique consists of applying a shock (heat, pressure or chemical) to a fertilized fish egg before the second polar body extrusion, blocking effectively meiosis II (JOHNSON et al., 2007). As a consequence, the fertilized egg has three haploid nuclei: one from the egg, one from the sperm and another from the second polar body and the three haploid nuclei fuse to form a triploid zygotic nucleus. The techniques for obtaining triploid fish have two main goals: growth improvement and sterility. Triploid fish have been raised in several species such as tilapia, catfish, trout and salmon (MOREIRA, 2001).

Obtaining triploid fish allows for stabilization in fish production (JOHNSON et al., 2007). According to the same authors, triploization increases phenotypical variation in growth and resistance traits in salmon. For better understanding of the cell events related to fish production, cytogenetics should be applied to fish breeding. According to Cruz et al. (2001), cytogenetics is a specialized area of genetics related to chromosomes and their association with genetic and phenotypic events. Many studies on the hybridization of species and sexual reversion have been performed in fish, contributing to a better understanding of the mechanisms of sexual determination, especially in Nile and other species of tilapia (LEONHARDT et al., 2006).

According to Boscolo et al. (2001), interest in tilapia farming has grown over the last eight years due to the use of sexual reversion technology, which consists of developing single sex populations to prevent fish reproduction in captivity. This technique is carried out by using feed with masculinizing hormones in new-born progenies until three to four weeks of age. The advantages of this technique include preventing energetic waste with reproductive activity, animal uniformity and reduction of sexual maturity effects in meat quality, among others (CESAR et al., 2009). Hybridization consists of crossing species using pure lines (RIBEIRO, 2001). Hybrid formation is potentially interesting in the evolution and reproduction of fish, as it has some advantages, such as rapid growth and greater resistance to diseases (LIU et al., 2007).

Characterization of genetic variability in flocks is important for establishing a breeding program. When conducting breeding in a population through traditional or molecular techniques, great initial genetic variability assures the existence of enough variability to achieve the desired improvements in successive generations (RESENDE et al., 2008). With regard to the genetic characterization of tilapia, there have been osses in the heterozygosis of flocks according to Melo et al. (2006).

This was supported by analyses conducted by these authors on six flocks of tilapia carried out in southeast Brazil; the Israel and Nile flocks were the most genetically similar, while the Chitralada and Taiwan flocks had the fewest genes in common. As Red tilapias (Oreochromis sp) are hybrid, they were the most genetically distinct of the six flocks.

Genetic variability reduction might occur in a single generation and therefore resistances to diseases as well as the capacity to adapt to new environments were also reduced because of reproductive management of fish, according to Lopes et al. (2008). According to Carneiro (2007), problems in aquaculture, such as inbreeding, might occur as a result of successive crossing that selects genes of specific interest through inbreeding, thus reducing genetic diversity. Building up a genetic data bank with frozen fish semen from several males can be used to fertilize eggs from females, significantly decreasing the risks of inbreeding without the need to maintain a large flock of reproducers. Carneiro (2007) further notes that this genetic material could also be used to recover fish population.

Genetic studies conducted with preserved material from a frozen sperm bank might help decrease the fishing stock limitations of different species by allowing them to adjust of fishing in a region (RESENDE et al., 2005). Techniques for inducing reproduction in fish using hormones have become essential in animal breeding for several species, according to Zaniboni Filho and Weingartner (2007). Hypophysation may be used for artificial procreation of fish, as it places special emphasis on efficiency, practicability and economy. This process consists of using collected fresh or preserved pituitary glands from donor fish that are then used to provoke final maturity of gametes in reproducers (FURUYA; FURUYA, 2001). Application of this process requires selection of reproducers, which is considered the most important phase of the egg-laying process. The criteria for this process are based on subjective traits, such as females with expanded soft abdomens and swollen and reddish genital papilla (ZANIBONI FILHO; WEINGARTNER, 2007).

CONCLUSION

Traditional breeding in fish farming has been conducted mainly through biometric analysis, which helps in choosing selection criteria, and assessment of genetic parameter such as heritability and correlations.

Practices for fish selection are often performed by the producers themselves and do not always achieve the desired results. Tilapia is the most farmed and studied specie in Brazil due to its adaptability to captivity. The use of biotechnology in fish breeding has been successful; it is an important strategy that allows breeding results to be achieved faster, although at a greater cost than traditional breeding. Fish breeding is still in its early stages, but it has become a field with strong growth expectations. There is still significant research to be conducted in this area beyond animal breeding and the different kinds of technologies that have been used in production. These studies have become essential for continue to advance the field and search for genetic progress in species.

REFERENCES

1. ARAGÃO, F. J. L. Organismos transgênicos: explicando e discutindo a tecnologia. Barueri: Manole. 2003. 116 p.
2. BLANCK, D. V.; GASPARINO, E.; RIBEIRO, R. P.; MARQUES, D. S. Polimorfismo no gene GH1-PstI associado a características corporais de linhagens de tilápia-do-nilo. Pesquisa Agropecuária Brasileira, Brasília, v.44, n.6, p.599-604, 2009.
3. BOSCOLO, W. R.; HAYASHI, C.; SOARES, C. M.; FURUYA, W. M.; MEURER, F. Desempenho e características de carcaça de machos revertidos de tilápias do Nilo (Oreochromis niloticus), linhagens tailandesa e comum, nas fases inicial e de crescimento. Revista Brasileira de Zootecnia, Viçosa, v. 30, n. 5 , p.1391-1396, 2001.
4. BOSWORTH, B. G.; HOLLAND, M.; BRAZIL, B. L. Evaluation of ultrasound imagery and body shape to predict carcass and fillet yield in farm-raised catfish. Journal of Animal Science, v. 79, n. 6, p. 1483-1490, 2001.
5. CAMARGO, S. G. O.; POUEY, J. L. O. F. Aquicultura - um mercado em expansão. Revista Brasileira de Agrociência, Pelotas, v. 11, n. 4, p. 393-396, 2005.
6. CARNEIRO, P. C. F. Tecnologia de produção e armazenamento de sêmen de peixes. Revista Brasileira de Produção Animal, Belo Horizonte, v.31, n. 3, p. 361-366, 2007.
7. CEREALI, S. S.; POMINI, E.; ROSA, R.; ZAWADZKI, C. H.;

FROEHLICH, O.; CAETANO, L. G. Karyotype description of two species of Hypostomus (Siluriformes, Loricariidae) of the Planalto da Bodoquena, Brazil. Genetics and Molecular Research, v. 7, n. 3, p. 538-591, 2008.

8. CESAR, M. P.; MURGAS, L. D. S.; ARAÚJO, R. V.; DRUMMOND, C. D. Métodos para obtenção de população monosexo na piscicultura. Boletim Agropecuário, Lavras, UFLA, n. 69, p.1-27, 2009.
9. CREPALDI, D. V.; TEIXEIRA, E. A.; FARIA, P. M.; RIBEIRO, L. P.; MELO, D. C.; OLIVEIRA, D. A. A.; TURRA, E. M.; QUEIROZ, B. M. Rendimento de carcaça em surubim (Pseudoplatystoma spp) avaliado por ultra-som. Revista Brasileira de Saúde e Produção Animal, Salvador, v. 9, n. 4, 813-824, 2008.
10. CRUZ, C. D.; VIANA, J. M. S.; CARNEIRO, P. C. Z. Genética. Viçosa: UFV, 2001. 476 p.
11. ESTRADA, M. P.; LUGO, J. M.; ACOSTA, J.; CARPIO, Y.; BORROTO, I.; MORERA, Y.; GONZÁLEZ, O.; RODRÍGUEZ, T.; RAMOS, L.; HUBERMAN, A. Effects of RNA interference on gene functions of aquatic organisms. Biotecnologia Aplicada, v.24, n.2, 2007.
12. FOOD AND AGRICULTURE ORGANIZATION - FAO. El estado mundial de la pésca e la acuicultura, 2008. Disponível em: . Acesso em: 12 Jul. 2009.
13. FUNK, W. C.; TYBURCZY, J. A.; KNUDSEN, L.; LINDNER, K. R.; ALLEDORF, F. W. Genetic basis of variation in morphological and life-history traits of a wild population of pink salmon. The journal of heredity, v. 96, n. 1, p. 24-31, 2005.
14. FURUYA, W. M.; FURUYA, V. R. B. Reprodução de peixes. In: MOREIRA, H. L. M.; VARGAS, L.; RIBEIRO, R. P.; ZIMMERMANN, S. (Ed.). Fundamentos da moderna aquicultura. Canoas: ULBRA, 2001. p.69-82.
15. GALL, G. A. Genetics and reproduction in fish culture. Journal of animal science, v. 69, n. 10, p. 4216-4220, 1991.
16. GONÇALVES, T. M.; ALMEIDA, A. J. L.; BORGES, E. E. S. Características de carcaça de tilápias do Nilo (Oreochromis niloticus) em quatro classes de peso ao abate. Acta Scientiarium, Animal Sciences, Maringá, v. 25, n. 1, p. 25-29, 2003.

17. GONZAGA, F.; ZANIBONI FILHO, E. Avaliação dos efeitos da indução de triploidização em Jundiá (Rhamdia quelen). 2006. Disponível em: . Acesso em: 3 jan. 2009.

18. GROENEN, M. A. M.; WAHLBERG, P.; FOGLIO, M.; CHENG, H. H.; MEGENS, H-J.; CROOIJMANS, R. P. M. A.; BESNIER, F.; LATHROP, M.; MUIR, W. M.; WONG, G.K-S.; GUT, I.; ANDERSSON, L. A high-density SNP-based linkage map of the chicken genome reveals sequence features correlated with recombination rate. Genome Research, v.19, p.510-519, 2008.

19. HEMMER-HANSEN, J.; NIELSEN, E.; MELDRUP, D.; MITTELHOLZER, C. Identification of single nucleotide polymorphisms in candidate genes for growth and reproduction in a nonmodel organism; the Atlantic cod, Gadus morhua. Molecular Ecology Resources, v.11, n.1, p.71-80, 2011.

20. HINRICHS, D.; WETTEN, M.; MEUWISSEN, T.H.E. An algorithm to compute optimal genetic contributions in selection programs with large numbers of candidates. Journal of Animal Science, v. 84, p. 3212-3218, 2006.

21. HOUSTON, R. D.; HALEY, C. S.; AMILTON, A.; GUY, D. R.; TINCH, A. E.; TAGGART, J. B.; MCANDREW, B. J.; BISHOP, S. C. Major quantitative trait loci affect resistance to infectious pancreatic necrosis in Atlantic salmon (Salmo salar). Genetics, v. 178, n. 2, p. 1109-1115, 2008.

22. HOWARD, R. D.; DEWOODY, J. A.; MUIR, W. M. Transgenic male mating advantage provides opportunity for Trojan gene effect in a fish. Proceedings of the National Academy of Sciences of the United States of America, v. 101, n. 9, p. 2934-2938, 2004.

23. IGARASHI, M. A.; MAGALHÃES NETO, E. O. Estratégias para o desenvolvimento da aquicultura no nordeste brasileiro. Revista Econômica do Nordeste, Fortaleza, v. 32, n. 2, p. 148-165, 2001.

24. JANGARELLI, M.; EUCLYDES, R. F.; CECON, P. R. Níveis de significância na identificação de marcadores moleculares no mapeamento genômico. Revista Brasileira de Zootecnia, Viçosa, v.40, n.2, p.308-313, 2011.

25. JOHNSON, R. M.; SHRIMPTON, J. M.; CHO, J. K.; HEATH, D. D. Dosage effects on eritability and maternal effects in diploid and

triploid Chinook salmon (Oncorhynchus tshawytscha). Heredity, v. 98, n. 5, p. 303-310, 2007.

26. KANG, J. H.; KIM, W. J.; LEE, W. J. Genetic Linkage Map of Oliver Flounder, Paralichthys olivaceus. International Journal of Biological Sciences, v.4, n.3, p.143-149, 2008.
27. KANG, J. H.; LEE, S. J.; PARK, R. S.; RYU, H. Y. DNA polymorphism in the growth hormone gene and its association with weight in olive flounder Paralichthys olivaceus. Fisheries Science, v.68, p.494-498, 2002.
28. KITAOKA, M.; WADA, T.; NISHIO, T.; GOTO, M. Fluorogenic Ribonuclease Protection (FRIP) Analysis of Single Nucleotide Polymorphisms (SNPs) in Japanese Rice (Oryza sativa L.) DNA for Cultivar Discrimination. Bioscience Biotechnology and Biochemistry, v.74, n.11, p.2189-2193, 2010.
29. KOCOUR, M.; KOHLMANN, K. Growth hormone gene polymorphisms in tench, Tinca tinca L. Aquaculture, v.310, p.298-304, 2011.
30. KONGCHUM, P.; PALTI, Y.; HALLERMAN, E. M.; HULATA, G.; DAVID, L. SNP discovery and development of genetic markers for mapping innate immune response genes in common carp (Cyprinus carpio). Fish & Shellfish Immunology, v.29, p.356-361, 2010.
31. KUBITZA, F.; ONO, E. A.; CAMPOS, J. L. Os caminhos da produção de peixes nativos no Brasil: uma análise da produção e obstáculos da piscicultura. Revista Panorama da Aquicultura, Laranjeiras, v. 17, p. 14-23, 2007.
32. LEE, C.Y.; HU, S. Y.; GONG, H. Y.; CHEN, M. H. C.; LU, J. K.; WU, J. L. Suppression of myostatin with vector-based RNA interference causes a double-muscle effect in transgenic zebrafish. Biochemical and Biophysical Research Communications, v.387, p. 766-771, 2009.
33. LEONHARDT, C. M. R.; CAETANO, L. G.; FENOCCHIO, A.; CAETANO FILHO, M.; LEONHARDT, J. H. Caracterização cariotípica de um estoque de tilápia do Nilo, Oreochromis niloticus, da Universidade Estadual de Londrina, mediante diversas técnicas de bandamento cromossômico. Semina: Ciências Agrárias, Londrina, v. 27, n. 4, p. 679-684, 2006.

34. LIU, Z.; KARSI, A.; LI, P.; CAO, D.; DUNHAM, R. An AFLP-Based genetic linkage map of channel Catfish (Ictalurus punctatus) Constructed by Using an Interspecific Hybrid Resource Family. Genetics, v. 165, n. 2, p. 687-694, 2003.

35. LIU, S.; QIN, Q.; XIAO, J.; LU, W.; SHEN, J.; LI, W.; LIU, J.; DUAN, W.; ZHANG, C.; TAO, M.; ZHAO, R.; YAN, J.; LIU, Y. The formation of the polyploid hybrids from different subfamily fish crossings and its evolutionary significance. Genetics Society of America, v. 176, p. 1023-1034, 2007.

36. LOPES, T. S.; RIBEIRO, R. P.; BARRERO, N. M. L.; SIROL, R. N.; POVH, J. A.; GOMES, P. C.; VARGAS, L. Caracterização genética de estoques de curimbatá (Prochilodus lineatus) utilizados em programas de repovoamento. Revista Brasileira de Saúde e Produção Animal, Salvador, v. 9, n. 4, p. 652-661, 2008.

37. MARKLUND, S.; CARLBORG, Ö. SNP detection and prediction of variability between chicken lines using genome resequencing of DNA pools. BMC Genomics, v.665, n.11, 2010.

38. MAREGONI, N. G.; MACHADO, M. R. F.; GASPARINO, E. Extração de DNA genômico em tecidos sólidos de peixes teleósteos. Semina: Ciências Agrárias, Londrina, v. 27, n. 1, p. 99-106, 2006.

39. MAREGONI, N. G.; BUENO, G. W.; GONÇALVES JÚNIOR, A. C.; OLIVEIRA, A. A. M. A. Desempenho produtivo e viabilidade econômica de juvenis de tilápia-do-Nilo cultivados na região oeste do Paraná sob diferentes densidades de estocagem. Revista Brasileira de Saúde e Produção Animal, Salvador, v. 9, n. 2, p. 341-349, 2008.

40. MELO, D. C.; OLIVEIRA, D. A. A.; RIBEIRO, L. P.; TEIXEIRA, C. S.; SOUSA, A. B.; COELHO, E. G. A.; CREPALDI, D. V.; TEIXEIRA, E. A. Caracterização genética de seis plantéis comerciais de tilápia (Oreochromis) utilizando marcadores microssatélites. Arquivo Brasileiro de Medicina Veterinária e Zootecnia, Belo Horizonte, v. 58, n. 1, p. 87-93, 2006.

41. MOEN, T.; FJALESTAD, K. T.; MUNCK, H.; GOMEZ-RAYA, L. A multistage testing strategy for detection of quantitative trait loci affecting disease resistance in Atlantic salmon. Genetics, v. 167, n. 2, p. 851-858, 2004.

42. MOREIRA, A. A.; HILSDORF, A. W. S.; SILVA, J. V.; SOUZA, V. R. Variabilidade genética de duas variedades de tilápia nilótica por meio de marcadores microssatélites. Pesquisa Agropecuária Brasileira, Brasília, v. 42, n. 4, p. 521-526, 2007.

43. MOREIRA, H. L. M. Genética e melhoramento de peixes. In: MOREIRA, H. L. M.; VARGAS, L.; RIBEIRO, R. P.; ZIMMERMANN, S. (Ed.). Fundamentos da moderna aquicultura. Canoas: ULBRA. 2001. p.135-148.

44. NINOV, K.; LEDUR, M. C.; NONES, K.; COLDEBELLA, A.; BERTOL, T. M.; CAETANO, A.; COUTINHO, L. L. Associação de polimorfismo de base única (SNP) no íntron 8 do gene do receptor da leptina em galinhas com rendimento de órgãos. In: REUNIÃO ANUAL DA SOCIEDADE BRASILEIRA DE ZOOTECNIA, 44., 2007, Jaboticabal. Anais... Jaboticabal: Sociedade Brasileira de Zootecnia, 2007.

45. ODEGARD, J.; BARANSKI, M.; GJERD, B.; GJEDREM, T. Methodology for genetic evaluation of disease resistance in aquaculture species: challenges and future prospects. Aquaculture Research, v.42, p.103-114, 2011.

46. PEREIRA, J. C. C. Melhoramento genético aplicado à produção animal. Belo Horizonte: FEP-MVZ. 1999. 493 p.

47. PIEDRAS, S. R. N.; MORAES, P. R. R.; POUEY, J. L. O. F. Crescimento de juvenis de jundiá (rhamdia quelen), de acordo com a temperatura da água. Boletim do Instituto de Pesca, São Paulo, v. 30, n. 2, p. 177-182, 2004.

48. PIEDRAS, S. R. N.; POUEY, J. L. O. F. Alimentação de alevinos de peixe-rei (Odontesthes bonariensis) com dietas naturais e artificiais. Ciência Rural, Santa Maria, v. 34, n. 4, p. 1303- 1206, 2004.

49. PEIXOTO, J. O.; FARIA, D. A.; SILVA, P.V .; FONSECA, I.; LOPES, P. S.; GUIMARÃES, S. E. F. Association between leptin gene single nucleotide polymorphisms and carcass traits in pigs. Revista Brasileira de Zootecnia, Viçosa, v.38, n.2, p.271-276, 2009.

50. QUEIROZ, J. F.; LOURENÇO, J. N. P.; KITAMURA, P. C. A Embrapa e a aquicultura - demandas e prioridades de pesquisa. Brasília: Embrapa Informação Tecnológica. 2002. 35p.

51. QUINN, N. L.; LEVENKOVA, N.; CHOW, W.; BOUFFARD,

P.; BOROEVICH, K. A.; KNIGHT, J. R.; JARVIE, T. P.; LUBIENIECKI, K. P.; DESANY, B.A.; KOOP, B.F.; HARKINS, T.T.; DAVIDSON, W. S. Assessing the feasibility of GS FLX Pyrosequencing for sequencing the Atlantic salmon genome. BMC Genomics, v.9, n.404, p.1-14, 2008.

52. QUINIOU, S. M.; WALDBIESER, G. C.; DUKE, M. V. A first generation BAC-based physical map of the channel catfish genome. BMC Genomics, v. 8, n. 40, p. 1-9, 2007.

53. REIDEL, A.; OLIVEIRA, L. G.; PIANA, P. A.; LEMAINSKI, D.; BOMBARDELLI, R. A.; BOSCOLO, W. R. Avaliação de rendimento e características morfométricas do curimbatá Prochilodus lineatus (VALENCIENNES, 1836), e do piavuçu Leporinus macrocephalus (GARAVELLO & BRITSKI, 1988), machos e fêmeas. Revista Varia Scientia, v. 4, n. 8, p. 71-78, 2004.

54. RIBEIRO, R. P. Espécies exóticas. In: MOREIRA, H.L. M.; VARGAS, L.; RIBEIRO, R. P.; ZIMMERMANN, S. (Ed.). Fundamentos da moderna aquicultura. Canoas: ULBRA. 2001. p.91-122.

55. RESENDE, E. K.; CATELLA, A. C.; MARQUES, D. K. S.; ROTTA, M. A. Banco de sêmen de peixes do Pantanal. Revista Eletrônica de Veterinária, v. 6, n. 10, p. 1-2, 2005.

56. RESENDE, E. K.; RIBEIRO, R. P.; LEGAT, A. P.; BENITES, C. Melhoramento genético em peixes – uma revolução na aquicultura do Brasil. 2008. Disponível em: . Acesso em: 6 jan. 2009.

57. REXROAD, C. E.; PALTI, Y.; GAHR, S. A.; VALLEJO, R. L. A second generation genetic map for rainbow trout (Oncorhynchus mykiss). BMC Genetics, v. 9, n. 74, p. 1-14, 2008.

58. ROCHA, M. A.; RIBEIRO, E. L. A.; MIZUBITI, I. Y.; SILVA, L. D. F. Parâmetros de crescimento e suas correlações em idades entre 60 a 240 dias na carpa húngara (Cyprinus carpio). Semina: Ciências Agrárias, Londrina, v. 23, n. 1, p. 29-34, 2002.

59. ROMAGOSA, E.; PIAVA, P.; GODINHO, H. M.; ANDRADE-TALMELLI, E. F. Características morfométricas e crescimento do cachara, Pseudoplatystoma fasciatum (Linnaeus, 1766), em cativeiro. Acta Scientiarium, Animal Sciences, Maringá, v. 25, n. 2, p. 277-283, 2003.

60. SANTOS, V. B.; FREITAS, R. T. F.; LOGATO, P. V. R.; FREATO, T. A.; ORFÃO, L. H.; MILLIOTI, L. C. Rendimento do processamento de linhagens de tilápias (Oreochromis niloticus) em função do peso corporal. Ciência e Agrotecnologia, Lavras, v. 31, n. 2, p. 554- 562, 2007.

61. SILVA, M.; MATOSO, D. A.; LUDWIG, L. A. M.; GOMES, E.; ALMEIDA, M. C.; VICARI, M.R.; ARTONI, R.F. Natural triploidy in Rhamdia quelen identified by cytogenetic monitoring in Iguaçu basin, southern Brazil. Environmental Biology Fish, v.91, n.3, p.361- 366, 2011.

62. SILVA R. F. Avaliação genética do crescimento de Tilápia do Nilo. Viçosa. 2007. 49 f. Dissertação (Mestrado em Zootecnia) - Universidade Federal de Viçosa , UFV, Viçosa, 2007.

63. VEHVILÄINEN, H.; KAUSE, A.; QUINTON, C.; KOSKINEN, H.; PAANANEN, T. Survival of the currently fittest: genetics of rainbow trout survival across time and space. Genetics, v. 180, n. 1, p. 507-516, 2008.

64. ZANIBONI FILHO, E.; WEINGARTNER, M. Técnicas da indução da reprodução de peixes migradores. Revista Brasileira de Reprodução Animal, Belo Horizonte, v. 31, n. 3, p. 367-373, 2007.

CHAPTER

4

GENETIC EVALUATION OF A SELECTIVE BREEDING PROGRAM FOR COMMON CARP CYPRINUS CARPIO CONDUCTED FROM 2004 TO 2014

Zaijie Dong[1], Nguyen Hong Nguyen[2] and Wenbin Zhu[1]

[1] Freshwater Fisheries Research Centre of Chinese Academy of Fishery Sciences, Key Laboratory of Freshwater Fisheries and Germplasm Resources Utilization, Ministry of Agriculture, Wuxi 214081, China.

[2] University of the Sunshine Coast, Locked Bag 4, Maroochydore, DC QLD 4558, Australia.

ABSTRACT

Background

The study evaluated genetic progress of a breeding program for common carp undergoing four generations of selection for increased harvest body weight from 2004 to 2014. The pedigree included 17,351 individual fish which were offspring of 342 sires and 352 dams. Genetic parameters for body weight at about two years of age and survival rate during grow-out period were also estimated using the residual maximum likelihood method applied to a two-trait linear mixed model. Direct response in body weight and correlated changes in survival were measured as the differences in: i) estimated breeding values (EBVs) between the two lines; and ii) EBVs of the selection line between successive generations.

Results

Direct gain in body weight ranged from 0.20 to 0.90 genetic standard deviation units and averaged 7 % of the base population per generation (two years per generation). Correlated changes in survival were negligible, indicating that the selection program for high growth did not have any adverse effect on this trait in the present population. The heritability for body weight was moderate (0.17, s.e. 0.05), whereas the estimate for survival was low (0.05–0.17) but significantly different from zero across linear mixed and threshold generalised statistical models. Our results predict that body weight or/and other growth related traits will continue to respond to selection and that there is potential to improve survival through direct genetic means. Correlated improvement in survival to selection for increased body weight was hardly achieved, as the genetic correlation between the two traits was not different from zero.

Conclusions

It is concluded that selection for increased harvest body weight resulted in significant improvement in growth performance of the present population of common carp *Cyprinus carpio*.

Keywords: Genetic improvement Selection response Heritability and common carp

BACKGROUND

Common carp (*Cyprinus carpio* L.) is one of the most economically important freshwater species for aquaculture in the world including China [1]. In China it is the third important cultured finfish, alongside grass carp (*Ctenopharyngodon idellus*), silver carp (*Hypophthalmichthys molitrix*) and bighead carp (*Hypophthalmichthys nobilis*). In 2011, the production of cultured common carp reached 2.71 million tonnes and accounted for 11.0 % of the total inland aquaculture production in China [2].

C. carpio is widely distributed throughout China and shows many morphological and genetic variations after local artificial breeding and natural selection. The most important wild and domesticated common carp populations in China include Huanghe carp (*C. carpio haematopterus* Temminck et Schlegel), Heilongjiang carp (*C. carpio haematopterus*), purse red carp (*C. carpio var. wuyuanensis*) and Xingguo red carp (*C. carpio var. xingguonensis*). China has a long history of culturing common carp and the species has played a significant socio-economic role in the society [3]. However, breeding this species in captivity only began in 1970's when some varieties were developed for aquaculture, mainly by crossbreeding technique [4]. Jian carp (*C. carpio* var. *jian*), which was bred by scientists in the Freshwater Fisheries Research Center (FFRC) of Chinese Academy of Fishery Sciences in 1990s [4], is the first variety of common carp that was produced through artificial breeding. The techniques used to produce this variety include hybridization, within-family selection and gynogenesis. The gross production of Jian carp is about 30 % higher than other varieties of common carp [4].

Due to the superiority in production characteristics of the Jian carp, a family selective breeding program to further improve quality and growth performance of this strain has been conducted at FFRC since 2004. We applied the method used for other aquatic animal species by establishing a synthetic base population from different sources and using mixed model methodology to rank selection candidates based on their estimated breeding values. This enabled us to select superior animals to become parents of the next generation [5]. This approach has been successfully applied to improve productivity of several terrestrial farmed animals [6] and aquaculture species [7]. Experimental studies with common carp report that selection for increased body weight also resulted in significant improvement in growth performance [8] and was correlated with increases in body traits (e.g. body length, depth or width) [9]. However, possible changes in fitness related traits, such as survival during grow-out, have not been reported.

The aim of this paper was to conduct genetic evaluation of the breeding program for common carp after four generations of selection in order to estimate genetic parameters and selection response for body weight and survival. Linear mixed and generalised threshold logistic models were used to estimate heritability for body weight and survival, respectively. The response (or genetic gain) was measured as the difference in estimated breeding values (EBVs) between the selection line and control group.

METHODS

Experimental Location

The breeding program for common carp was conducted at the Freshwater Fisheries Research Centre (FFRC) of the Chinese Academy of Fishery Sciences (CAFS), following The Guide for the Care and Use of Experimental Animals of China.

FFRC is located in Wuxi city besides the third largest freshwater lake (Lake Tai) and is about 30 km to the south of the Yangtze River. The annual temperature ranges between −7 and 38 °C (average 15.5 °C) and the average annual rainfall is approximately 1000 mm. Topographically, FFRC is located in low plain areas (200 m above sea level) along the lower reaches of the Yangtze River. The freshwater pond, pH 7.2–8.5, has a salinity level below 0.2 ppt.

Origin of the common carp population

The synthetic base population was formed from a complete diallel cross involving the Jian carp and Huanghe carp strains produced in April 2004. A total of 78 families were produced, tagged and reared for 20 months. Measurements of body weight, length and height were made on all individual fish. Data from this base population were analysed to estimate breeding values. These were used to carry out during the first round of selection. In 2006, a mating protocol to form the selection and control lines was designed based on the analysis of growth data of the 78 families produced in 2004. A total of 81 pairs of fish spawned successfully for the selection line, whereas 19 pairs of fish spawned for the control line. Their progeny were contemporaneously reared by family under the same pond conditions. In 2008, the third strain of Heilongjiang carp (20 females, 1.85 kg and 20 males, 1.5 kg) was incorporated into the population and 66 selection families were established.

Family production and rearing

Mating design and family production

Parental breeders in each generation were paired on the basis of their estimated breeding values (EBV) and their genetic relationship with other individuals in the pedigree. Sexual maturity of the breeders was examined externally as common carp normally reach maturation in the spawning season (April-May). Induced spawning was applied to produce full- and half-sib families. A hormone injection was given to parental fish: females were injected with 500 IU HCG + 4 μg LRH-A2 per kg; males 250 IU HCG + 2 μg LRH-A2 per kg. Spawning began around 12 h later.

Each pair of the injected fish was released into an individual hapa ($1 \times 1 \times 1$ m^3, 16 mesh per cm) installed in the two earthen ponds. The fish spawned naturally (without outside intervention) in the hapas.

Twenty four hours after the hormone injections, the cages and eggs were transferred to the egg-hatching and fry-rearing site in an earthen pond (0.35 ha, 2 m deep). The hatching hapas were of $1 \times 1 \times 1$ m^3 (20 mesh per cm) installed in indoor cement tanks (70 m^3). Twenty hapas were installed in each tank. Eggs hatched after 48 ± 2 h. Eggs from each family were separately hatched in different hapas.

Family rearing procedures

Each full sib group was reared in the same hapa ($1 \times 1 \times 1$ m^3) for 7d before being transferred to a cage with larger mesh size (8 mesh per cm). The stocking density in the cage was 1000 larvae per m^3. Fry were not counted immediately after hatching due to high fecundity of the species. Soybean milk and fine granular formulated diet were used as the feed for the fry and provided twice a day.

When fish larvae reached the length of approximately 3 mm (after about 2 months after hatching), they were transferred to cages $1 \times 1 \times 1$ m^3 with larger mesh size (3 mm). The stocking density was reduced to 100 pieces per m^3. For each family, only 100 randomly sampled individuals were retained and restocked in the new cage. The fish were reared in the same cage for about two months and were fed twice a day with pellet feed until tagging. The feeding rate was from 2 to 4 % of the body weight, adjusted according to ambient temperature. The diet had 32 % protein and 3 % lipid.

Offspring of each family were reared in separate cages until tagging was completed. All the cages were installed in one pond to ensure uniform rearing conditions. The cages were cleaned and checked for netting once a week during the rearing period.

Individual tagging

The fish were tagged after 2 months of rearing at an average body weight of about 20 g. The tagging of all the families was completed within two days. In each family, a random sample of 50 fingerlings were tagged using Passive Integrated Transponder (PIT) for individual identification.

All the tagged fish were kept in indoor cement tanks (72 m^2) for 1–2 weeks to monitor mortality before releasing them into an earthen pond for further rearing. The stocking density was 20 fingerlings per m^2. Normal feeding was resumed one day after tagging. During this period a small number (~1 %) of fish died or lost their tag (about 0.1 %) and these were replaced by their siblings from the same family.

Communal Testing

Communal testing of all families was conducted at two different stages over a period of 11.5–13.5 months. In the first period/year, all the tagged fish were released in one earthen pond of 0.16 ha and a depth of 1.8 m, after the temporary holding. The fish were daily fed with pellet feed (28 % protein and 3 % lipid). Feeding was practiced twice a day, during early morning and late afternoon. The stocking density in this period was about 18,000 fish per ha. The water temperature was below 25 °C.

In the second period, the fish were reared in the new pond (0.34 ha) with a water depth of 2.0 m for 6 months until the second sampling. Feed (28 % protein) was provided twice a day, once in the early morning and once in the late afternoon throughout the culture period. The feeding rate was between 1 and 4 % of the estimated body mass. Some mortality occurred in the hot season. The dead fish were collected and their tags were recovered.

The stocking density in this period was 15,000 tagged fish per ha, together with 2000 silver carp per ha and 1000 bighead carp per ha. Stocking of silver carp and bighead carp aimed to regulate the water fertility. The water temperature during this period (January to December) ranged from 4 to 32 °C.

Harvesting

Harvesting and measurement were conducted every twelve months of culture and after about one and a half years from birth (482 d in G2 and G3 and 557d in G4). The fish were harvested through total drainage of the pond because common carp are difficult to catch by netting. All the harvested fish were scanned and measured, including body weight, body length and body height as well as their tag identification and sex. Only a small proportion of the tagged fish (0.009 % or 27 fish) were not identified because: (i) fish had lost their tags during the rearing; (ii) tags were defective; or (iii) fish died during the rearing period and were not collected before sinking to the bottom and decomposing.

Selection Procedures

Linear animal mixed model analyses were performed each generation to estimate the breeding value (EBV) for body weight of individual fish. Based on individual and family rankings by EBV, best (highest EBV) fish were selected to become parents in the selection line, whereas the control group was selected based on the EBV mean of the population. Two to three times more fish than the actual number of breeders were selected as candidates for the selection line and control group. The mating protocol was designed with a restriction on the number of fish per family contributed to the next generation. The mating of close relatives was also avoided. These fish were kept in an earthen pond and provided a high quality diet.

In each generation, the mating pairs consisted of 90 for the selection and 20 for the control. The selected male and female breeders were kept in separate earthen ponds to avoid any possible natural breeding. Feeding, management and induced breeding were practised as details above. The majority of the mating pairs spawned 48 h after the injection. A small number of pairs (3–10 %) failed to spawn across the generations. Some of them had high EBV ranking. They were held in indoor tanks (males separated from females) and fed a high quality brooder diet. Following the second hormone injection, about 60 % of the pairs that initially failed spawned. High mortality also occurred at hatching in 2–4 families each generation (less than 100 fry survived out of five to 10 thousands), probably due to poor egg quality.

In total, 66–84 selection and 16–19 control families were successfully produced across generations. Fry nursing/rearing, tagging, communal grow-out and harvest data were recorded as described in the above sections. Genetic evaluation was then conducted and a new breeding cycle was repeated every two years (i.e. generation). The same selection procedures, animal husbandry and management regimes were practised in all generations.

Table 1: The number of sires, dams and offspring for the selection line and control group in four generations of selection

Generation	Year	Line	Sire	Dam	Offspring
1	2004 – 2005	Base population	78	78	2911
2	2006 – 2007	Selection	84	84	4266
		Control	19	19	949
3	2008 - 2009	Selection	66	73	3886
		Control	14	13	817
4	2012 - 2013	Selection	72	72	3683
		Control	16	16	813
Total	All years		342	352	17,351

Statistical Analysis

Genetic parameters for body weight were estimated using linear mixed model [Eq 1] that included the fixed effects of generation (G_i =4), line (L_j =2, selection and control), sex (S_k=2, female and male) and their two-way interactions between these factors. Stocking weight (W_l) within line and generation was also fitted as a linear covariate in the model. The random terms were the additive genetics of individual fish (a_m) and common full-sibs (c_n).

$$y_{ijklmn} = \mu + G_i + L_j + S_k + G \times L_{ij} + G \times S_{ik} + L \times S_{jk} + W_l(G, L) + a_m + c_n + e_{ijklmn} \quad (1)$$

where y_{ijklmn} is the trait observation, G_i, L_j, S_k, W_l, a_m and c_n are as defined above and e_{ijklmn} is the error term

As survival was recorded as a binary expression (fish that survived at harvest were coded as 1, n =11966 and those absent at harvest were coded as 0, n =3765), this trait was analysed using the threshold generalised logistic sire model [10]. Under the logistic model, calculation of heritability for survival assumed that residual variance was corrected by $\pi^2/3$ (3.289) factor. In addition, a standard animal mixed model was used to estimate heritability for this trait.

Phenotypic and genetic correlations were obtained from a bivariate animal mixed model. All the analyses were conducted using ASReml version 3.0 [11].

Selection responses for both body weight and survival were measured as the difference in estimated breeding values (EBVs) between the selection line and control group or between successive generations. The direct genetic gain for body weight and correlated changes in survival were expressed in actual units (gram for body weight and % for survival), genetic standard deviation unit (SD_A) and percentage of the base population. The statistical model used to estimate EBVs for body weight and survival were the same as those used to estimate the heritability.

RESULTS

Characteristics of the population and data

Over four generations of selection from 2004 to 2014, a total of 17,351 offspring produced from 342 sires and 352 dams were performance tested in earthen pond over/during an average grow-out period of 383 days. The number of offspring and their parents (sires and dams) in each generation is given in Table 1.

At final harvest (383 days), the number of fish with the data for body weight and survival are shown in Table 2together with basic statistic parameters for these traits. The average body weight of the population at final harvest was 0.9 kg and survival rate during grow-out was 74 %. This population showed a large variation in body weight as shown by the high coefficient of variation of 70 %.

Table 2: Basic statistics for traits studied

Traits	Unit	N	Mean	SD	CV (%)
Weight	g	11680	944.7	657.4	69.6
Survival	%	17,351	73.9	43.9	59.4

The analysis of variance using the general linear model showed that the main effects of generation, line and sex were statistically significant for body weight. The two-way interactions among these factors (except for line × sex), as well as the linear covariate of stocking weight within line and generation, were also significant for this trait, i.e. body weight ($P < 0.001$). However, none of these effects was significant for survival when the generalised linear model was used.

Least squares means for body weight and survival of the selection line and control

Least squares means (LSMs) for body weight and survival were obtained using the linear mixed model and the threshold generalised logistic model, respectively. They are presented for each generation in Table 3. The selection line had significantly greater body weight than that of the control in all generations from 2004 to 2014 ($P < 0.05$ to 0.01). However, the difference in LSMs for survival between the selection line and control group was not significant ($P > 0.05$).

Sexual size dimorphism

Sexual size dimorphism (SSD) occurred in the early phase of growth. The difference in body weight of fingerlings at stocking was small (2 %) but significant ($P < 0.05$). After 1 year of culture from hatching, female common carp had 34.4 % greater body weight than that of male ($P < 0.01$). At final harvest (one and a half to 2 years from birth) the between-sex difference in body weight was statistically significant ($P < 0.001$) in all generations (Fig. 1). In contrast to body weight, survival during grow-out did not differ between female and male common carp during the course of the selection program.

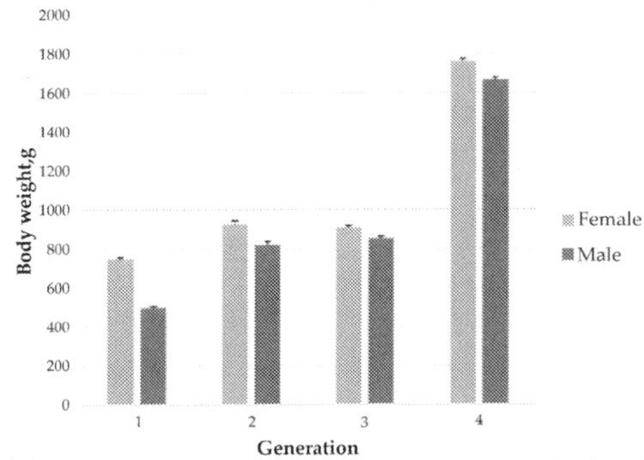

Figure 1: Least squares means of body weight by sex in four generations of selection.

Heritability, common environmental effects and correlations

Estimates of heritability for body weight and survival and the phenotypic and genetic correlations between the two traits are given in Table 4. Body weight is moderately heritable ($h^2 = 0.17 \pm 0.05$) and the common full-sib effect (c^2) accounted for 16 % of the total phenotypic variation ($c^2 = 0.16$). The estimates of heritability for survival, estimated from linear animal mixed model and threshold logistic sire, were low (0.05 ± 0.01) and moderate (0.17 ± 0.03), respectively. However, both estimates were significant, based on their low standard errors.

Table 3: Least squares means (± se) for body weight (g) and survival (%) in the selection line and control group

Trait	Generation	Control	Selection	% difference
Weight	G2	575.7 ± 9.1	590.8 ± 4.3	2.62
	G3	701.8 ± 9.5	747.2 ± 4.5	6.47
	G4	2085.1 ± 11.2	2154.2 ± 5.3	3.31
Survival	G2	71.0 ± 1.47	69.1 ± 0.71	−2.89
	G3	74.4 ± 1.53	71.3 ± 0.73	−4.22
	G4	58.7 ± 1.73	60.0 ± 0.81	2.32

Selection Response

Genetic response to selection was measured as the difference either in estimated breeding values (EBVs) between the selection line and control (Table 5) or those between successive generations (Table 6). For each method, the results were presented in actual unit of measurements (g for body weight and % for survival), genetic standard deviation unit (SD_A) and percentage of the base population. Regardless of expression units, the estimates of direct genetic gain for body weight and correlated response for survival were consistent between the two methods. For body weight, selection achieved a cumulative direct genetic gain by 1.2 SD_A or 28 %, averaging 7 % per generation. Correlated genetic changes in survival were small (−0.09 to 0.264 SD_A).

Table 4: Heritability, maternal and common environmental effects for body weight and survival and phenotypic (r_p) and genetic (r_g) correlations between the two traits

Traits	Heritability	Full-sib effects	r_g	r_p
Weight	0.17 ± 0.05	0.16 ± 0.02	0.10 ± 0.08	−0.03 ± 0.01
Survival	0.05 ± 0.01[a]			
	0.17 ± 0.03[b]			

[a] Estimated from animal mixed model, [b] estimated from threshold logistic model

Table 5: Genetic gain calculated as the difference in breeding values between the selection line and control in actual unit of measurements or expressed in genetic standard deviation (SD_A) unit

Generation	Weight		Survival	
	Actual unit	Genetic SD	Actual unit	Genetic SD
G1	0.07	0.0008	0.0468	0.485
G2	46.61	0.495	0.0016	0.017
G3	84.31	0.895	0.0127	0.132
G4	18.23	0.194	0.0198	0.206

Table 6: Genetic gain calculated as the difference in breeding values of the selection line between successive generations in actual unit of measurements or expressed in genetic standard deviation unit

Generation	Weight		Survival	
	Actual unit	Genetic SD	Actual unit	Genetic SD
G1	0.07	0.0008	0.0027	0.027
G2 – G1	39.53	0.420	0.0180	0.186
G3 – G2	34.33	0.365	0.0254	0.264
G4 – G3	7.43	0.079	0.0006	0.006

DISCUSSION

Selection Response and Future Improvement

Genetic evaluation of 17,351 animals produced from 342 sires and 352 dams over four generations between 2004 and 2014 demonstrated that the selection program resulted in a remarkable improvement in growth performance in the common carp population. The average genetic gain was approximately 7 % per generation (2 years per generation). Our results are in good agreement with those reported recently for common carp by Ninh et al. [9] and Vandeputte et al. [8] as well as other aquatic animal species such as tilapia [12] and giant freshwater prawn [13]. Across farmed aquaculture species, direct response to selection for increased body weight or high growth range from 5–15 % per generation [5, 7]. Estimates of genetic gain depend on statistical methods used or whether a comparison is made with a control (fish having breeding values close to the population mean) or with wild stocks. In the present study, a separate control group of the same genetic origin as the selection line was maintained and contemporaneously produced in each generation. The genetic gain was estimated using the difference in: i) EBVs between the two lines (method 1); and ii) EBV between successive generations (method 2). The magnitude of genetic gain obtained from methods one and two are similar. Hung et al. [13] reported that the genetic gain slightly differed by the two methods. The EBV is a measure of the genetic superiority in a trait of interest of an animal as compared to its contemporaries and is calculated from the phenotypes of the individual and pedigree data. Thus, an analysis of the genetic trend calculated using EBV provided a more accurate indication of the amount of genetic progress attained in selected populations for traits studied as compared to using the phenotypic means [14]. Our results, when considered with those reported by Hung et al. [13] and Hamzah et al. [12], suggest that genetic response can be estimated with a minimum bias by having a contemporary control in parallel with the selection line in all generations. However, in commercial breeding programs, where resources are not available to maintain a separate control, assessment of genetic progress in selection populations could be estimated as the differences in EBV between consecutive generations.

For aquaculture enterprises, body weight and survival rate are the two most important economic traits. We investigated these changes associated with the selection program for common carp. Using two estimation methods and four generations, changes in survival rate during grow-out were negligible. This is consistent with the absence of genetic correlation between body weight and survival (Table 4). Correlated changes in survival to selection for increased body weight of common carp have not been reported previously but the non-significant changes in survival during grow-out in the present population are in agreement with reports in other freshwater fish, e.g. tilapia [15, 16]. There is a growing concern about negative changes in fitness related traits in artificial selection programs [5]. However, the non-significant correlated response in survival shown in our research indicates that selection for high growth did not have detrimental effect on fitness in the common carp population over a period of ten years.

Although survival was not improved from the selection program for high growth, this trait has a heritable additive genetic component (heritability = 0.05 to 0.17), indicating that direct selection to improve survival, albeit at a slow rate, could be possible in practical genetic improvement programs for common carp. A number of studies report that index selection, combining growth and survival, resulted in the improvement in both traits in Pacific white leg shrimp *Liptopenaus vannamei* [17], tilapia *Oreochromis aureus* [18] and abalone *Haliotis diversicolor* [19]. Expanding the breeding objectives for common carp by including new traits (especially survival rate or disease resistance) is currently under investigation in our breeding program.

The heritability of body weight was shown to be moderate and significant in our study and indicates that this population of common carp will continue to show a positive response to future selection. Our estimate of heritability for body weight is consistent with studies reported recently in common carp [20, 21, 22, 23] and other fishes [12] as well as in crustacean species such as shrimp [24, 25] and mollusc, e.g. abalone [19]. Vandeputte [26] reported that the heritability range for body weight in common carp was 0.00 to 0.75. An evaluation of the literature, together with the heritability estimate obtained from our present study, suggest that selective breeding is an effective way of improving growth related traits.

Common Full-Sib Effects

It is not unsurprising that the common full-sib effect (c^2) accounted for about 17 % of the total phenotypic variance for body weight. This is mainly a result of the separate early rearing of each family for 2–3 months before tagging. Under separate family rearing in a different selected population of common carp, Ninh et al. [20] reported that the c^2 effect ranged from 0.11 to 0.30. A large proportion of common full-sib variance was also reported for body traits in other finfishes such as tilapia [27], Atlantic salmon [28], rainbow trout [29], and giant freshwater prawn [25, 30]. Early communal rearing of all families soon after birth can reduce the c^2effect as reported across aquaculture species, including common carp [20] and marine yellowtail kingfish [31].

On Farm Testing of the Improved Strain

On farm testings of the improved common carp strain were carried out in several different geography and climate locations covering east China (Jiangsu and Shandong Provinces), southwest China (Sichuan Province and Guizhou Province), northwest China (Gansu Province and Ningxia Province) and northeast China (Liaoning Province). Compared with local common carp varieties, our improved strain had greater growth performance (20.1–39.2 %), higher survival (1.0–8.9 %) and lower food conversion rate (8.5–22.8 %) across all the regions. These results are in good agreement with those reported in other Asian countries, showing that improved carp strains are superior to local stocks across different farming systems in Bangladesh, Thailand and Vietnam [32]. In India,

Mahapatra et al. [33] reported that the improved rohu carp had 96 % greater body weight than the stock of farmers. The superiority of our improved common carp strain, under both selection and production environments, demonstrate that the genetic progress achieved in the nucleus is also expressed effectively under practical conditions and that the use of the improved strain can help farmers/producers to accelerate commercial production.

Existing Challenges

Further to the successful outcome from the selective breeding program for common carp in the present population, there are challenges for the long-term success in genetic improvement for this species. For instance, induced breeding at the first sexual maturation of the females did not result in 100 % success rate of spawning. A second injection was needed for females that failed to spawn the first time, causing a delay of one to two weeks compared with the normal reproduction for culture production. In all generations, offspring from all families were reared separately in hapas (net cages). The use of outdoor net cages installed in a pond for family rearing has proven successful, but the growth and survival rates of the fish in the early stages of development, were lower than for those reared under normal pond conditions. The fine mesh (8–16 mesh per cm) of the cage restricted water exchange and supply of natural food into the cage, which had an adverse impact on the growth and survival of the fish. It was also difficult to maintain a uniform culture environment for different families contained in separate cages, and to ensure a normal rate of growth and development when there are limitations with the physical rearing facilities. However, using individual earthen ponds for family rearing was not feasible due to the large number of families and cages were used for the family rearing in all generations. One option to overcome these limitations is to apply DNA markers for parentage assignment to enable the early communal rearing of all families after birth, as demonstrated in common carp [9]. It is also necessary to record growth and survival traits during the early phase of rearing in order to allow a formal genetic evaluation for these characters in efforts to broaden the breeding objectives in the present common carp population.

CONCLUSION

Four generations of selection for increased harvest body weight achieved an average direct genetic gain of approximately 7 % per generation. The selection program did not have a negative impact on survival rate of the animal over the ten year research period. The moderate heritability for body weight indicates that the population will respond to future selection. The significant additive genetic component for survival provides the possibility of further improving this characteristic in the future selective breeding program for *C. carpio*.

DECLARATIONS

Acknowledgement

This project was funded by the Chinese Earmarked Fund for Modern Agro-Industry Technology Research System (CARS-46) (2008 – present), the Key Projects in the National Science & Technology Pillar Program during the Eleventh Five-Year Plan Period (2006BAD01A1208) (2006–2010) and partially funded by Asian Development Bank ADB-RECTA 6136 (2001–2007). We extend our thanks to Dr Raul Ponzoni formerly from the WorldFish Centre for his technical advice during the early planning stage of the project. Colleagues and technicians at FFRC, and students at Wuxi Fisheries College are greatly appreciated for their assistance in the management of the animal and data collection. Prof Richard Burns at the University of the Sunshine Coast kindly helped to review language of the manuscript.

Authors' Contributions

ZD and NHN conceived and designed the experiment. ZD and WZ contributed the research materials and ran the breeding program. NHN and ZD performed the statistical analysis and drafted the manuscript. All authors (ZD, NHN and WZ) read and approved the final manuscript.

REFERENCES

1. FAO. The state of world fisheries and aquaculture. Rome: Food and Agriculture Organization of the United Nations; 2012.

2. TFBMA. China yearbook of fishery statistics. The Fishery Bureau of Ministry of Agriculture. Beijing, Chinese Agriculture Press. 2012.
3. Balon EK. Origin and domestication of the wild carp, Cyprinus carpio: from Roman gourmets to the swimming flowers. Aquaculture. 1995;129(1):3–48.
4. Dong Z, Yuan X. The utilizations of heterosis in common carp in China. Aquaculture Asia. 2002;7(2):14-15.
5. Nguyen HN. Genetic improvement for important farmed aquaculture species with a reference to carp, tilapia and prawns in Asia: achievements, lessons and challenges. Fish and Fisheries 2015:https://doi.org/10.1111/faf.12122.
6. Gianola D, Rosa GJ. One hundred years of statistical developments in animal breeding. Annu Rev Anim Biosci. 2015;3:19–56.
7. Gjedrem T. Genetic improvement for the development of efficient global aquaculture: a personal opinion review. Aquaculture. 2012;344:12–22.
8. Vandeputte M, Kocour M, Mauger S, Rodina M, Launay A, Gela D, et al. Genetic variation for growth at one and two summers of age in the common carp (Cyprinus carpio L.): Heritability estimates and response to selection. Aquaculture. 2008;277:7–13.
9. Ninh NH, Ponzoni RW, Nguyen NH, Woolliams JA, Taggart JB, McAndrew BJ, et al. A comparison of communal and separate rearing of families in selective breeding of common carp (*Cyprinus carpio*): Responses to selection. Aquaculture. 2013;408–409:152–9.
10. Nguyen N, Whatmore P, Miller A, Knibb W. Quantitative genetic properties of four measures of deformity in yellowtail kingfish *Seriola lalandi* Valenciennes, 1833. J Fish Dis. 2015. doi:https://doi.org/10.1111/jfd.12348.
11. Gilmour AR, Gogel B, Cullis B, Thompson R, Butler D. ASReml user guide release 3.0. Hemel Hempstead, UK: VSN International Ltd; 2009.
12. Hamzah A, Ponzoni RW, Nguyen NH, Khaw HL, Yee HY, Nor SAM. Genetic evaluation of the Genetically Improved Farmed Tilapia (GIFT) strain over ten generations of selection in Malaysia. J Trop Agr Sci. 2014;37(4):411–29.

13. Hung D, Vu NT, Nguyen NH, Ponzoni RW, Hurwood DA, Mather PB. Genetic response to combined family selection for improved mean harvest weight in giant freshwater prawn (*Macrobrachium rosenbergii*) in Vietnam. Aquaculture. 2013;412–413:70–3.
14. Nicholas FW. Introduction to veterinary genetics: John Wiley & Sons; 2009.
15. Hamzah A, Mekkawy W, Khaw HL, Nguyen NH, Yee HY, Bakar KRA, et al. Genetic parameters for survival during the grow out period in the GIFT strain and correlated response to selection for body weight. West Sussex, United Kingdom: Aquaculture Research 2015; accepted.
16. Ninh NH, Thoa NP, Knibb W, Nguyen NH. Selection for enhanced growth performance of Nile tilapia (*Oreochromis niloticus*) in brackish water (15–20 ppt) in Vietnam. Aquaculture. 2014;428–429:1–6.
17. Gitterle T, Johansen H, Erazo C, Lozano C, Cock J, Salazar M, et al. Response to multi-trait selection for harvest weight, overall survival, and resistance to white spot syndrome virus (WSSV) in Penaeus (Litopenaeus) vannamei. Aquaculture. 2007;272:S262.
18. Thodesen J, Rye M, Wang Y-X, Li S-J, Bentsen HB, Gjedrem T. Genetic improvement of tilapias in China: Genetic parameters and selection responses in growth, pond survival and cold-water tolerance of blue tilapia (Oreochromis aureus) after four generations of multi-trait selection. Aquaculture. 2013;396–399:32–42.
19. Liu J, Lai Z, Fu X, Wu Y, Bao X, Hu Z, et al. Genetic parameters and selection responses for growth and survival of the small abalone Haliotis diversicolor after four generations of successive selection. Aquaculture. 2015;436:58–64.
20. Ninh NH, Ponzoni RW, Nguyen NH, Woolliams JA, Taggart JB, McAndrew BJ, et al. A comparison of communal and separate rearing of families in selective breeding of common carp (*Cyprinus carpio*): estimation of genetic parameters. Aquaculture. 2011;322–323:39–46.
21. Vandeputte M, Kocour M, Mauger S, Dupont-Nivet M, De Guerry D, Rodina M, et al. Heritability estimates for growth-related traits using microsatellite parentage assignment in juvenile common carp

(Cyprinus carpio L.). Aquaculture. 2004;235(1):223–36.

22. Kocour M, Mauger S, Rodina M, Gela D, Linhart O, Vandeputte M. Heritability estimates for processing and quality traits in common carp (Cyprinus carpio L.) using a molecular pedigree. Aquaculture. 2007;270(1):43–50.
23. Nielsen HM, Ødegård J, Olesen I, Gjerde B, Ardo L, Jeney G, et al. Genetic analysis of common carp (*Cyprinus carpio*) strains: I: Genetic parameters and heterosis for growth traits and survival. Aquaculture. 2010;304(1):14–21.
24. Nguyen NH, Quinn J, Powell D, Elizur A, Thoa NP, Nocillado J, et al. Heritability for body colour and its genetic association with morphometric traits in Banana shrimp (Fenneropenaeus merguiensis). BMC Genet. 2014;15(1):132.
25. Hung D, Nguyen NH, Hurwood DA, Mather PB. Quantitative genetic parameters for body traits at different ages in a cultured stock of giant freshwater prawn (*Macrobrachium rosenbergii*) selected for fast growth. Mar Freshw Res. 2014;65(3):198–205.
26. Vandeputte M. Selective breeding of quantitative traits in the common carp (Cyprinus carpio): a review. Aquat Living Resour. 2003;16(05):399–407.
27. Nguyen NH, Ponzoni RW, Abu-Bakar KR, Hamzah A, Khaw HL, Yee HY. Correlated response in fillet weight and yield to selection for increased harvest weight in genetically improved farmed tilapia (GIFT strain), Oreochromis niloticus. Aquaculture. 2010;305(1–4):1–5.
28. Yáñez JM, Lhorente JP, Bassini LN, Oyarzún M, Neira R, Newman S. Genetic co-variation between resistance against both Caligus rogercresseyi and Piscirickettsia salmonis, and body weight in Atlantic salmon (Salmo salar). Aquaculture. 2014;433:295–8.
29. Haffray P, Bugeon J, Rivard Q, Quittet B, Puyo S, Allamelou JM, et al. Reprint of: Genetic parameters of in-vivo prediction of carcass, head and fillet yields by internal ultrasound and 2D external imagery in large rainbow trout (Oncorhynchus mykiss). Aquaculture. 2014;420:S134–42.

30. Hung D, Nguyen HN. Genetic inheritance of female and male morphotypes in giant freshwater prawn *Macrobrachium rosenbergii*. PLoS One. 2014;9(2):e90142.

31. Whatmore P, Nguyen NH, Miller A, Lamont R, Powell D, D'Antignana T, et al. Genetic parameters for economically important traits in yellowtail kingfish *Seriola lalandi*. Aquaculture. 2013;400–401:77–84.

32. Nguyen NH, Ponzoni RW. Genetic improvement of carp reduces poverty and hunger in Asia. Global Aquaculture Advocate 2008:76–78.

33. Mahapatra KD, Saha J, Sarangi N, Jana R, Gjerde B, Nguyen N, et al. Genetic improvement and dissemination of Rohu (Labeo rohita, Ham.) in India. In: Genetic improvement: making it happen. Proceedings of the Seventeenth Conference of the Association for the Advancement of Animal Breeding and Genetics: 23rd-26th September. Armidale, N.S.W. Australia Armidale, New South Wales, Australia: Association for the Advancement of Animal Breeding and Genetics; 2007;17:37–40.

CHAPTER

5

AQUACULTURE GENOMICS, GENETICS AND BREEDING IN THE UNITED STATES: CURRENT STATUS, CHALLENGES, AND PRIORITIES FOR FUTURE RESEARCH

Hisham Abdelrahman[1], Mohamed ElHady[2], Acacia Alcivar-Warren[3], Standish Allen[4], Rafet Al-Tobasei[5], Lisui Bao[1], Ben Beck[6], Harvey Blackburn[7], Brian Bosworth[8], John Buchanan[9], Jesse Chappell[1], William Daniels[1], Sheng Dong[1], Rex Dunham[1], Evan Durland[10], Ahmed Elaswad[1], Marta Gomez-Chiarri[11], Kamal Gosh[1], Ximing Guo[12], Perry Hackett[13], Terry Hanson[1], Dennis Hedgecock[14], Tiffany Howard[1], Leigh Holland[1], Molly Jackson[15], Yulin Jin[1], Karim Khalil[1], Thomas Kocher[16], Tim Leeds[17], Ning Li[1], Lauren Lindsey[1], Shikai Liu[1], Zhanjiang Liu[1], Kyle Martin[18], Romi Novriadi[1], Ramjie Odin[1], Yniv Palti[17], Eric Peatman[1], Dina Proestou[19], Guyu Qin[1], Benjamin Reading[20], Caird Rexroad[21], Steven Roberts[22], Mohamed Salem[5], Andrew Severin[23], Huitong Shi[1], Craig Shoemaker[6], Sheila Stiles[24], Suxu Tan[1], Kathy F. J. Tang[25], Wilawan Thongda[1], Terrence Tiersch[26], Joseph Tomasso[1], Wendy Tri Prabowo[1], Roger Vallejo[17], Hein van der Steen[27], Khoi Vo[1], Geoff Waldbieser[8], Hanping Wang[28], Xiaozhu Wang[1], Jianhai Xiang[29], Yujia Yang[1], Roger Yant[30], Zihao Yuan[1], Qi-fan Zeng[1] and Tao Zhou[1]

[1] School of Fisheries, Aquaculture and Aquatic Sciences, Auburn University, Auburn, AL 36849, USA.

[2] Department of Biological Sciences, Auburn University, Auburn, AL 36849, USA.

[3] Environmental Genomics Inc., Southborough, MA 01772-1801, USA.

[4] Aquaculture Genetics & Breeding Technology Center, Virginia Institute of Marine Science, Gloucester Point, VA 23062, USA.

[5] Department of Biology, Middle Tennessee State University, Murfreesboro, TN 37132, USA.

[6] Aquatic Animal Health Research Unit, USDA-ARS, 990 Wire Road, Auburn, AL 36832, USA.

[7] USDA-ARS-NL Wheat & Corn Collections at a Glance GRP, National Animal Germplasm Program, 1111 S. Mason St., Fort Collins, CO 80521-4500, USA.

[8] USDA-ARS/CGRU, 141 Experimental Station Road, Stoneville, MS 38701, USA.

[9] Center for Aquaculture Technologies, 8395 Camino Santa Fe, Suite E, San Diego, CA 92121, USA.

[10]Department of Fisheries and Wildlife, Oregon State University, Corvallis, OR 97331, USA.

[11]Department of Fisheries, Animal & Veterinary Science, 134 Woodward Hall, 9 East Alumni Avenue, Kingston, RI 02881, USA.

[12]Haskin Shellfish Research Laboratory, Department of Marine and Coastal Sciences, Rutgers University, 6959 Miller Avenue, Port Norris, NJ 08349, USA.

[13]Department of Genetics, Cell Biology and Development, 5-108 MCB, 420 Washington Avenue SE, Minneapolis, MN 55455, USA.

[14]Department of Biological Sciences, University of Southern California, Los Angeles, CA 90089-0371, USA.

[15]Taylor Shellfish Farms, 130 SE Lynch RD, Shelton, WA 98584, USA.

[16]Department of Biology, University of Maryland, 2132 Biosciences Research Building, College Park, MD 20742, USA.

[17]National Center for Cool and Cold Water Aquaculture, Agricultural Research Service, United States Department of Agriculture, Kearneysville, WV 25430, USA.

[18]Troutlodge, 27090 Us Highway 12, Naches, WA 98937USA.

[19]USDA ARS NEA NCWMAC Shellfish Genetics at the University Rhode Island, 469

CBLS, 120 Flagg Road, Kingston, RI 02881, USA.

[20]Department of Applied Ecology, North Carolina State University, Raleigh, NC 27695-7617, USA.

[21]USDA ARS Office of National Programs, George Washington Carver Center Room 4-2106, 5601 Sunnyside Avenue, Beltsville, MD 20705, USA.

[22]School of Aquatic and Fishery Sciences, University of Washington, Seattle, WA 98105, USA.

[23]Genome Informatics Facility, Office of Biotechnology, Iowa State University, Ames, IA 50011, USA.

[24]USDOC/NOAA, National Marine Fisheries Service, NEFSC, Milford Laboratory, Milford, Connectcut 06460, USA.

[25]School of Animal and Comparative Biomedical Sciences, University of Arizona, Tucson, AZ 85721, USA.

[26]Aquatic Germplasm and Genetic Resources Center, School of Renewable Natural Resources, Louisiana State University Agricultural Center, Baton Rouge, LA 70820, USA.

[27]Stonebridge breeding Ltd, Gate House, Abbotswood, Evesham WR11 4NS, UK.

[28]Aquaculture Genetics and Breeding Laboratory, The Ohio State University South Centers, Piketon, OH 45661, USA.

[29]Key Laboratory of Experimental Marine Biology, Institute of Oceanology, Chinese Academy of Sciences, Qingdao 266071, China.

[30]Hybrid Catfish Company, 1233 Montgomery Drive, Inverness, MS 38753, USA.

ABSTRACT

Advancing the production efficiency and profitability of aquaculture is dependent upon the ability to utilize a diverse array of genetic resources. The ultimate goals of aquaculture genomics, genetics and breeding research are to enhance aquaculture production efficiency, sustainability, product quality, and profitability in support of the commercial sector and for the benefit of consumers. In order to achieve these goals, it is important to understand the genomic structure and organization of aquaculture species, and their genomic and phenomic variations, as well as the genetic basis of traits and their interrelationships. In addition, it is also important to understand the mechanisms of regulation and evolutionary conservation at the levels of genome, transcriptome, proteome, epigenome, and systems biology. With genomic information and information between the genomes and phenomes, technologies for marker/causal mutation-assisted selection, genome selection, and genome editing can be developed for applications in aquaculture. A set of genomic tools and resources must be made available including reference genome sequences and their annotations (including coding and non-coding regulatory elements), genome-wide polymorphic markers, efficient genotyping platforms, high-density and high-resolution linkage maps, and transcriptome resources including non-coding transcripts. Genomic and genetic control of important performance and production traits, such as disease resistance, feed conversion efficiency, growth rate, processing yield, behaviour, reproductive characteristics, and tolerance to environmental stressors like low dissolved oxygen, high or low water temperature and salinity, must be understood. QTL need to be identified, validated across strains, lines and populations, and their mechanisms of control understood. Causal gene(s) need to be identified. Genetic and epigenetic regulation of important aquaculture traits need to be determined, and technologies for marker-assisted selection, causal gene/mutation-assisted selection, genome selection, and genome editing using CRISPR and other technologies must be developed, demonstrated with applicability, and application to aquaculture industries.

Major progress has been made in aquaculture genomics for dozens of fish and shellfish species including the development of genetic linkage maps, physical maps, microarrays, single nucleotide polymorphism (SNP) arrays, transcriptome databases and various stages of genome reference sequences. This paper provides a general review of the current status, challenges and future research needs of aquaculture genomics, genetics, and breeding, with a focus on major aquaculture species in the United States: catfish, rainbow trout, Atlantic salmon, tilapia, striped bass, oysters, and shrimp. While the overall research priorities and the practical goals are similar across various aquaculture species, the current status in each species should dictate the next priority areas within the species. This paper is an output of the USDA Workshop for Aquaculture Genomics, Genetics, and Breeding held in late March 2016 in Auburn, Alabama, with participants from all parts of the United States.

Keywords: Aquaculture Genetic resources Genome Transcriptome QTL RNA-Seq SNP Fish Shellfish

Abbreviations

ANN	Artificial neural networks
CRISPR	Clustered regulatory interspaced short palindromic repeats
ESD	Environmental sex determination
EST	Expressed sequence tag
FAANG	Functional Annotation of Animal Genomes
GMO	Genetically modified organisms
GS	Genome selection
GSD	Genetic sex determination
GWAS	Genome-wide association studies
IPN	Infectious pancreatic necrosis
IPNV	Infectious pancreatic necrosis virus
MAS	Marker-assisted selection
NRSP-8	National Research Support Project 8
QTL	Quantitative trait loci
SNP	Single nucleotide polymorphism
SPF	Specific pathogen-free
SSD	Sexual size dimorphism
SVM	Support vector machines
TALEN	Transcription activator-like effector nucleases
ZFN	Zinc finger nuclease

BACKGROUND

The major goals of research programs having components related to aquaculture genomics, genetics and breeding are to enhance aquaculture production efficiency, sustainability, product quality and profitability in support of the commercial sector and for the benefit of U.S. consumers. Progress towards achieving these goals includes genetic improvement of production, performance and animal welfare/fitness traits, and this progress is predicated upon the access and utilization of an array of genetic resources within each species group. To this end, various genetic stock enhancement approaches are currently being studied by the aquaculture research community, and major progress has been made since the start of aquaculture genomics research 20 years ago [1]. Such progress includes advances in traditional selection, intraspecific crossbreeding, interspecific hybridization, genome-enabled selection (e.g., marker/causal mutation-assisted selection and/or genomic selection), polypoidy, sex reversal and breeding, xenogenesis, gene transfer, and genome editing. Some of the most important traits studied for genetic improvement in U.S. aquaculture species include disease resistance, feed conversion efficiency, growth rate, behaviour, processing yield, reproductive characteristics and tolerance to environmental stressors like low dissolved oxygen, high or low water temperature and salinity, body composition, and flesh quality. Traditionally, genetic improvement in the commercial aquaculture sector relied on phenotypes and pedigree information, but recently leading international breeding companies have begun to implement genome technologies into their breeding programs for some of the species where advanced genomic resources and tools are available (e.g., [2, 3, 4]).

Genomic information provides powerful tools to enhance physiological research, the results of which may be used for optimization of husbandry practices, feeding and feed formulations, breeding technologies, or non-genetic selection or screening (e.g., epigenetics, proteomics, and metabolomics). Whole genome sequences, in various states of assembly, are now available for many aquaculture species, enabling the identification of genomic variations such as insertions/deletions, single nucleotide polymorphisms (SNPs), copy number variations, and differentially methylated regions. However, this information is only useful when used to predict phenotypes that have a positive impact on production or product quality. For this reason, genetic mapping, quantitative trait loci (QTL) analysis, genome-wide association studies (GWAS), expression profiling, and bioinformatic analysis can be used to identify genotypic variants associated with particular phenotypic traits, which can then be exploited in breeding programs. For some aquaculture species we have reached the point where genome-based technologies such as marker-assisted and whole genome selection can be applied to enhance aquaculture traits and research is beginning to shift toward understanding functional polymorphisms and the gene regulatory networks underlying commercially important traits. A more complete understanding of the gene networks underlying growth, reproduction, and disease resistance will provide the knowledge-base for developing more robust and productive genetic stocks for the aquaculture industry.

The degree to which genome enabled technologies and genomic information have been or can be applied in genetic improvement programs varies across aquaculture species. Private sector investment in research and development for the implementation of new technologies is dependent on unique industry structure (e.g., overall size of the industry, size of individual companies) and the level of vertical integration. In addition, the approach used for germplasm improvement and status of existing breeding programs dictates whether and which genome enabled technologies are suitable for a given industry. Industries with centralized breeding, such as rainbow trout and salmon, have greater potential to benefit from new technologies compared to industries where breeding activities are widely distributed. Finally, the current demand for species-specific genomic tools (such as high through-put genotyping assays) among the diverse aquaculture industry sectors is low, rendering them commercially unaffordable. This forces some industries interested in genetic improvement to rely on the public sector for resources that enable application of state-of-the-art genomic technologies.

Here we review the development of genomic tools and application of genome enabled technologies for the genetic improvement of aquaculture species. Specifically, we review the status of genome mapping and sequencing, identify gaps in our current knowledge, and highlight the need to implement new technologies in aquaculture. We then propose a set of priorities for future research in aquaculture genomics, genetics, and breeding.

Whole Genome Sequencing and Assembly

The genomes of several major aquaculture species in the United States, especially those under the USDA National Research Support Project 8 (NRSP-8), have been sequenced or are being sequenced (Table 1), including catfish [5], Atlantic salmon [6], rainbow trout [7], tilapia [8], striped bass (Reading, personal communication), Pacific oyster [9], eastern oyster (Gomez-Chiarri, personal communication), and Pacific white shrimp (Xiang, personal communication) as well as yellow perch and bluegill sunfish (Wang, personal communication). These accomplishments were achieved through support of USDA, NOAA, and other U.S. funding agencies. National Institute of Food and Agriculture (NIFA) AFRI programs, especially the Animal Genomics, Genetics and Breeding program, were central to the historical achievements of generating the reference genome sequences for these fish and shellfish species. Strong international collaboration was also important for the achievements. For instance, the genome project for the Pacific oyster was led by scientists from China and the U.S. [9]; the Atlantic salmon project was led by scientists from Norway, Canada, and Chile [6]; the rainbow trout project was led by scientists from France and currently is a collaborative international effort primarily between the U.S. and Norway; the genome project for Pacific white shrimp is led by Chinese scientists (Xiang, personal communication), and a reference genome for the original specific pathogen-free (SPF) broodstocks developed by the U.S. Marine Shrimp Farming Program in Oahu, HI is being generated (Alcivar-Warren, personal communication).

Table 1: Some examples of whole genome sequencing of aquatic and aquaculture species

Species	References
***Ictalurus punctatus*(Channel catfish)**	**Liu et al. 2016** [5]
***Ictalurus furcatus*(Blue catfish)**	**Waldbieser and Liu, unpublished data**
Oncorhynchus mykiss (Rainbow trout)	**Berthelot et al. 2014** [7]
***Salmo salar*(Atlantic salmon)**	**Lien et al. 2016** [6]
***Oreochromis niloticus*(Nile tilapia)**	**Brawand et al. 2014** [8]
***Crassostrea virginica*(Eastern oyster)**	**Gomez-Chiarri et al. 2015** [15] **& personal communication**
***Crassostrea gigas*(Pacific oyster)**	**Zhang et al. 2012** [9]
***Penaeus/Litopenaeus vannamei*(Pacific white shrimp)**	**Xiang, 2016, personal communication**
***Penaeus monodon*(Giant tiger prawn)**	**Warren, personal communication**
Atlantic Cod	Star et al. 2011 [142]
Bluegill sunfish	Wang, personal communication

California yellowtail	Severin, Purcell, Hyde, personal communication
Cavefish	McGaugh et al. 2014 [143]
Coelacanth	Amemiya et al. 2013 [144]
Common carp	Xu et al. 2014b [145]
Indian catfish	Das, personal communication
Japanse flounder	Chen, Yellow Sea Fisheries Institute, China, personal communication
Grass carp	Wang et al. 2015 [146]
Lamprey	Smith et al. 2013 [147]
Medaka	Kasahara et al. 2007 [148]
Pacific abalone	Severin, Purcell, Hyde, personal communication
Pearl oyster	Du, personal communication
Platyfish	Schartl et al. 2013 [149]
Rohu carp	Das, personal communication
Sea bass	Tine et al. 2014 [150]
Scallops	Bao, Ocean University of China, personal communication
Sea cucumber	Xiang, Chinese Academy of Sciences, China, personal communication
Shark	Venkatesh et al. 2014 [151]
Sole	Chen et al. 2014 [152]
Stickleback	Jones et al. 2012 [153]
Striped bass	Reading, 2016, personal communication
Tetraodon	Jaillon et al. 2004 [154]
Turbot	Figueras et al. 2016 [155]
White bass	Reading, 2016, personal communication
Yellow croaker	Wu et al. 2014 [156]
Yellow perch	Wang, personal communication
Zebrafish	Howe et al. 2013 [157]

Bold data are the species initially included in the NRSP-8 Project (Alcivar-Warren et al. 1997)

Various technologies have been used for the generation of the whole genome sequence of aquaculture species. However, the Illumina and PacBio platforms have contributed the most to the progress of aquaculture genome sequencing. Illumina sequencing generates accurate, but short reads at a relatively low cost, while PacBio sequencing generates longer, but less accurate reads at a higher cost. The proportion of sequences generated using these two platforms varies depending on the species and the status of the sequencing technology when the genome sequencing projects were initiated. A significant decrease in the cost of PacBio sequencing has generally led to increased use of this technology and the enhancement of contig/scaffold lengths in sequence assemblies.

Like the selection of sequencing technologies, various sequencing templates were used for the generation of the whole genome sequence assemblies in aquaculture species. These included mixtures of outbred individuals, single diploid males or females, individuals from inbred lines, and completely homozygous doubled haploids. The use of more homozygous templates greatly simplifies the computation of genome assemblies, which are complicated by the high levels of heterozygosity and sequence polymorphism characteristic of several aquaculture species [6, 7, 10]. The choice of sequencing templates has largely been dictated by the availability of the preferred homozygous templates. For instance, doubled haploids, produced through gynogenesis or androgenesis, are the preferred sequencing template for most teleost sequencing projects. However, the generation of doubled haploids is not generally feasible in many shellfish species, due to the unequal first cleavage that is sensitive to manipulation [11]. For some species, multiple individuals must be used because the DNA extracted from a single individual is not sufficient for the sequencing process.

While the mechanics of generating a large number of sequence reads is no longer difficult, calculation of a high quality of sequence assembly remains a challenging task. Four specific metrics are generally used to evaluate the quality of whole genome sequence assemblies including 1) *Contiguity*, as reflected in contig numbers and distribution of contig sizes; 2) *Connectivity*, as reflected in the number of scaffolds and distribution of scaffold sizes; 3) *Completeness*, as reflected in the total size of the genome assemblies and the percentage of coverage of the whole genome; and 4) *Accuracy*, as validated by at least one additional methodology such as genetic linkage mapping, physical mapping, or optical mapping. In addition, the integration of the whole genome sequences with genetic linkage maps is important for genetic studies.

The quality of current whole genome sequences as measured by these four metrics, varies among species. Quality measurements of the whole genome sequence assemblies of the aquaculture species are summarized in Table 2. In general, sequence assemblies of fish species are of higher quality than those for shellfish species. This is in part because the genomes of the shellfish species are highly heterozygous and contain a high level of repetitive elements. For instance, the oysters are among the most polymorphic animals; SNP density was estimated at 1.22 SNPs per 100 bp for the Pacific oyster [9] and either 1.85 SNPs [12] or 4.2 SNPs per 100 bp [10] at population levels for the eastern oyster. Moreover, repetitive elements account for over 80% of the shrimp genome (Xiang, personal communications).

Table 2: Status of whole genome sequencing and assembly of major aquaculture species in the United States, listed in the order of scaffold N50 sizes

Species	**Contig N50 [141]**	**Scaffold N50 (Mb)**	**Scaffolds**	**% on chromosome**	**Sequencing platform**	**Total size (Mb)**	**References**
Catfish	77.2	7.73	9974	97.2	Illumina, PacBio	783	Liu et al. 2016 [5]
				99.1			Zeng et al. 2017 [13]
Atlantic Salmon	57.6	2.97	843,055	75.4	Sanger, Illumina, PacBio	2970	Lien et al. 2016 [6]
Tilapia	29.3	2.80	-	70.9	Illumina, PacBio	928	Brawand et al. 2014 [8]
	3090	-		86.9		1010	Conte et al. 2016, PC
Eastern oyster	1.59	2.50	849	In progress	PacBio, Illumina	819	Wes Warren, PC
Rainbow trout	7.7	0.38		54.0	Illumina	1900	Brawand et al. 2014 [8] Palti and Gao, PC
	13.9	1.72		82.0		2178	
Zebrafish	25.0	1.55		96.5	Sanger, Illumina	1410	Howe et al. 2013 [157]
California yellowtail	139.3	1.49	4439	-	Illumina	685	Andrew Severin, PC
					PacBio		

Pacific white shrimp (Litopenaeus vannamei)	57.1	0.66	6007	71.6	Illumina PacBio	1779	Jianhai Xiang, 2016, PC
Pacific oyster	19.4	0.4	11,969	-	Illumina	559	Zhang et al. 2012b [9]
Striped bass	20.9	0.03	35,010	-	Illumina PacBio	585	Benjamin Reading, 2016, PC
White bass (male/female)	In process	In process	56,818/ 57,533	-	Illumina	644/643	Benjamin Reading, 2016, PC
Pacific abalone	In process	In process	-	-	Illumina	2000	Severin, Purcell, Hyde, PC
Yellow perch (male/female)	In process	In process	-	-	Illumina	1380/1240	Haping Wang, PC

Zebrafish is included as a reference. PC: personal communications

For species under the NRSP-8 program, reference genome sequence assemblies for catfish, tilapia, Atlantic salmon, and rainbow trout are of good quality. For catfish, 50% of the genome sequence is included in only 31 of the largest scaffolds; 90, 95, and 98% of the genome is included in 185, 314, and 594 scaffolds, respectively. The catfish reference genome sequence was assessed to be nearly complete as 99.7% of re-sequencing reads were mapped to the reference genome sequence. In addition, the number of complete genes included in the reference genome sequence is larger than that of any of the sequenced diploid fish species, including zebrafish [5]. The catfish reference genome sequence assembly was validated by genetic mapping. The positions of 253,744 genetically mapped SNPs were fully concordant with those on the reference genome sequence with four exceptions [13]. The vast majority of the reference genome sequence (99.1%) has been anchored to chromosomes [13].

The reference genome assembly of Atlantic salmon is also of high quality [6]. The genome was sequenced with Sanger and Illumina technologies. It is complete as 2.97 Gb reference genome sequences were assembled, with the unassembled sequences being just repetitive elements. The largest 9447 scaffolds accounted for 2.24 Gb of the 2.97 Gb genome sequence. This is a remarkable achievement considering the very complex nature of the genome. The Atlantic salmon genome is largely tetraploid due to a recent genome duplication. It also has a high repeat content (58–60%); the dispersed Tc1 transposons represented

12.89% of the genome [6]. Similarly, the assembly of the rainbow trout genome is of good quality [7]. Since the publication of the genome paper, the reference genome sequence of rainbow trout has been further improved. The contig N50 has increased from 7.7 Kb to 13.9 Kb, and the scaffold N50 has increased from 380 Kb to 1.72 Mb. More importantly, over 82% of the genome sequence has been mapped to chromosomes (Palti, personal communication).

The published tilapia genome sequence [8] was already of good quality, but the recent use of PacBio long sequencing technology allowed a new high quality assembly (Matthew Conte, personal communication). The contig L50 length reached 3.09 Mb, and 50% of the genome is included in the largest 93 contigs. Importantly, over 86.9% of the reference genome sequence is anchored to chromosomes, enhancing the utility of the reference genome sequence for genetic analyses. The whole genome sequences of striped bass, white bass, yellow perch, and bluegill sunfish are at the stage of draft assemblies.

A published genome sequence exists for the Pacific oyster, but the assembly is highly fragmented [9]. Efforts are ongoing to improve the genome assembly and contiguity, completeness, and accuracy are significantly better now (Zhang, personal communication). Linkage analyses were conducted to validate the genome sequence assembly [14]. The whole genome assembly of eastern oysters is at the draft sequence stage. Several strategies were employed to address challenges encountered in the assembly of the Pacific oyster genome. A single, highly inbred individual, produced through multiple generations of inbreeding and one generation of meiotic gynogenesis (Guo, personal communication) was used as a template. PacBio sequencing was used to provide 50x genome coverage in addition to Illumina sequencing ([15]; Gomez-Chiarri, personal communication). Initial statistics suggest this assembly is of much higher quality than that of the Pacific oyster. The draft assembly (Table 2) is now being validated using high-density linkage maps generated by the Guo laboratory.

The Pacific shrimp genome has been sequenced and assembled, but is not yet published. As shown in Table 2, the genome assembly is of high quality, with a contig N50 of 57.1 kb. The whole genome is included in 6007 scaffolds. Importantly, 71.6% of the genome sequence is anchored to chromosomes through linkage mapping (Xiang, personal communications). Of all the aquaculture genomes, the shrimp genome is perhaps the hardest to deal with because of the difficulty in isolating high molecular weight DNA due to enhanced DNase activity, the large chromosome number, and high levels of heterozygosity and repetitive elements. Physical mapping has been hindered by the lack of BAC libraries with very large inserts. The only BAC library of shrimp, pECBAC1, has an average insert size of approximately 101 kb [16].

For XY heterogametic species, often only the homogametic gender was used as sequencing template, and so information on the sex chromosomes is lacking. For instance, the catfish genome sequence was produced using a doubled haploid female produced through gynogenesis, and therefore the Y sex chromosome was not sequenced. Similarly, the Atlantic salmon genome was produced by using DNA template from a single double-haploid female produced by mitotic androgenesis. Therefore, the Y chromosome is not included in the reference genome. The rainbow trout genome was sequenced using a YY doubled haploid. While it provided Y chromosome information, the X chromosome was not covered in the reference genome sequence. Furthermore, sex determination in some fish and shellfish is complicated by having multifactorial sex determining mechanisms, including genetic sex determination (GSD), environmental sex determination (ESD) and their interactions. With WZ heterogametic species like some of tilapia species, sequencing a single representative of each gender may not be sufficient if there is a polygenic sex determination.

The first genome sequence is a historical milestone for any aquaculture species. However, in order to enable the utility of a reference sequence, additional work is required. For all aquaculture species, further refinement of the reference genome sequence, including improvements in contiguity, completion, and accuracy, as well as anchoring the reference genome sequence to chromosomes and obtaining sex chromosome sequences, is a priority (Table 3). Integration of genome sequence and linkage maps is also very important for genetic and breeding work, and can be accomplished relatively quickly. Sequencing of the Y or X chromosome is essential to study sex determining mechanisms, and sex-related traits, such as sexual dimorphism in growth or sexual size dimorphism (SSD). For instance, with tilapia and bluegill, males grow much faster and bigger than females. In contrast, females grow faster and bigger with yellow perch (Hanping Wang, personal communication). Such differences can be exploited as excellent natural models for the analysis of the genomic basis for sexual bimorphisms.

Table 3: Examples of additional work to enhance the utility of the whole genome reference sequences of major aquaculture species in the United States

Species	Contiguity, completion, and accuracy	Anchoring sequence to chromosomes	Sex chromosome sequencing
Catfish	+	+	Y chromosome need to be sequenced
Atlantic salmon	++	++	Y chromosome need to be sequenced
Tilapia	+	+	
Rainbow trout	++	+++	
California yellowtail	++	+++++	
Pacific oyster	+++	+++	
Striped bass	++++	+++++	
White bass	++++	+++++	
Eastern oyster	+++	++++	
Shrimp	+++	+++	
Pacific abalone	+++	+++++	

+ indicate some additional work required, and additional "+" signs indicate the level of additional work required; additional "+" signs indicate larger amount of improvements are needed

Genomic Variations, Polymorphic Markers, and Genotyping Platforms

Catalogues of genome variations and efficient genotyping platforms are essential to fully exploit whole genome sequences. One of the most useful by-products of whole genome sequencing is the development of thousands of DNA markers. In the first decade of aquaculture genome research, major effort was focused on developing polymorphic markers [17]. As whole genome sequencing projects were conducted, large numbers of polymorphic markers were identified. Whole genome sequencing with diploid sequencing templates allows identification of both microsatellites and SNPs. Analysis of SNPs between the two alleles of the sequenced individual also allow a rough assessment of the level of heterozygosity of the species.

In addition to whole genome sequencing, SNPs can be identified through genome re-sequencing or RNA-Seq projects. For instance, genome re-sequencing projects have identified more than 8.3 and 9.7 million putative SNPs in channel catfish [18] and Atlantic salmon [19] respectively. Large numbers of SNPs have been identified in most major aquaculture species, with those for the species under the NRSP-8 summarized in Table 4. SNP markers are a much-needed resource for genetic and genomic studies, the construction of high-density SNP arrays, and the development of high-density linkage maps. Validation and testing of these SNPs using SNP arrays will form the material basis for GWAS and whole genome-based selection.

Table 4: Some examples of SNPs identified from the aquaculture species under NRSP-8

Species	SNPs from genome sequencing	Numbers of SNPs	Method of identification	Reference
Catfish	None	8.3 million	Genome re-sequencing, transcriptome sequencing	Sun et al. 2014 [18]
				Liu et al. 2012 [158]
Rainbow trout	None	145,168	RAD sequencing	Palti et al. 2014 [159]
		5052	RNA-Seq	Christensen et al. 2013 [160], Al-Tobasei et al. 2016 [161]
		50,000	RNA-Seq	Palti et al. 2015 [23]
		1.8 million	Genome re-sequencing	
Atlantic salmon	None	9.7 million	Genome re-sequencing	Yáñez et al. 2016 [19]
Tilapia	Yes	3569	Genome re-sequencing	Van Bers et al. 2012 [162]
Striped bass	Yes	-	RNA-Seq	Li et al. 2014 [163]

Pacific oyster	Yes	3.8 million	Genome re-sequencing	Zhang et al., 2012 [9]
		4122	RNA-Seq	Hedgecock et al. 2015 [14]
Pacific white shrimp	Yes	96,040	RNA-Seq	Yu et al. 2014 [164]

Those SNPs identified from genome sequencing are not included here

A key advantage of SNP over microsatellite markers is the potential for rapid, low-cost genotyping. For many aquaculture species, the identification of large numbers of SNPs led to the development of efficient genotyping platforms. Available high-density SNP arrays for aquaculture species are listed in Table 5 and include the 15, 286, and 930 K Atlantic salmon arrays [6, 20, 21], the 250 and 690 K catfish arrays [13, 22], the 57 K rainbow trout array [23], and the 250 K common carp array [24]. The SNP arrays for each of the four aforementioned species have high marker densities and good genome coverage. SNP arrays need to be developed for tilapia, striped bass, oysters, and shrimp. As with genome assembly, the development of SNP arrays for some species (e.g., oysters and shrimp) is complicated by extremely high levels of polymorphism.

Table 5: Development of high density SNP arrays in aquaculture species, PC: personal communications

Species	**SNP array technology**	**SNP array density**	**References**
Atlantic salmon	Illumina iSelect technology	15 K	Gidskehaug et al. 2011 [20]
	Affymetrix Axiom technology	286 K	Houston et al. 2014 [21]
	Affymetrix Axiom technology	930 K	Lien et al. 2016 [6]
Catfish	Affymetrix Axiom technology	250 K	Liu et al. 2014 [22]
	Affymetrix Axiom technology	690 K	Zeng et al. 2017 [13]
Common carp	Affymetrix Axiom technology	250 K	Xu et al. 2014 [147]

Rainbow trout	Affymetrix Axiom technology	57 K	Palti et al. 2015 [23]
	Affymetrix Axiom technology	50 K	Salem et al. PC

Linkage Mapping and Physical Mapping

Ultimately, genomic information must be translated into genetic terms to facilitate genetic enhancement in aquaculture. Genetic linkage maps derived from genetic analysis of recombination during meiosis are important for the assembly of chromosome-scale sequence scaffolds. Mapping of sequence-tagged genetic markers derived from the reference genome allows sequence contigs to be arranged in an order that corresponds to the linkage group or chromosome. In addition, linkage mapping is a good method for validating reference genome assemblies.

Linkage maps have been constructed for most of the major aquaculture species (Table 6). For the species under NRSP-8, high density maps exist for catfish, Atlantic salmon, rainbow trout, tilapia, oysters, and shrimp. The linkage maps for catfish and salmonids have the highest marker densities, with the latest catfish linkage map ordering 253,087 markers [13], and the Atlantic salmon linkage map ordering 565,887 markers [6]. The latest linkage maps for the Pacific and eastern oysters have 3367 and 4316 markers, respectively [25] (Guo, personal communication). A large proportion of the genome sequence has been anchored to linkage maps in catfish (99.1%), tilapia (86.9%), Atlantic salmon (75.4%), Pacific shrimp (71.6%), and rainbow trout (54%).

Table 6: Examples of genetic linkage maps in aquaculture species, with the species under the NRSP-8 in bold

Species	Number and type of markers	Mapping population	Unique map positions	References
Asian seabass	790 microsatellites and SNPs	93 fish from two families	501	Wang et al. 2011 [165]
Atlantic salmon	5650 SNPs	3297 fish from 143 families	2894 in female genetic map, 1009 in male specific map	Lien et al. 2011[166]

Brown trout	288 micro-satellites, 13 allozymes	93 fish from 4 families	-	Gharbi et al. 2006 [167]
Catfish	54,342 SNPs	576 fish from three channel catfish families	15,598	Li et al. 2015[168]
	26,239 SNPs	288 interspecific backcross progenies	12,776	Liu et al. 2016[169]
	253,087 SNPs	465 fish from four channel catfish families	30,591	Zeng et al. 2017[13]
Common carp	28,194 SNPs	108 fish from one yellow river carp family	14,146	Peng et al. 2016 [170]
Eastern oyster	4607 SNPs	112 progenies from one family	4136	Guo, personal communication
European seabass	190 microsatellites, 176 AFLP, 2 SNP	50 fish from one Venezia Fbis family	-	Chistiakov et al. 2008 [171]
Grass carp	279 microsatellites and SNPs	192 progenies from two families	245	Xia et al. 2010 [172]
Japanese flounder	1268 microsatellites, 105 SNPs, 2 genes	45 offspring from one family	235 in male genetic map, 184 in female genetic map	Castaño-Sánchez et al. 2010 [173]
Pacific oyster	1172 SNPs and microsatellites	336 progenies from five families	1172 unique markers mapped	Hedgecock et al. 2015 [14]
			424 in consensus linkage map	
Rainbow trout	2226 microsatellites and SNPs	120 individuals from two unrelated doubled haploid lines	1366 in synthetic map	Guyomard et al. 2012 [174]
	47,939 SNPs	5716 fish	47,939 mapped to genome sequence scaffolds	Gonzalez-Pena et al. 2016 [26]; Palti personal communication

Scallop	3806 SNPs	96 progenies from one Farrer's scallop family	2983	Jiao et al. 2013 [175]
Sea bream	321 microsatellites, ESTs, and SNPs	50 individuals from one family	229	Tsigenopoulos et al. 2014 [176]
Giant tiger prawn	3959 SNPs	1024 offspring from seven black tiger shrimp family	-	Baranski et al. 2014 [177]
Pacific white shrimp	429 AFLP, 22 microsatellites	F2 cross of slow and fast growth parents, 43 shrimp	-	Andriantahina et al. 2013 [178]
	6146 SNPs	205 progenies from one Pacific white shrimp family	4650	Yu et al. 2015[179]
Tilapia	525 microsatellites, 20 genes	70 individuals from one family	435	Lee et al. 2005[180]
	401 microsatellites	95 individuals from two families	352	Liu et al. 2013[181]
Yellowtail	217 microsatellites	90 progenies from one family	105 in female genetic map, 83 in male genetic map	Ohara et al. 2005 [182]
	1480 microsatellites and 601 SNPs	94 offspring of one family	-	Aoki et al. 2015 [183]
	6275 SNPs	460 individuals from five wild families	-	Ozaki et al. 2016 [184]

The major issue for linkage maps of aquaculture species is resolution. While the number of markers on the high density SNP arrays is large, map resolution has been limited by the size of the mapping populations. In most cases, the number of samples used for genetic mapping was not very large, leading to a high level of marker stacking. The exceptions are the Atlantic salmon and rainbow trout where over 2000 and 5000 individuals, respectively, were used for linkage analysis, leading to a very high resolution of the linkage map [6, 26]. While the high fecundity of fish and shellfish species makes it possible to generate large mapping families, the major limitation for high resolution linkage mapping is funding, as genotyping costs are directly proportional to the sample sizes in linkage analysis.

Physical maps have been constructed for only a few aquaculture species (Table 7) including Atlantic salmon [27], tilapia [28], catfish [29, 30], rainbow trout [31], common carp [32], Asian seabass [33], Pacific oyster [34] and scallop [35]. Over time, BAC-based physical mapping has been replaced in favour of next generation sequencing and optical mapping technologies [36]. The existing physical maps and related BAC resources, however, are still useful for validation of reference genome sequences.

Table 7: Examples of physical maps constructed from aquaculture species

Species with physical maps	References
Atlantic salmon	Ng et al. 2005 [27]
Tilapia	Katagiri et al. 2005 [28]
Channel catfish	Xu et al. 2007 [30]
	Quiniou et al. 2007 [29]
Pacific oyster	Gaffney, 2008 [34]
Rainbow trout	Palti et al. 2009 [31]
Common carp	Xu et al. 2011 [32]
Pacific white shrimp	Yu et al. 2015 [179]
Asian seabass	Xia et al. 2010 [33]
Scallop	Zhang et al. 2011 [35]

Transcriptome Resources

Proper annotation of the genome sequences presents a challenge that can be at least partially overcome with transcriptome information. Specifically, gene models and gene structures need to be supported by experimental data; exon-intron borders need to be defined; alternatively spliced and differentially polyadenylated transcripts need to be identified and their translated proteins verified; and expression and function of the genes need to be studied. In addition to protein-coding genes, non-coding RNAs need to be identified and mechanisms of their target interactions need to be understood.

Large numbers of expressed sequence tag (EST) resources exist for major aquaculture species. As summarized in Table 8, almost a half million

ESTs were generated for Atlantic salmon, over 350,000 for channel catfish, and almost 290,000 for rainbow trout. These EST resources are useful for the assembly of full length transcripts for genome annotation; however, with the advent of low-cost next generation sequencing technologies, transcriptomes are now more efficiently characterized with RNA-Seq.

Table 8: EST resources of selected aquaculture species (with >10,000 ESTs)

Species	Number of ESTs
Danio rerio (zebrafish)	1,488,275
Ciona intestinalis	1,205,674
Xenopus laevis (African clawed frog)	677,911
Oryzias latipes (Japanese medaka)	666,891
Salmo salar (Atlantic salmon)	498,245
Ictalurus punctatus (channel catfish)	354,516
Oncorhynchus mykiss (rainbow trout)	287,564
Morone saxatilis (striped bass)	230,151
Crassostrea gigas	206,388
Litopenaeus vannamei	161,248
Ictalurus furcatus	139,475
Oreochromis niloticus (Nile tilapia)	120,991
Petromyzon marinus (sea lamprey)	120,731
Sparus aurata	79,216

Zebrafish is included as a reference

Large RNA-Seq datasets have been generated by various institutions for important aquaculture species in the United States (https://www.ncbi.nlm.nih.gov/sra) to characterize differentially expressed genes in response to disease or stress in catfish [37, 38, 39, 40], disease in salmon [41, 42] and to identify markers associated with growth, heat stress, and disease and tissue specificity in rainbow trout [43, 44, 45, 46]. In striped bass, RNA-Seq studies focused on reproduction traits and egg quality [47, 48, 49], while in tilapia, they were conducted to identify genes responsive to alkalinity stress [50], salinity adaptation [51], and adaptation to low or high fat diets [52]. In yellow perch and bluegill, RNA sequencing of neo-males (perch), neo-females (bluegill), regular males and regular females is being conducted to investigate epigenomic modification of SSD and sex determination in fish (Wang, personal communication). RNA-Seq studies have also been conducted to characterize the Pacific oyster response to environmental stress (e.g., temperature, salinity, air exposure and heavy metals) [9, 53, 54, 55] and Ostreid herpesvirus [52]. In eastern oysters, RNA-Seq studies identified genes associated with osmoregulation [12], characterized the transcriptomic response to a bacterial pathogen [56], and revealed extensive expansion of gene families associated with innate immunity [15, 57]. In shrimp, genes associated with early development [58] and resistance to Taura syndrome virus (TSV) [59] have been identified via RNA-Seq analysis, and improved shrimp transcriptome were reported [60]. When coupled with genetic analysis such as bulk segregant analysis (e.g., [38, 43]), transcriptome analyses using RNA-Seq will enable the identification of candidate genes for important aquaculture traits.

Transcriptome resources also empower proteomics analysis [48, 61, 62]. Proteomics offers great promise for advancing our understanding of the functions of genes that underlie important production traits, however these methods rely on existing homologous protein-coding sequence databases, which remain incomplete for many non-model organisms, including important aquaculture species. Tandem mass spectrometry approaches in proteomics use these databases to identify protein fragments by mass spectrometry and thus require amino acid (or protein-coding nucleic acid) sequence information, optimally from the research organism under investigation. Thousands of different proteins have already been identified and measured with tandem mass spectrometry approaches to answer important questions about reproduction in striped bass and the closely related white perch, which serves as a research model [61, 62, 63, 64]. A similar proteomic approach identified important proteins related to muscle atrophy in rainbow trout [65].

Non-Coding Transcripts, Regulation of Genome Expression, And Epigenomics

Despite their importance in regulating gene expression, non-coding transcripts are much less understood than protein-coding transcripts in aquaculture species. Limited work has been conducted in this relatively new area of research. Among aquaculture species, most of the work on non-coding RNAs was conducted in rainbow trout. A few studies were devoted to identification of microRNAs and long non-coding RNAs [66, 67, 68, 69, 70, 71]. In a number of cases, microRNAs were found to be associated with performance traits. For instance, a large number of microRNAs were differentially expressed between sexually mature and immature fish; in association with egg quality and muscle growth and quality [72, 73, 74]. In addition, differential expression of long non-coding RNAs studied in three genetic lines of rainbow trout identified important long-coding RNAs in response to infection with *Flavobacterium psychrophilum* [75].

In Atlantic salmon, several studies were conducted to characterize the microRNA repertoire. In one study, Bekaert et al. [76] identified 888 microRNA genes. In another study, Andreassen et al. [77] identified a total 180 distinct mature microRNAs, and found that many microRNAs were conserved across species, and a few microRNAs were expressed in a tissue-specific fashion. In another study, Kure et al. [78] found that 18 microRNAs were differentially expressed upon exposure to acidic aluminium-rich water.

Research on non-coding RNAs in catfish, striped bass, tilapia, oysters, and shrimp is limited. For instance, residue microRNA profiling was reported in catfish [79, 80, 81], tilapia [82], oysters [83, 84], and shrimp [85, 86]. However, now with the high quality reference genome sequences, it is expected that large numbers of projects will be conducted with aquaculture species in this area. This aligns very well with the FAANG (Functional Annotation of Animal Genomes) Project. As the importance and detailed operational protocols are well discussed in the white paper published in Genome Biology [87], we will not repeat them here, but this will be an important area for future research with aquaculture species as well, especially those with a well assembled reference genome sequence.

Genome scale analysis of epigenetic regulation have been conducted with oysters [88, 89, 90, 91, 92, 93], Atlantic salmon [94], rainbow trout [95, 96], and tilapia [97], yellow perch, bluegill (Wang, personal communication) and additional projects are being initiated in several other major aquaculture species. Apparently, this is an area of active research, and functional annotation of non-protein coding genome elements is an important area. Again, this aligns well with those objectives of the FAANG Project [87].

Performance Traits, Phenotypic Variations, and QTL Analysis

The practical purpose of aquaculture genomics and genetics studies is to reveal the genetic basis of performance and production traits, and to use such information for genetic enhancement programs. Domestication of most aquaculture species is still in the early stages, occurring over the last few decades, compared to other food animals and crops which have been domesticated over hundreds or even thousands of years. Because of this short history of domestication, aquaculture species still segregate considerable genetic variation among strains, lines, families and individuals.

Many aquaculture phenotypes are complex and quantitative in nature. Therefore, a major goal of aquaculture genetics research is to leverage genome information to predict complex phenotypes. In aquaculture species, QTL mapping and GWAS analysis are well-established procedures for correlating genetic and phenotypic variation; however additional work is required to identify specific genetic variants responsible for phenotypic variations. The identification of the causal SNPs or the genes underlining the performance traits is not only important for aquaculture applications, but also important for understanding the molecular mechanisms of phenotypic expression.

Progress with QTL/GWAS analysis has been greatly accelerated by the application of SNP arrays. Some examples of QTL mapping and GWAS analysis in aquaculture species are listed in Table 9. Most of the work has focused on disease resistance, growth traits, tolerance to stresses, and development or sexual maturity. Some of the best examples of QTL studies are from salmon research. For instance, the resistance against infectious pancreatic necrosis (IPN) virus was mapped to a major QTL that account for vast majority of phenotypic variance [98, 99], and further analysis identified the causal gene as epithelial cadherin [100, 101]. In catfish, QTL have been identified for a number of traits including disease resistance [38, 102, 103], heat stress [104], hypoxia tolerance [105], and head size [106]. In most of these cases, QTL were mapped within a region smaller than one million base pairs, allowing speculation of candidate genes, but fine mapping will be required to identify the specific causal genes. An interesting finding of these studies is the identification of functional hubs [102, 106] linking genes with roles in the same pathway. In addition, there appears to be a high level of evolutionary conservation of genes responsible for a number of traits in various species ranging across mammals, amphibians, and fishes. For instance, genes involved in the small GTPase pathway were found to affect head size and shape in catfish, frogs, mouse, and dogs [107]. Such discoveries open the possibility of comparative quantitative genomics.

Table 9: QTL studies in selected aquaculture species with major US aquaculture species in bold

Species	Traits	Reference
Arctic charr	Body weight and sexual maturation; Salinity tolerance	Küttner et al. 2011 [185]
		Norman et al. 2011 [186]
Asian seabass	Resistance against viral nervous necrosis disease	Liu et al. 2016 [187]
	Growth-related traits	Wang et al. 2006 [188]
	Omega-3 fatty acids	Xia et al. 2014 [189]
Atlantic salmon	Growth traits and flesh colour	Baranski et al. 2010 [190]; Tsai et al. 2014 [191]; 2015 [192]; Moen et al. 2009 [99]; 2015
	Resistance against IPN	[101]; Houston et al. 2008 [98]; 2010 [100]
	Late sexual maturation	Gutierrez et al. 2014 [193]
	Resistance to pancreas disease	Gonen et al. 2015 [194]
Catfish	Columnaris disease resistance	Geng et al. 2015 [102]
	ESC disease resistance	Wang et al. 2013 [38]; Zhou et al. 2017 [103]
	Hypoxia tolerance	Wang et al. 2016 [105];
	Heat stress	Jin et al. 2016 [104]
	Head size	Geng et al. 2016 [106]

Common carp	Muscle fiber traits	Zhang et al. 2011 [195]
	Morphometric traits	Boulton et al. 2011 [196]
	Swimming ability	Laghari et al. 2014 [197]
Eastern oyster	Disease resistance	Yu and Guo, 2006 [110]
European sea-bass	Growth, body weight	Louro et al. 2016 [198],
	Morphometric traits and stress response	Massault et al. 2010 [199]
Pacific white shrimp	Growth parameters	Andriantahina et al. 2013 [178]
Giant tiger prawn	Disease resistance and sex determination	Robinson et al. 2014 [200]
Japanese flounder	Vibrio anguillarum resistance	Wang et al. 2014 [201]
Pacific oyster	Growth	Guo et al. 2012 [112]
	Resistance against summer mortality	Sauvage et al. 2010 [202]
	Viability	Plough & Hedgecock, 2011 [111]; Plough et al. 2016 [113]
Gilthead seabream	Skeletal deformities	Negrín-Báez et al. 2015 [203]
	Sex determination and body growth	Loukovitis et al. 2011 [204]
	Resistance to fish pasteurellosis	Massault et al. 2011 [205]

Rainbow trout	Growth related traits	Kocmarek et al. 2015 [206]; Wringe at al., 2010[207]; Leder at al., 2006 [208]; Easton et al. 2011[209]; Miller et al. 2012 [210]
	Spawning time; development rate	
	Upper thermal tolerance	Perry et al. 2005 [211]
	Whirling disease resistance	Baerwald et al. 2011 [212]
	Bacterial cold water disease resistance	Vallejo et al. 2014 [107]; Palti et al. 2015 [108]; Liu et al. 2015 [109]; Campbell et al. 2014 [213]
	IHNV disease resistance	Rodriguez et al. 2004 [214]; Campbell et al. 2014[213]
	Fillet yield	Gonzalez-Pena et al. 2016 [26]
	Osmoregulation capacity	Le Bras et al. 2011 [215]
	Response to crowding stress	Rexroad et al. 2013 [216]; Liu et al. 2015 [217]
Turbot	Growth traits	Sánchez-Molano et al. 2011 [218]
	Aeromonas resistance	Rodríguez-Ramilo et al. 2011 [219]
	Resistance against Philasterides	Rodríguez-Ramilo et al. 2013 [220]
	Resistance to viral haemorrhagic septicaemia	Rodríguez-Ramilo et al. 2014 [221]
Tilapia	Growth traits	Liu et al. 2014 [222]; Wang et al. 2015 [223]
	Sex	Palaiokostas et al. 2015 [224]

Similarly, QTL have been identified for growth and reproductive traits, upper thermal tolerance, osmoregulation capacity, stress responses, and disease resistance in rainbow trout. Significant efforts have been devoted to the analysis of resistance to bacterial cold water disease (BCWD) [107, 108, 109]. QTL analysis and genome selection for BCWD resistance are facilitating significant genetic improvement for this trait in rainbow trout [3, 4].

Although QTL have been identified for disease resistance, viability and growth-related traits in eastern and Pacific oysters [110, 111, 112, 113], low marker density limits QTL resolution. Candidate gene-based studies have led to the identification of variation in a serine protease inhibitor associated with *Perkinsus marinus*-resistance in the eastern oyster [114]. QTL analysis in tilapia, striped bass, and shrimp are at the early stages, but with the efficient genotyping systems, rapid progress is expected.

Aside from lack of genetic and genomic resources (e.g., inbred lines/families, sequenced genomes, efficient genotyping platforms) in some aquaculture species, several additional challenges face aquaculture researchers. First, unlike many livestock species where phenotypic and genotypic data can be collected on a large proportion of the cultured animals, phenotypic and genotypic data collection on the entire population of an aquaculture species is impossible. It is therefore essential that aquaculture geneticists understand QTL in all strains used in the industry, because a QTL present in one population may not be present in another. Second, fish and shellfish are outbred species with extremely large numbers of founders. Their high fecundities make QTL analysis within families extremely efficient, but whether the identified QTL are conserved across families, strains, and populations are unknown.

Genome-Based Technologies and Regulatory Framework

A number of technologies, including polyploidization, gynogenesis, androgenesis, sex reversal, gamete cryopreservation, and gene transfer, are still very useful for aquaculture breeding programs. There are opportunities for enhancing these technologies by using genomic information. At the same time, genomic research has generated new technologies that can be used for genetic enhancement of aquaculture species, including marker-assisted selection (MAS), genome selection (GS), and genome editing.

Marker-assisted selection has been successfully used in aquaculture. The best example of MAS in an aquaculture species is selection for disease resistance in Japanese flounder. A microsatellite locus, Poli9-8TUF, was mapped near the major QTL for resistance to lymphocystis disease. Additional analysis indicated that the disease resistance was controlled by a single gene, and that the resistance allele was dominant. Based on the marker linkage information, Fuji et al. [115] developed a new population of Japanese flounder using MAS with the marker Poli9-8TUF. They selected a female homozygous for the favourable allele (B-favourable) and a male with a higher growth rate and good body shape, but without the resistant allele as parents. All the progeny are heterozygotes with the resistance allele and entirely resistant to lymphocystis disease, while the control group without B-favourable alleles showed incidences of 4.5 and 6.3% of mortality due to lymphocystis disease. These results clearly demonstrate that MAS is an efficient strategy for breeding [116].

Another good example of MAS is the selection of IPN resistance in Atlantic salmon. One major QTL was mapped to linkage group 21, which accounts for 29% and 83% of the phenotypic and genetic variances, respectively. Three microsatellite markers were tightly linked to the QTL, and these markers have been used for the selection of IPN resistance [99]. Recently, the gene responsible for IPN resistance was identified as a cadherin expressed in the epithelium where the protein binds to IPNV virions [101]. Marker-assisted selection allowed production of IPN-resistant salmon, leading to a 75% reduction in the number of IPN outbreaks in the salmon farming industry [101].

Sex identification using sex markers is a special case of MAS. Sex markers have been developed and used in quite a few aquaculture species, including common carp [117], tilapia [118], catfish [119], zhikong scallop [120], half-smooth tongue sole [121], white shrimp [122], kuruma prawn [123], yellowtail [124] and rainbow trout [125]. These sex-linked markers have been useful for the identification of sex without phenotypic data.

Recent advances in genome analysis including the availability of a large number of polymorphic markers, highly efficient genotyping platforms such as SNP arrays, and the application of next generation sequencing technologies, allowed mapping of dense markers across the entire genome, which in turn enables an estimation of the genetic merit of every chromosome fragment contributing variation in a population with phenotypic observations. Not only can the merit of every chromosomal segment be estimated, but also all the traits of interest can be estimated simultaneously. Whole genome selection is based on estimating the value of every chromosomal fragment contributing variation in a population with phenotypic observations (Training), and then the results of training are used to predict the merit of new animals (Testing) that are not included in the training dataset.

Genome selection was first proposed by Meuwissen et al. [126]. Since then it has gained tremendous attention in the animal genetics community. Compared with MAS, genomic selection uses the estimated effect of many loci across the entire genome at once, not just the small number of linked loci as done with MAS. Although genome selection has been successfully used in dairy cow and beef cattle and other livestock species [127], its use in aquaculture species has been limited to just a few species [128, 129]. In rainbow trout, genome selection was carried out for the selection of bacterial cold water disease [3]. In Atlantic salmon, genome selection was used to predict breeding values for resistance to sea lice [130]. Although demonstrated to be effective, genome selection has not been commercially applied in aquaculture species primarily due to financial limitations.

Supervised machine learning is similar in concept to whole genome selection using Training and Testing datasets, and includes Support Vector Machines (SVMs) and Artificial Neural Networks (ANNs). These are systems that can be trained to recognize certain data input patterns and then can be used to predict outcomes or classify data. Machine learning has been used to classify transcriptome and proteome data by pattern recognition (expression "fingerprinting") in an analytical bioinformatics approach [49]. Expression patterns of genes and proteins can be modelled to identify the most important ones contributing to a trait or response. Machine learning ANNs have been used to analyze tens of thousands of expressed genes in microarray and RNA-Seq studies to show that the collective changes in the expression of 233 ovary genes (less than 2% of the genes measured) explained over 90% of the variation in striped bass embryo survival [47, 49]. These trained ANNs also predict, with a correct classification rate over 80%, which female striped bass will produce fertile or infertile eggs based on gene expression profiles of ovary tissues sampled prior to ovulation. Additionally, SVMs have been used to model the striped bass ovary proteome (355 proteins) and this system can predict the specific ovary growth stage with 83% accuracy based on quantitative tandem mass spectrometry data [61]. A portion of the plasma proteome (94 proteins) also has been similarly modelled to accurately predict gender of white perch [131]. Therefore, machine learning additionally poses a potential use as a diagnostic tool, for example in identifying those females that will produce poor quality eggs, or determining reproductive state or gender. Future applications of machine learning could include modelling genomic markers, such as SNPs, to identify those most important to a particular trait and then to predict the future performance of an individual based on the presence or absence of those SNP markers.

Genome editing refers to the ability to make specific changes at targeted genomic sites [132]. With the initial zinc finger nuclease (ZFN) technology developed in 1996, genome editing technologies have evolved and become more and more efficient, with the development of TALEN (transcription activator-like effector nucleases) and CRISPR/Cas9 (clustered regulatory interspaced short palindromic repeats). These new genome editing technologies overcome the disadvantages of ZFN technology and they have become very efficient for the modification of genomes. CRISPR/Cas9 has been demonstrated to be very efficient in zebrafish [133, 134], tilapia [135] and catfish (Liu, unpublished data). Mutation rates of 70–100% can be achieved with very low levels of mosacism in channel catfish [136].

Genome editing technologies can be used to introduce an immediate improvement in a phenotype in a single generation; hence, these technologies hold great promise for improving aquaculture. However, genetically modified organisms (GMO) have encountered low public acceptance, especially with aquaculture species. As demonstrated with the lengthy approval process of AquAdvantage transgenic Atlantic salmon, decades of time and millions of dollars were spent in coping with the regulatory issues ([137]; Hackett, 2016, personal communications). It could be argued that genome editing technologies differ from traditional gene transfer technologies because no foreign DNA is introduced. The scientific community must be proactive of research in the area of regulatory issues and public perception. Escaping the GMO label is possible with genome editing. For example, recently, the USDA decided that a CRISPR-modified mushroom can be cultured and sold without passing through the agency's regulatory process [138].

Leveraging Investments in Genomics through Integration with Germplasm Repositories

Rapid development and adaptation of genomic tools among various aquatic species is a double-edged sword in terms of how such tools may create genetic diversity and thereby limit industry options in the future. Evidence of such contractions have been demonstrated with livestock and in particular the Holstein cow and how gene banks can facilitate the alleviation of genetic bottlenecks [139, 140]. While genomic research continues to rapidly proceed among various aquaculture species, there are some major technological gaps preventing the aquaculture sector from securing and utilizing improved genetic resources. As shown in other life forms such as livestock species, there is a critical need to understand and acquire genetically diverse samples from all major aquatic species, cryopreserve those samples, and to present them in publically available databases for viewing of information about the sampled populations. Such information would include phenotypes, management system descriptors, environmental conditions, locality data, and comprehensive genomic information. The Animal-GRIN information system operated by USDA/ARS is designed in this manner and is publically accessible via the internet. Acquiring and integrating this wide range of data not only serves to make germplasm and tissue samples more useful in the present, it will also allow researchers to perform studies not foreseen today and to respond to future challenges such as disease outbreaks or losses of critical genetic diversity in cultured lines. Viewing of genetic resources (via germplasm) and its associated detailed information as a public resource serves to speed innovation, as well as to leverage the considerable investments being made in genomic research. In essence this affords us new and more cost-effective approaches to produce, maintain, and distribute genetic improvement across the breadth of cultured aquatic species [141].

To respond to these needs, there is a requirement to collect and cryogenically store gametes and tissues from a wide range of species that can be used by industry members and public researchers alike. Coupled with these samples should be the ability to store genomic information from publicly funded research, as well as from industry. Such an information system would link samples with genomic, phenotypic, locality (GIS-based), and environmental descriptors and make this information publically available through a user interface via the internet. Aquatic species researchers could use this resource for varied experimental purposes (e.g., of crossing spring and fall spawning populations) and for corrective mating. As such, the collection and curation of germplasm or tissue samples has value, just as does the determination of genomic information. It is the purposeful integration of these genetic and informational resources that provides a synergistic leveraging or expansion of value and potential utility. Indeed, the value of information or germplasm samples is directly magnified by their coupling or association in a comprehensive repository system.

Future Research Priorities

Economically important aquaculture species are a diverse group of organisms and research priorities vary depending on the unique biology of each species. Although fishes are the most diverse vertebrate group, aquacultured teleosts are similar enough phylogenetically and biologically that they can follow a similar research program. Invertebrates are not as uniform, and may each have special properties that require different approaches.

The research tasks needed to develop a program of genetic enhancement in aquaculture species can be divided into two phases (Fig. 1). The first phase, development of species-specific genomic resources, is the one that has been pursued for the major aquaculture species over the past 20 years. It includes the development of genetic and physical maps, annotated genome sequences, and platforms for high-throughput genotyping. Some species (e.g., catfish, tilapia, rainbow trout, and salmon) may be nearing the completion of Phase I. Other species are just beginning this phase, hopefully benefitting from the experience of other species, and taking shortcuts available with new technologies. From the perspective of genetics and breeding, generation of complete sets of heritabilities and genetic correlations is also needed.

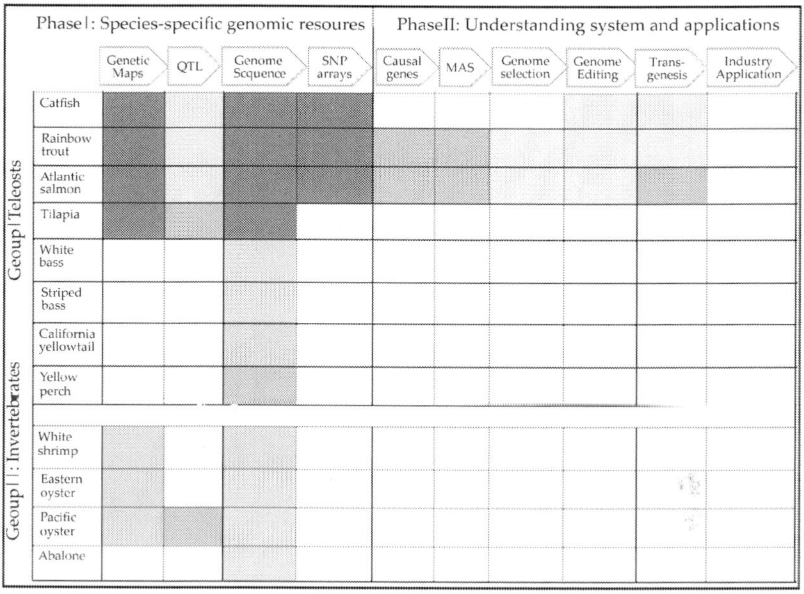

Figure 1: Schematic presentation of the goals and current status of aquaculture genomics and genetics research. The major aquaculture species in the United States are grouped into teleost fish and invertebrate species, with the species names listed in the first column. Major milestones of research goals are listed in the first row, while current status for each species is indicated in the appropriate cells with various colors: *Dark green*: good status; *light green*, outstanding progress has been made, but additional work still needed; *dark yellow*: significant progress has been made, but significant amount of additional work still needed; *light yellow*, some progress has been made.

Within Phase I, we can make a distinction between development of resources and application of resources to commercial aquaculture. For instance, genetic maps enable QTL/MAS, and genome sequences enable genomic selection. Application of MAS/genomic selection is generally beyond the scope and funding of the academic laboratories that have participated in the development of the genomic tools that enable it, although not beyond the scope and mission of government laboratories and academic laboratories that take on genetic stock enhancement for smaller or regional aquaculture species.

In Phase II, the genomic resources developed in Phase I can be used to develop a functional understanding of animal systems. As an example, a number of laboratories are working to develop an understanding of the gene regulatory network underlying sex determination in fishes. The research program involves not only RNA-Seq to characterize patterns of gene expression in the developing gonad, but also CRISPR modifications to test linkages in the gene regulatory network model. It should be possible to develop an understanding of the gene network underlying sex determination that will be broadly applicable among aquaculture species. Similar research programs are underway to understand the genetic basis of other important traits including growth, disease resistance, etc. In each case, the goal is to develop an understanding of animal systems that can be easily transferred to related species.

When this more detailed understanding of animal systems is complete, it will become possible to make specific genetic modifications (e.g., using CRISPR) to improve animals for commercial production. The safety and effectiveness of such modifications are important topics for research, but commercial application of these technologies will require stable business models and well established regulatory frameworks that ensure the safe application of these technologies to the species being targeted, the public, and the environment.

The current status of breeding technologies in US commercial aquaculture is summarized in Table 10. With significant differences in the structures of the aquaculture industries among species, practical strategies suitable to specific situations must be developed. In addition, development of comprehensive germplasm repositories will ensure protection of valuable genetic resources of aquaculture species and the investments made in developing them.

Table 10: Current status of breeding technologies in U.S. commercial aquaculture

Species	Status
Catfish	Private sector efforts to conduct genetic enhancement programs appear to have been successful, but the private sector has not made a great effort in genetics and breeding. Currently, some on-farm selection is practiced, but not in a very controlled manner. Genetic improvement is primarily conducted by public sector research programs, which has resulted in 7 releases to the industry of varying impacts. Most of these fish populations were developed by mass selection and in some cases family selection with the most emphasis on growth rate. Advanced genomic tools and technologies are available but have yet to be implemented by industry.
	The industry has widely adopted the channel female x blue male interspecific catfish hybrid which demonstrates significantly greater performance for numerous traits in comparison to the traditionally grown channel catfish with hybrids now comprising 60–70% of the industry. The vast majority of hybrids are produced with a single line of blue catfish.
Atlantic salmon	Private sector breeding is integrated with a publicly funded research program. Genetic improvement is based on quantitative genetics to improve growth, fillet quality and disease traits. Due to international interest in this species advanced genome tools and technologies are widely available, their implementation in the U.S. was recently initiated in a public/private partnership with efforts to incorporate MAS for sea lice resistance.
	In 2015 the AquAdvantage Salmon was approved for sale in the U.S. by FDA, however it is expected to reach the marketplace in 2017.
Rainbow trout	Public sector breeding programs utilize quantitative genetics to select for growth performance and disease resistance in all-female populations. Chromosome set manipulation is used to provide all-female triploids for net pen operations that require sterile fish; they are also valued for their superior growth characteristics at larger sizes.
	Publically funded research programs have released germplasm improved for growth and disease resistance characteristics. Advanced genome tools and technologies are widely available and have been implemented into the private sector. Proof of concept studies for genomic selection for disease resistance in a research population have motivated initial implementation in a commercial breeding population.

Tilapia	Private sector family based breeding for Nile tilapia for improved growth, yield and disease resistance is enhanced through publicly funded research programs. Although genome tools and technologies are available, they have not yet been implemented by the private sector.
Striped bass	Private sector fingerling producers incorporate germplasm from wild caught and captive (domestic) populations. Significant genetic improvement has been achieved through the production of hybrids created primarily by crossing domestic striped bass males x domestic or wild caught white bass females, with parental species improvement achieved primarily via mass selection techniques. Genomic technologies are under development and have not yet incorporated into commercial breeding, although domestic striped bass and white bass are available through a publically funded research program.
Oysters	The Pacific oyster industry is supported through public and private programs for ploidy manipulation, family-based selection and crossbreeding. Polyploid and improved broodstocks are widely used by the U.S. West Coast industry. Genetic improvement of the eastern oyster is publically funded. For much of the past 40 years, improvements in eastern oyster growth and survival have been realized using mass-selection techniques; however, there has been a recent shift toward applying quantitative genetics and ploidy manipulation to enhance production traits. Broodstock from these breeding programs are widely used by the private sector in the Northeast and Mid-Atlantic. Genome tools for both oyster species are coming online, but have not yet been implemented.
Shrimp	Shrimp breeders in the public and private sector selectively breed to produce specific pathogen resistant shrimp.

CONCLUSIONS

Based on the current status, trends, and industry needs of aquaculture genomics, genetics and breeding research, the following areas of research need to be priorities:

Phase I goals for each species

- Highly contiguous and complete genome sequence
- Full annotation of the genome sequence, including functional (genome to phenome) studies
- Identification of genetic variants in different broodstocks, and their relationship to performance traits
- Development of systems for high-throughput genotyping

- Anchoring of the genome sequence to genetic maps
- Identification of QTL for performance and production traits
- Bioinformatic capabilities to manage these data
- Training the next generation of aquaculture breeders
- Establishment of high-throughput cryopreservation protocols and pathways for aquaculture species

Phase II goals for each group of species

- Proof of concept demonstrations which apply genome technologies to improve production efficiency, production sustainability, animal welfare and/or product quality in the commercial sector
- Development of standardized measures of organismal phenotypes
- Understanding epigenetic effects that contribute to variation in gene expression
- Validation of QTL, identification of the causative genetic variants underlying variations in performance, and determine the mechanisms of actions
- Determine general and specific combining abilities in both intraspecific and interspecific systems
- Marker-assisted selection and genome selection for production traits
- Characterization of the gene regulatory networks underlying phenotypic traits important to commercial aquaculture production
- Determine the genomic basis of heterosis and genomic predictors of heterosis
- Identification of conserved regulatory mechanisms and pathways for growth, feed conversion efficiency, disease resistance, stress tolerance, sex and other traits among aquaculture species
- Development and application of gene editing technologies and the associated regulatory frameworks, first for basic research, and eventually for commercial production

- Development of tools that can be easily used by the industry
- Industry applications of genome technologies
- Establishment of a comprehensive germplasm repository system to protect, maintain and distribute genetic resources developed through genomic technologies

NOTES

Acknowledgements

We thank Dr. Lakshmi Matukumalli and Dr. Gene Kim, national program leaders for Animal Genomics, Genetics and Breeding program, and Aquaculture Program, respectively, for their participation in the workshop, and their encouragement for writing this paper. We are grateful to the staff assistance provided by the College of Agriculture and the Office of the Vice President for Research and Economic Development at Auburn University during the workshop.

Funding

The Aquaculture Genomics, Genetics and Breeding Workshop held in Auburn, Alabama, USA was supported partially by a part of a competitive grant from the Animal Genomics, Genetics and Breeding Program of the USDA National Institute of Food and Agriculture (#2015-67015-22907), and partially by USDA NRSP-8 Aquaculture Coordinator's funds.

Authors' Contributions

All authors are signatories of the Aquaculture Genomics, Genetics and Breeding Workshop; they have contributed to its conception and in drafting of the manuscript. ZL is the organizer of the workshop and provided the very initial draft. MG-C, XG, TK, YP, DP, BR, CR, SR, and MS wrote the first draft and contributed to initial revisions. All authors were involved in subsequent rounds of revisions, and read and approved the final manuscript. Authors are listed alphabetically based on last names.

REFERENCES

1. Alcivar-Warren A, Dunham R, Gaffney P. First aquaculture species genome mapping workshop. Anim Genet. 1997;28(6):451–2.
2. Tsai H-Y, Hamilton A, Tinch AE, Guy DR, Bron JE, Taggart JB, Gharbi K, Stear M, Matika O, Pong-Wong R. Genomic prediction of host resistance to sea lice in farmed Atlantic salmon populations. Genet Sel Evol. 2016;48:47.
3. Vallejo RL, Leeds TD, Fragomeni BO, Gao G, Hernandez AG, Misztal I, Welch TJ, Wiens GD, Palti Y. Evaluation of genome-enabled selection for bacterial cold water disease resistance using progeny performance data in rainbow trout: Insights on genotyping methods and genomic prediction models. Front Genet. 2016;7:96.
4. Vallejo RL, Leeds TD, Gao G, Parsons JE, Martin KE, Evenhuis JP, Fragomeni BO, Wiens GD, Palti Y. Genomic selection models double the accuracy of predicted breeding values for bacterial cold water disease resistance compared to a traditional pedigree-based model in rainbow trout aquaculture. Genet Select Evol. 2017;49(1):17.
5. Liu Z, Liu S, Yao J, Bao L, Zhang J, Li Y, Jiang C, Sun L, Zhang Y, Zhou T et al. The channel catfish genome sequence provides insights into the evolution of scale formation in teleosts. Nat Commun. 2016;7:11757.
6. Lien S, Koop BF, Sandve SR, Miller JR, Kent MP, Nome T, Hvidsten TR, Leong JS, Minkley DR, Zimin A, et al. The Atlantic salmon genome provides insights into rediploidization. Nature. 2016;533(7602):200–5.
7. Berthelot C, Brunet F, Chalopin D, Juanchich A, Bernard M, Noël B, Bento P, Da Silva C, Labadie K, Alberti A. The rainbow trout genome provides novel insights into evolution after whole-genome duplication in vertebrates. Nat Commun. 2014;5:3657.
8. Brawand D, Wagner CE, Li YI, Malinsky M, Keller I, Fan S, Simakov O, Ng AY, Lim ZW, Bezault E, et al. The genomic substrate for adaptive radiation in African cichlid fish. Nature. 2014;513(7518):375–81.
9. Zhang G, Fang X, Guo X, Li L, Luo R, Xu F, Yang P, Zhang L, Wang X, Qi H, et al. The oyster genome reveals stress adaptation and complexity of shell formation. Nature. 2012;490(7418):49–54.

10. Zhang L, Guo X. Development and validation of single nucleotide polymorphism markers in the eastern oyster *Crassostrea virginica* Gmelin by mining ESTs and resequencing. Aquaculture. 2010;302(1):124–9.
11. Guo X, Wang Y, Xu Z, Yang H. Chromosome set manipulation in shellfish, New Technologies in Aquaculture: Improving Production Efficiency, Quality and Environmental Management. 2009. p. 165–94.
12. Eierman LE, Hare MP. Transcriptomic analysis of candidate osmoregulatory genes in the eastern oyster *Crassostrea virginica*. BMC Genomics. 2014;15(1):1.
13. Zeng Q, Fu Q, Li Y, Liu S, Waldbieser G, Yang Y, Bao L, Yuan Z, Li N, Liu Z. Development of 690 K SNP arrays for whole genome mapping and genetic studies in catfish. Sci Rep. 2017;7:40347.
14. Hedgecock D, Shin G, Gracey AY, Den Berg DV, Samanta MP. Second-Generation Linkage Maps for the Pacific Oyster *Crassostrea gigas* Reveal Errors in Assembly of Genome Scaffolds. G3: Genes/Genomes/Genetics. 2015;5(10):2007–19.
15. Gomez-Chiarri M, Warren WC, Guo X, Proestou D. Developing tools for the study of molluscan immunity: the sequencing of the genome of the eastern oyster, *Crassostrea virginica*. Fish Shellfish Immunol. 2015;46(1):2–4.
16. Zhang X, Zhang Y, Scheuring C, Zhang H-B, Huan P, Wang B, Liu C, Li F, Liu B, Xiang J. Construction and characterization of a bacterial artificial chromosome (BAC) library of Pacific white shrimp, *Litopenaeus vannamei*. Mar Biotechnol. 2010;12(2):141–9.
17. Liu Z, Cordes J. DNA marker technologies and their applications in aquaculture genetics. Aquaculture. 2004;238(1):1–37.
18. Sun L, Liu S, Wang R, Jiang Y, Zhang Y, Zhang J, Bao L, Kaltenboeck L, Dunham R, Waldbieser G. Identification and analysis of genome-wide SNPs provide insight into signatures of selection and domestication in channel catfish (*Ictalurus punctatus*). PLoS One. 2014;9(10):e109666.
19. Yanez JM, Naswa S, Lopez ME, Bassini L, Correa K, Gilbey J, Bernatchez L, Norris A, Neira R, Lhorente JP, et al. Genomewide single nucleotide polymorphism discovery in Atlantic salmon

(*Salmo salar*): validation in wild and farmed American and European populations. Mol Ecol Resour. 2016;16(4):1002–11.

20. Gidskehaug L, Kent M, Hayes BJ, Lien S. Genotype calling and mapping of multisite variants using an Atlantic salmon iSelect SNP array. Bioinformatics (Oxford, England). 2011;27(3):303–10.
21. Houston RD, Taggart JB, Cézard T, Bekaert M, Lowe NR, Downing A, Talbot R, Bishop SC, Archibald AL, Bron JE. Development and validation of a high density SNP genotyping array for Atlantic salmon (*Salmo salar*). BMC Genomics. 2014;15:90.
22. Liu S, Sun L, Li Y, Sun F, Jiang Y, Zhang Y, Zhang J, Feng J, Kaltenboeck L, Kucuktas H. Development of the catfish 250 K SNP array for genome-wide association studies. BMC Res Notes. 2014;7:135.
23. Palti Y, Gao G, Liu S, Kent M, Lien S, Miller M, Rexroad C, Moen T. The development and characterization of a 57 K single nucleotide polymorphism array for rainbow trout. Mol Ecol Resour. 2015;15(3):662–72.
24. Xu J, Zhao Z, Zhang X, Zheng X, Li J, Jiang Y, Kuang Y, Zhang Y, Feng J, Li C, et al. Development and evaluation of the first high-throughput SNP array for common carp (*Cyprinus carpio*). BMC Genomics. 2014;15:307.
25. Wang J, Li L, Zhang G. A High-Density SNP Genetic Linkage Map and QTL Analysis of Growth-Related Traits in a Hybrid Family of Oysters (*Crassostrea gigas* × *Crassostrea angulata*) Using Genotyping-by-Sequencing. G3: Genes/Genomes/Genetics. 2016;6(5):1417–26.
26. Gonzalez-Pena D, Gao G, Baranski M, Moen T, Cleveland BM, Kenney PB, Vallejo RL, Palti Y, Leeds TD. Genome-Wide Association Study for Identifying Loci that Affect Fillet Yield, Carcass, and Body Weight Traits in Rainbow Trout (*Oncorhynchus mykiss*). Front Genet. 2016;7:203.
27. Ng SHS, Artieri CG, Bosdet IE, Chiu R, Danzmann RG, Davidson WS, Ferguson MM, Fjell CD, Hoyheim B, Jones SJM, et al. A physical map of the genome of Atlantic salmon. Salmo Salar Genomics. 2005;86(4):396–404.
28. Katagiri T, Kidd C, Tomasino E, Davis JT, Wishon C, Stern JE,

Carleton KL, Howe AE, Kocher TD. A BAC-based physical map of the Nile tilapia genome. BMC Genomics. 2005;6:89.

29. Quiniou SM, Waldbieser GC, Duke MV. A first generation BAC-based physical map of the channel catfish genome. BMC Genomics. 2007;8:40.
30. Xu P, Wang S, Liu L, Thorsen J, Kucuktas H, Liu Z. A BAC-based physical map of the channel catfish genome. Genomics. 2007;90(3):380–8.
31. Palti Y, Luo M-C, Hu Y, Genet C, You FM, Vallejo RL, Thorgaard GH, Wheeler PA, Rexroad CE. A first generation BAC-based physical map of the rainbow trout genome. BMC Genomics. 2009;10:462.
32. Xu P, Wang J, Wang J, Cui R, Li Y, Zhao Z, Ji P, Zhang Y, Li J, Sun X. Generation of the first BAC-based physical map of the common carp genome. BMC Genomics. 2011;12:537.
33. Xia JH, Feng F, Lin G, Wang CM, Yue GH. A First Generation BAC-Based Physical Map of the Asian Seabass (*Lates calcarifer*). PLoS One. 2010;5(8):e11974.
34. Gaffney PM. A BAC-based physical map for the Pacific oyster genome. J Shellfish Res. 2008;27(4):1009.
35. Zhang X, Zhao C, Huang C, Duan H, Huan P, Liu C, Zhang X, Zhang Y, Li F, Zhang H-B, et al. A BAC-Based Physical Map of Zhikong Scallop (*Chlamys farreri* Jones et Preston). PLoS One. 2011;6(11):e27612.
36. Schwartz D, Li X, Hernandez L, Ramnarain S, Huff E, Wang Y. Ordered restriction maps of *Saccharomyces cerevisiae* chromosomes constructed by optical mapping. Science. 1993;262(5130):110–4.
37. Rajendran K, Zhang J, Liu S, Peatman E, Kucuktas H, Wang X, Liu H, Wood T, Terhune J, Liu Z. Pathogen recognition receptors in channel catfish: II. Identification, phylogeny and expression of retinoic acid-inducible gene I (RIG-I)-like receptors (RLRs). Dev Comp Immunol. 2012;37(3):381–9.
38. Wang R, Sun L, Bao L, Zhang J, Jiang Y, Yao J, Song L, Feng J, Liu S, Liu Z. Bulk segregant RNA-seq reveals expression and positional candidate genes and allele-specific expression for disease resistance against enteric septicemia of catfish. BMC Genomics. 2013;14:929.

39. Sun F, Peatman E, Li C, Liu S, Jiang Y, Zhou Z, Liu Z. Transcriptomic signatures of attachment, NF-κB suppression and IFN stimulation in the catfish gill following columnaris bacterial infection. Dev Comp Immunol. 2012;38(1):169–80.
40. Liu S, Wang X, Sun F, Zhang J, Feng J, Liu H, Rajendran K, Sun L, Zhang Y, Jiang Y. RNA-Seq reveals expression signatures of genes involved in oxygen transport, protein synthesis, folding, and degradation in response to heat stress in catfish. Physiol Genomics. 2013;45(12):462–76.
41. Valenzuela-Miranda D, Boltana S, Cabrejos ME, Yanez JM, Gallardo-Escárate C. High-throughput transcriptome analysis of ISAV-infected Atlantic salmon *Salmo salar* unravels divergent immune responses associated to head-kidney, liver and gills tissues. Fish Shellfish Immunol. 2015;45(2):367–77.
42. Xu C, Evensen Ø, Munang'andu HM. De novo assembly and transcriptome analysis of Atlantic salmon macrophage/dendritic-like TO cells following type I IFN treatment and Salmonid alphavirus subtype-3 infection. BMC Genomics. 2015;16:96.
43. Salem M, Vallejo RL, Leeds TD, Palti Y, Liu S, Sabbagh A, Rexroad III CE, Yao J. RNA-Seq identifies SNP markers for growth traits in rainbow trout. PLoS One. 2012;7(5):e36264.
44. Narum SR, Campbell NR. Transcriptomic response to heat stress among ecologically divergent populations of redband trout. BMC Genomics. 2015;16:103.
45. Cano I, Verner-Jeffreys DW, van Aerle R, Paley RK, Peeler EJ, Green M, Rimmer GS, Savage J, Joiner CL, Bayley AE. Puffy Skin Disease Is an Emerging Transmissible Condition in Rainbow Trout *Oncorhynchus mykiss* Walbaum. PLoS One. 2016;11(7):e0158151.
46. Marancik D, Gao G, Paneru B, Ma H, Hernandez AG, Salem M, Yao J, Palti Y, Wiens GD. Whole-body transcriptome of selectively bred, resistant-, control-, and susceptible-line rainbow trout following experimental challenge with *Flavobacterium psychrophilum*. Front Genet. 2015;5:453.
47. Chapman RW, Reading BJ, Sullivan CV. Ovary transcriptome profiling via artificial intelligence reveals a transcriptomic fingerprint

predicting egg quality in striped bass, *Morone saxatilis*. PloS One. 2014;9(5):e96818.

48. Reading BJ, Chapman RW, Schaff JE, Scholl EH, Opperman CH, Sullivan CV. An ovary transcriptome for all maturational stages of the striped bass (*Morone saxatilis*), a highly advanced perciform fish. BMC Res Notes. 2012;5:111.
49. Sullivan CV, Chapman RW, Reading BJ, Anderson PE. Transcriptomics of mRNA and egg quality in farmed fish: Some recent developments and future directions. Gen Comp Endocrinol. 2015;221:23–30.
50. Zhao Y, Wang J, Thammaratsuntorn J, Wu J, Wei J, Wang Y, Xu J, Zhao J. Comparative transcriptome analysis of Nile tilapia (*Oreochromis niloticus*) in response to alkalinity stress. Genet Mol Res. 2015;14(4):17916–26.
51. Xu Z, Gan L, Li T, Xu C, Chen K, Wang X, Qin JG, Chen L, Li E. Transcriptome profiling and molecular pathway analysis of genes in association with salinity adaptation in nile tilapia *Oreochromis niloticus*. PLoS One. 2015;10(8):e0136506.
52. He AY, Ning LJ, Chen LQ, Chen YL, Xing Q, Li JM, Qiao F, Li DL, Zhang ML, Du ZY. Systemic adaptation of lipid metabolism in response to low-and high-fat diet in Nile tilapia (*Oreochromis niloticus*). Physiol Rep. 2015;3(8):e12485.
53. Clark MS, Thorne MA, Amaral A, Vieira F, Batista FM, Reis J, Power DM. Identification of molecular and physiological responses to chronic environmental challenge in an invasive species: the Pacific oyster, *Crassostrea gigas*. Ecol Evol. 2013;3(10):3283–97.
54. Zhao X, Yu H, Kong L, Liu S, Li Q. Comparative transcriptome analysis of two oysters, *Crassostrea gigas* and *Crassostrea hongkongensis* provides insights into adaptation to hypo-osmotic conditions. PLoS one. 2014;9(11):e111915.
55. Zhao X, Yu H, Kong L, Li Q. Transcriptomic responses to salinity stress in the Pacific oyster *Crassostrea gigas*. PLoS One. 2012;7(9):e46244.
56. McDowell IC, Modak TH, Lane CE, Gomez-Chiarri M. Multi-species protein similarity clustering reveals novel expanded immune gene families in the eastern oyster Crassostrea virginica.

Fish Shellfish Immunol. 2016;53:13–23.

57. Zhang L, Li L, Zhu Y, Zhang G, Guo X. Transcriptome analysis reveals a rich gene set related to innate immunity in the Eastern oyster (*Crassostrea virginica*). Mar Biotechnol. 2014;16(1):17–33.
58. Wei J, Zhang X, Yu Y, Huang H, Li F, Xiang J. Comparative transcriptomic characterization of the early development in Pacific white shrimp *Litopenaeus vannamei*. PLoS One. 2014;9(9):e106201.
59. Sookruksawong S, Sun F, Liu Z, Tassanakajon A. RNA-Seq analysis reveals genes associated with resistance to Taura syndrome virus (TSV) in the Pacific white shrimp *Litopenaeus vannamei*. Dev Comp Immunol. 2013;41(4):523–33.
60. Ghaffari N, Sanchez-Flores A, Doan R, Garcia-Orozco KD, Chen PL, Ochoa-Leyva A, Lopez-Zavala AA, Carrasco JS, Hong C, Brieba LG. Novel transcriptome assembly and improved annotation of the whiteleg shrimp (*Litopenaeus vannamei*), a dominant crustacean in global seafood mariculture. Sci Rep. 2014;4:7081.
61. Reading BJ, Williams VN, Chapman RW, Williams TI, Sullivan CV. Dynamics of the striped bass (*Morone saxatilis*) ovary proteome reveal a complex network of the translasome. J Proteome Res. 2013;12(4):1691–9.
62. Schilling J, Nepomuceno A, Schaff JE, Muddiman DC, Daniels HV, Reading BJ. Compartment proteomics analysis of white perch (*Morone americana*) ovary using support vector machines. J Proteome Res. 2014;13(3):1515–26.
63. Schilling J, Loziuk PL, Muddiman DC, Daniels HV, Reading BJ. Mechanisms of egg yolk formation and implications on early life history of white perch (*Morone americana*). PLoS One. 2015;10(11):e0143225.
64. Williams VN, Reading BJ, Amano H, Hiramatsu N, Schilling J, Salger SA, Islam Williams T, Gross K, Sullivan CV. Proportional accumulation of yolk proteins derived from multiple vitellogenins is precisely regulated during vitellogenesis in striped bass (*Morone saxatilis*). J Exp Zool A Ecol Genet Physiol. 2014;321(6):301–15.
65. Salem M, Kenney PB, Rexroad CE, Yao J. Proteomic signature of muscle atrophy in rainbow trout. J Proteome. 2010;73(4):778–89.
66. Mennigen JA, Panserat S, Larquier M, Plagnes-Juan E, Medale

F, Seiliez I, Skiba-Cassy S. Postprandial regulation of hepatic microRNAs predicted to target the insulin pathway in rainbow trout. PLoS One. 2012;7(6):e38604.

67. Mennigen JA, Skiba-Cassy S, Panserat S. Ontogenetic expression of metabolic genes and microRNAs in rainbow trout alevins during the transition from the endogenous to the exogenous feeding period. J Exp Biol. 2013;216(9):1597–608.

68. Juanchich A, Bardou P, Rue O, Gabillard JC, Gaspin C, Bobe J, Guiguen Y. Characterization of an extensive rainbow trout miRNA transcriptome by next generation sequencing. BMC Genomics. 2016;17:164.

69. Mennigen JA, Zhang D. MicroTrout: A comprehensive, genome-wide miRNA target prediction framework for rainbow trout, *Oncorhynchus mykiss*. Comp Biochem Physiol Part D Genomics Proteomics. 2016;20:19–26.

70. Al-Tobasei R, Paneru B, Salem M. Genome-wide discovery of long non-coding RNAs in rainbow trout. PLoS One. 2016;11(2):e0148940.

71. Wang Z, Schwacke R, Kunze R. DNA damage-induced transcription of transposable elements and long non-coding RNAs in Arabidopsis is rare and ATM-dependent. Mol Plant. 2016;9(8):1142–55.

72. Farlora R, Valenzuela-Miranda D, Alarcón-Matus P, Gallardo-Escárate C. Identification of microRNAs associated with sexual maturity in rainbow trout brain and testis through small RNA deep sequencing. Mol Reprod Dev. 2015;82(9):651–62.

73. Ma H, Weber GM, Hostuttler MA, Wei H, Wang L, Yao J. MicroRNA expression profiles from eggs of different qualities associated with post-ovulatory ageing in rainbow trout (*Oncorhynchus mykiss*). BMC Genomics. 2015;16:201.

74. Paneru B, Al-Tobasei R, Leeds T, Kenney B, Salem M: MicroRNAs associated with muscle growth and fillet quality in rainbow trout. In: PAG-XXV Plant & Animal Genomes Conference: Plant and Animal Genome. San Diego; 2017.

75. Paneru B, Al-Tobasei R, Palti Y, Wiens GD, Salem M. Differential expression of long non-coding RNAs in three genetic lines of rainbow trout in response to infection with *Flavobacterium psychrophilum*. Sci Rep. 2016;6(36032):1–14.

76. Bekaert M, Lowe NR, Bishop SC, Bron JE, Taggart JB, Houston RD. Sequencing and Characterisation of an Extensive Atlantic Salmon (*Salmo salar* L.) MicroRNA Repertoire. PLoS One. 2013;8(7):e70136.
77. Andreassen R, Worren MM, Høyheim B. Discovery and characterization of miRNA genes in atlantic salmon (*Salmo salar*) by use of a deep sequencing approach. BMC Genomics. 2013;14:482.
78. Kure EH, Saebo M, Stangeland AM, Hamfjord J, Hytterod S, Heggenes J, Lydersen E. Molecular responses to toxicological stressors: profiling microRNAs in wild Atlantic salmon (*Salmo salar*) exposed to acidic aluminum-rich water. Aquat Toxicol. 2013;138–139:98–104.
79. Barozai MY, Baloch IA, Din M. Identification of MicroRNAs and their targets in Helianthus. Mol Biol Rep. 2012;39(3):2523–32.
80. Xu Z, Qin Q, Ge J, Pan J, Xu X. Bioinformatic identification and validation of conservative microRNAs in *Ictalurus punctatus*. Mol Biol Rep. 2012;39(12):10395–405.
81. Xu Z, Chen J, Li X, Ge J, Pan J, Xu X. Identification and characterization of microRNAs in channel catfish (*Ictalurus punctatus*) by using Solexa sequencing technology. PLoS One. 2013;8(1):e54174.
82. Tao W, Sun L, Shi H, Cheng Y, Jiang D, Fu B, Conte MA, Gammerdinger WJ, Kocher TD, Wang D. Integrated analysis of miRNA and mRNA expression profiles in tilapia gonads at an early stage of sex differentiation. BMC Genomics. 2016;17:328.
83. Xu F, Wang X, Feng Y, Huang W, Wang W, Li L, Fang X, Que H, Zhang G. Identification of conserved and novel microRNAs in the pacific oyster *Crassostrea gigas* by deep sequencing. PLoS One. 2014;9(8):e104371.
84. Zhou Z, Wang L, Song L, Liu R, Zhang H, Huang M, Chen H. The identification and Characteristics of immune-related microRNAs in hemocytes of oyster *Crassostrea gigas*. PLoS One. 2014;9(2):e88397.
85. Zeng D, Chen X, Xie D, Zhao Y, Yang Q, Wang H, Li Y, Chen X. Identification of highly expressed host microRNAs that respond to white spot syndrome virus infection in the Pacific

white shrimp *Litopenaeus vannamei* (Penaeidae). Genet Mol Res. 2015;14(2):4818–28.

86. Xi Q, Xiong Y, Wang Y, Cheng X, Qi Q, Shu G, Wang S, Wang L, Gao P, Zhu X, et al. Genome-wide discovery of novel and conserved microRNAs in white shrimp (*Litopenaeus vannamei*). Mol Biol Rep. 2015;42(1):61–9.
87. Andersson L, Archibald AL, Bottema CD, Brauning R, Burgess SC, Burt DW, Casas E, Cheng HH, Clarke L, Couldrey C, et al. Coordinated international action to accelerate genome-to-phenome with FAANG, the Functional Annotation of Animal Genomes project. Genome Biol. 2015;16:57.
88. Fellous A, Favrel P, Guo X, Riviere G. The Jumonji gene family in *Crassostrea gigas*suggests evolutionary conservation of Jmj-C histone demethylases orthologues in the oyster gametogenesis and development. Gene. 2014;538(1):164–75.
89. Fellous A, Favrel P, Riviere G. Temperature influences histone methylation and mRNA expression of the Jmj-C histone-demethylase orthologues during the early development of the oyster *Crassostrea gigas*. Mar Genomics. 2015;19:23–30.
90. Riviere G. Epigenetic features in the oyster *Crassostrea gigas* suggestive of functionally relevant promoter DNA methylation in invertebrates. Front Physiol. 2014;5(129):1–7.
91. Olson CE, Roberts SB. Genome-wide profiling of DNA methylation and gene expression in *Crassostrea gigas* male gametes. Front Physiol. 2014;5:224.
92. Wang X, Li Q, Lian J, Li L, Jin L, Cai H, Xu F, Qi H, Zhang L, Wu F. Genome-wide and single-base resolution DNA methylomes of the Pacific oyster *Crassostrea gigas* provide insight into the evolution of invertebrate CpG methylation. BMC Genomics. 2014;15:1119.
93. Saint-Carlier E, Riviere G. Regulation of *Hox* orthologues in the oyster *Crassostrea gigas*evidences a functional role for promoter DNA methylation in an invertebrate. FEBS Lett. 2015;589(13):1459–66.
94. Morán P, Pérez-Figueroa A. Methylation changes associated with early maturation stages in the Atlantic salmon. BMC Genet. 2011;12:86.

95. Baerwald MR, Meek MH, Stephens MR, Nagarajan RP, Goodbla AM, Tomalty KMH, Thorgaard GH, May B, Nichols KM. Migration-related phenotypic divergence is associated with epigenetic modifications in rainbow trout. Mol Ecol. 2016;25(8):1785–800.

96. Marandel L, Lepais O, Arbenoits E, Véron V, Dias K, Zion M, Panserat S. Remodelling of the hepatic epigenetic landscape of glucose-intolerant rainbow trout (*Oncorhynchus mykiss*) by nutritional status and dietary carbohydrates. Sci Rep. 2016;6:32187.

97. Kratochwil CF, Meyer A. Mapping active promoters by ChIP-seq profiling of H3K4me3 in cichlid fish- a first step to uncover cis-regulatory elements in ecological model teleosts. Mol Ecol Resour. 2015;15(4):761–71.

98. Houston RD, Haley CS, Hamilton A, Guy DR, Tinch AE, Taggart JB, McAndrew BJ, Bishop SC. Major quantitative trait loci affect resistance to infectious pancreatic necrosis in Atlantic salmon (*Salmo salar*). Genetics. 2008;178(2):1109–15.

99. Moen T, Baranski M, Sonesson AK, Kjøglum S. Confirmation and fine-mapping of a major QTL for resistance to infectious pancreatic necrosis in Atlantic salmon (*Salmo salar*): population-level associations between markers and trait. BMC Genomics. 2009;10:368.

100. Houston RD, Haley CS, Hamilton A, Guy DR, Mota-Velasco JC, Gheyas AA, Tinch AE, Taggart J, Bron J, Starkey W. The susceptibility of Atlantic salmon fry to freshwater infectious pancreatic necrosis is largely explained by a major QTL. Heredity. 2010;105(3):318–27.

101. Moen T, Torgersen J, Santi N, Davidson WS, Baranski M, Ødegård J, Kjøglum S, Velle B, Kent M, Lubieniecki KP. Epithelial cadherin determines resistance to infectious pancreatic necrosis virus in Atlantic salmon. Genetics. 2015;200(4):1313–26.

102. Geng X, Sha J, Liu S, Bao L, Zhang J, Wang R, Yao J, Li C, Feng J, Sun F. A genome-wide association study in catfish reveals the presence of functional hubs of related genes within QTLs for columnaris disease resistance. BMC Genomics. 2015;16:196.

103. Zhou T, Liu S, Geng X, Jin Y, Jiang C, Bao L, Yao J, Zhang Y, Zhang J, Sun L et al. QTLs for ESC disease resistance of catfish as

revealed by a genome wide association study. Mol Genet Genomics. 2017;292:231–42.

104. Jin Y, Zhou T, Geng X, Liu S, Chen A, Yao J, Jiang C, Tan S, Su B, Liu Z. A genome-wide association study of heat stress-associated SNPs in catfish. Anim Genet. 2016. doi:10.1111/age.12482.

105. Wang X, Liu S, Jiang C, Geng X, Zhou T, Li N, Bao L, Li Y, Yao J, Yang Y et al. Multiple across-strain and within-strain QTLs suggest highly complex genetic architecture for hypoxia tolerance in channel catfish. Mol Gen Genomics. 2016. doi:10.1007/s00438-016-1256-2.

106. Geng X, Liu S, Yao J, Bao L, Zhang J, Li C, Wang R, Sha J, Zeng P, Zhi D, et al. A Genome Wide Association Study Identifies Multiple Regions Associated with Head Size in Catfish. G3: Genes/Genomes/Genetics. 2016;6(10):3389–98.

107. Vallejo RL, Palti Y, Liu S, Marancik DP, Wiens GD. Validation of linked QTL for bacterial cold water disease resistance and spleen size on rainbow trout chromosome Omy19. Aquaculture. 2014;432:139–43.

108. Palti Y, Vallejo RL, Gao G, Liu S, Hernandez AG, Rexroad III CE, Wiens GD. Detection and validation of QTL affecting bacterial cold water disease resistance in rainbow trout using restriction-site associated DNA sequencing. PLoS One. 2015;10(9):e0138435.

109. Liu S, Vallejo RL, Palti Y, Gao G, Marancik DP, Hernandez AG, Wiens GD. Identification of single nucleotide polymorphism markers associated with bacterial cold water disease resistance and spleen size in rainbow trout. Front Genet. 2015;6:298.

110. Yu Z, Guo X. Identification and mapping of disease-resistance QTLs in the eastern oyster, *Crassostrea virginica* Gmelin. Aquaculture. 2006;254(1):160–70.

111. Plough LV, Hedgecock D. Quantitative trait locus analysis of stage-specific inbreeding depression in the Pacific oyster *Crassostrea gigas*. Genetics. 2011;189(4):1473–86.

112. Guo X, Li Q, Wang QZ, Kong LF. Genetic mapping and QTL analysis of growth-related traits in the Pacific oyster. Mar Biotechnol. 2012;14(2):218–26.

113. Plough L, Shin G, Hedgecock D. Genetic inviability is a major

driver of type-III survivorship in experimental families of a highly fecund marine bivalve. Mol Ecol. 2016;25(4):895–910.

114. He Y, Yu H, Bao Z, Zhang Q, Guo X. Mutation in promoter region of a serine protease inhibitor confers Perkinsus marinus resistance in the eastern oyster (*Crassostrea virginica*). Fish Shellfish Immunol. 2012;33(2):411–7.

115. Fuji K, Hasegawa O, Honda K, Kumasaka K, Sakamoto T, Okamoto N. Marker-assisted breeding of a lymphocystis disease-resistant Japanese flounder (*Paralichthys olivaceus*). Aquaculture. 2007;272(1–4):291–5.

116. Ozaki A, Araki K, Okamoto H, Okauchi M, Mushiake K, Yoshida K, Tsuzaki T, Fuji K, Sakamoto T, Okamoto N. Progress of DNA marker-assisted breeding in maricultured finfish. Bull Fisheries Res Agency. 2012;35:31–7.

117. Chen J, Wang Y, Yue Y, Xia X, Du Q, Chang Z. A novel male-specific DNA sequence in the common carp, *Cyprinus carpio*. Mol Cell Probes. 2009;23(5):235–9.

118. Lee BY, Penman DJ, Kocher TD. Identification of a sex-determining region in Nile tilapia (*Oreochromis niloticus*) using bulked segregant analysis. Anim Genet. 2003;34(5):379–83.

119. Ninwichian P, Peatman E, Liu H, Kucuktas H, Somridhivej B, Liu S, Li P, Jiang Y, Sha Z, Kaltenboeck L. Second-generation genetic linkage map of catfish and its integration with the BAC-based physical map. G3: Genes| Genomes| Genetics. 2012;2(10):1233–41.

120. Li L, Xiang J, Liu X, Zhang Y, Dong B, Zhang X. Construction of AFLP-based genetic linkage map for Zhikong scallop, *Chlamys farreri* Jones et Preston and mapping of sex-linked markers. Aquaculture. 2005;245(1–4):63–73.

121. Chen S, Li J, Deng S, Tian Y, Wang Q, Zhuang Z, Sha Z, Xu J. Isolation of Female-Specific AFLP Markers and Molecular Identification of Genetic Sex in Half-Smooth Tongue Sole (*Cynoglossus semilaevis*). Mar Biotechnol. 2007;9(2):273–80.

122. Pérez F, Erazo C, Zhinaula M, Volckaert F, Calderón J. A sex-specific linkage map of the white shrimp *Penaeus (Litopenaeus) vannamei* based on AFLP markers. Aquaculture. 2004;242(1–4):105–18.

123. Li Y, Byrne K, Miggiano E, Whan V, Moore S, Keys S, Crocos P, Preston N, Lehnert S. Genetic mapping of the kuruma prawn *Penaeus japonicus* using AFLP markers. Aquaculture. 2003;219(1–4):143–56.

124. Fuji K, Yoshida K, Hattori K, Ozaki A, Araki K, Okauchi M, Kubota S, Okamoto N, Sakamoto T. Identification of the sex-linked locus in yellowtail, Seriola quinqueradiata. Aquaculture. 2010;308:S51–5.

125. Felip A, Young WP, Wheeler PA, Thorgaard GH. An AFLP-based approach for the identification of sex-linked markers in rainbow trout (*Oncorhynchus mykiss*). Aquaculture. 2005;247(1–4):35–43.

126. Meuwissen THE, Hayes BJ, Goddard ME. Prediction of Total Genetic Value Using Genome-Wide Dense Marker Maps. Genetics. 2001;157(4):1819–29.

127. Hayes BJ, Bowman PJ, Chamberlain AJ, Goddard ME. Invited review: Genomic selection in dairy cattle: progress and challenges. J Dairy Sci. 2009;92(2):433–43.

128. Ragavendran A, Muir WM. Genomic Selection in Aquaculture: Methods and Practical Considerations. In: Next Generation Sequencing and Whole Genome Selection in Aquaculture. Ames: Wiley-Blackwell; 2011. p. 165–183.

129. Sonesson AK. Genomic Selection for Aquaculture: Principles and Procedures. In: Next Generation Sequencing and Whole Genome Selection in Aquaculture. Ames: Wiley-Blackwell; 2011. p. 151–163.

130. Ødegård J, Moen T, Santi N, Korsvoll SA, Kjøglum S, Meuwissen THE. Genomic prediction in an admixed population of Atlantic salmon (*Salmo salar*). Front Genet. 2014;5:402.

131. Schilling J, Nepomuceno AI, Planchart A, Yoder JA, Kelly RM, Muddiman DC, Daniels HV, Hiramatsu N, Reading BJ. Machine learning reveals sex-specific 17β-estradiol-responsive expression patterns in white perch (*Morone americana*) plasma proteins. Proteomics. 2015;15(15):2678–90.

132. Nemudryi AA, Valetdinova KR, Medvedev SP, Zakian SM. TALEN and CRISPR/Cas Genome Editing Systems: Tools of Discovery. Acta Nat. 2014;6(3):19–40.

133. Hwang WY, Fu Y, Reyon D, Maeder ML, Tsai SQ, Sander JD,

Peterson RT, Yeh JRJ, Joung JK. Efficient genome editing in zebrafish using a CRISPR-Cas system. Nat Biotech. 2013;31(3):227–9.

134. Jao LE, Wente SR, Chen W. Efficient multiplex biallelic zebrafish genome editing using a CRISPR nuclease system. Proc Natl Acad Sci. 2013;110(34):13904–9.

135. Li M, Yang H, Zhao J, Fang L, Shi H, Li M, Sun Y, Zhang X, Jiang D, Zhou L. Efficient and heritable gene targeting in tilapia by CRISPR/Cas9. Genetics. 2014;197(2):591–9.

136. Elaswad A. Genetic Technologies for Disease Resistance Research and Enhancement in Catfish. PhD Dissertation. Auburn: Auburn University; 2016.

137. Buchanan J. Regulatory Approval of Genetically Engineered AquAdvantage Salmon. In: Plant and Animal Genome XXIV Conference: Plant and Animal Genome. San Diego; 2016.

138. Waltz E. Gene-edited CRISPR mushroom escapes US regulation. Nat News. 2016;532:293.

139. Danchin-Burge C, Hiemstra S, Blackburn H. Ex situ conservation of holstein-friesian cattle: comparing the Dutch, French, and US germplasm collections. J Dairy Sci. 2011;94(8):4100–8.

140. Yue XP, Dechow C, Liu WS. A limited number of Y chromosome lineages is present in North American Holsteins. J Dairy Sci. 2015;98(4):2738–45.

141. Tiersch TR, Green CC. Cryopreservation in aquatic species. 2nd ed. Baton Rouge: World Aquaculture Society; 2011.

142. Star B, Nederbragt AJ, Jentoft S, Grimholt U, Malmstrøm M, Gregers TF, Rounge TB, Paulsen J, Solbakken MH, Sharma A. The genome sequence of Atlantic cod reveals a unique immune system. Nature. 2011;477(7363):207–10.

143. McGaugh SE, Gross JB, Aken B, Blin M, Borowsky R, Chalopin D, Hinaux H, Jeffery WR, Keene A, Ma L, et al. The cavefish genome reveals candidate genes for eye loss. Nat Commun. 2014;5:5307.

144. Amemiya CT, Alfoldi J, Lee AP, Fan S, Philippe H, MacCallum I, Braasch I, Manousaki T, Schneider I, Rohner N, et al. The African coelacanth genome provides insights into tetrapod evolution. Nature. 2013;496(7445):311–6.

145. Xu P, Zhang X, Wang X, Li J, Liu G, Kuang Y, Xu J, Zheng X, Ren L, Wang G. Genome sequence and genetic diversity of the common carp, *Cyprinus carpio*. Nat Genet. 2014;46(11):1212–9.
146. Wang Y, Lu Y, Zhang Y, Ning Z, Li Y, Zhao Q, Lu H, Huang R, Xia X, Feng Q. The draft genome of the grass carp (*Ctenopharyngodon idellus*) provides insights into its evolution and vegetarian adaptation. Nat Genet. 2015;47(6):625–31.
147. Smith JJ, Kuraku S, Holt C, Sauka-Spengler T, Jiang N, Campbell MS, Yandell MD, Manousaki T, Meyer A, Bloom OE, et al. Sequencing of the sea lamprey (*Petromyzon marinus*) genome provides insights into vertebrate evolution. Nat Genet. 2013;45(4):415–21.
148. Kasahara M, Naruse K, Sasaki S, Nakatani Y, Qu W, Ahsan B, Yamada T, Nagayasu Y, Doi K, Kasai Y, et al. The medaka draft genome and insights into vertebrate genome evolution. Nature. 2007;447(7145):714–9.
149. Schartl M, Walter RB, Shen Y, Garcia T, Catchen J, Amores A, Braasch I, Chalopin D, Volff J-N, Lesch K-P, et al. The genome of the platyfish, *Xiphophorus maculatus*, provides insights into evolutionary adaptation and several complex traits. Nat Genet. 2013;45(5):567–72.
150. Tine M, Kuhl H, Gagnaire P-A, Louro B, Desmarais E, Martins RS, Hecht J, Knaust F, Belkhir K, Klages S. European sea bass genome and its variation provide insights into adaptation to euryhalinity and speciation. Nat Commun. 2014;5:5770.
151. Venkatesh B, Lee AP, Ravi V, Maurya AK, Lian MM, Swann JB, Ohta Y, Flajnik MF, Sutoh Y, Kasahara M, et al. Elephant shark genome provides unique insights into gnathostome evolution. Nature. 2014;505(7482):174–9.
152. Chen S, Zhang G, Shao C, Huang Q, Liu G, Zhang P, Song W, An N, Chalopin D, Volff J-N. Whole-genome sequence of a flatfish provides insights into ZW sex chromosome evolution and adaptation to a benthic lifestyle. Nat Genet. 2014;46(3):253–60.
153. Jones FC, Grabherr MG, Chan YF, Russell P, Mauceli E, Johnson J, Swofford R, Pirun M, Zody MC, White S, et al. The genomic basis of adaptive evolution in threespine sticklebacks. Nature. 2012;484(7392):55–61.

154. Jaillon O, Aury J-M, Brunet F, Petit J-L, Stange-Thomann N, Mauceli E, Bouneau L, Fischer C, Ozouf-Costaz C, Bernot A, et al. Genome duplication in the teleost fish *Tetraodon nigroviridis* reveals the early vertebrate proto-karyotype. Nature. 2004;431(7011):946–57.

155. Figueras A, Robledo D, Corvelo A, Hermida M, Pereiro P, Rubiolo JA, Gómez-Garrido J, Carreté L, Bello X, Gut M et al. Whole Genome Sequencing of Turbot (Scophthalmus maximus; Pleuronectiformes): A Fish Adapted to Demersal Life. DNA Res. 2016;23(3):181–192.

156. Wu C, Zhang D, Kan M, Lv Z, Zhu A, Su Y, Zhou D, Zhang J, Zhang Z, Xu M. The draft genome of the large yellow croaker reveals well-developed innate immunity. Nat Commun. 2014;5:5227.

157. Howe K, Clark MD, Torroja CF, Torrance J, Berthelot C, Muffato M, Collins JE, Humphray S, McLaren K, Matthews L, et al. The zebrafish reference genome sequence and its relationship to the human genome. Nature. 2013;496(7446):498–503.

158. Liu S, Zhang Y, Zhou Z, Waldbieser G, Sun F, Lu J, Zhang J, Jiang Y, Zhang H, Wang X. Efficient assembly and annotation of the transcriptome of catfish by RNA-Seq analysis of a doubled haploid homozygote. BMC Genomics. 2012;13:595.

159. Palti Y, Gao G, Miller MR, Vallejo RL, Wheeler PA, Quillet E, Yao J, Thorgaard GH, Salem M, Rexroad CE. A resource of single-nucleotide polymorphisms for rainbow trout generated by restriction-site associated DNA sequencing of doubled haploids. Mol Ecol Resour. 2014;14(3):588–96.

160. Christensen KA, Brunelli JP, Lambert MJ, DeKoning J, Phillips RB, Thorgaard GH. Identification of single nucleotide polymorphisms from the transcriptome of an organism with a whole genome duplication. BMC Bioinformatics. 2013;14:325.

161. Al-Tobasei R: Allelic-imbalance analysis in pooled RNA-Seq samples identifies muscle-associated genetic markers in rainbow trout: improved bioinformatics practices. In: Plant and Animal Genome XXIV Conference: Plant and Animal Genome. San Diego; 2016.

162. Van Bers N, Crooijmans R, Groenen M, Dibbits B, Komen J. SNP marker detection and genotyping in tilapia. Mol Ecol Resour.

2012;12(5):932–41.

163. Li C, Beck BH, Fuller SA, Peatman E. Transcriptome annotation and marker discovery in white bass (*Morone chrysops*) and striped bass (*Morone saxatilis*). Anim Genet. 2014;45(6):885–7.

164. Yu Y, Wei J, Zhang X, Liu J, Liu C, Li F, Xiang J. SNP discovery in the transcriptome of white Pacific shrimp *Litopenaeus vannamei* by next generation sequencing. PLoS One. 2014;9(1):e87218.

165. Wang CM, Bai ZY, He XP, Lin G, Xia JH, Sun F, Lo LC, Feng F, Zhu ZY, Yue GH. A high-resolution linkage map for comparative genome analysis and QTL fine mapping in Asian seabass, Lates calcarifer. BMC Genomics. 2011;12:174.

166. Lien S, Gidskehaug L, Moen T, Hayes BJ, Berg PR, Davidson WS, Omholt SW, Kent MP. A dense SNP-based linkage map for Atlantic salmon (*Salmo salar*) reveals extended chromosome homeologies and striking differences in sex-specific recombination patterns. BMC Genomics. 2011;12:615.

167. Gharbi K, Gautier A, Danzmann RG, Gharbi S, Sakamoto T, Høyheim B, Taggart JB, Cairney M, Powell R, Krieg F, et al. A linkage map for brown trout (*Salmo trutta*): chromosome homeologies and comparative genome organization with other salmonid fish. Genetics. 2006;172(4):2405–19.

168. Li Y, Liu S, Qin Z, Waldbieser G, Wang R, Sun L, Bao L, Danzmann RG, Dunham R, Liu Z. Construction of a high-density, high-resolution genetic map and its integration with BAC-based physical map in channel catfish. DNA Res. 2015;22(1):39–52.

169. Liu S, Li Y, Qin Z, Geng X, Bao L, Kaltenboeck L, Kucuktas H, Dunham R, Liu Z. High-density interspecific genetic linkage mapping provides insights into genomic incompatibility between channel catfish and blue catfish. Anim Genet. 2016;47(1):81–90.

170. Peng W, Xu J, Zhang Y, Feng J, Dong C, Jiang L, Feng J, Chen B, Gong Y, Chen L, et al. An ultra-high density linkage map and QTL mapping for sex and growth-related traits of common carp (*Cyprinus carpio*). Sci Rep. 2016;6:26693.

171. Chistiakov DA, Tsigenopoulos CS, Lagnel J, Guo YM, Hellemans B, Haley CS, Volckaert FAM, Kotoulas G. A combined AFLP and microsatellite linkage map and pilot comparative genomic

analysis of European sea bass *Dicentrarchus labrax* L. Anim Genet. 2008;39(6):623–34.

172. Xia JH, Liu F, Zhu ZY, Fu J, Feng J, Li J, Yue GH. A consensus linkage map of the grass carp (*Ctenopharyngodon idella*) based on microsatellites and SNPs. BMC Genomics. 2010;11(1):554.

173. Castaño-Sánchez C, Fuji K, Ozaki A, Hasegawa O, Sakamoto T, Morishima K, Nakayama I, Fujiwara A, Masaoka T, Okamoto H, et al. A second generation genetic linkage map of Japanese flounder (*Paralichthys olivaceus*). BMC Genomics. 2010;11(1):554.

174. Guyomard R, Boussaha M, Krieg F, Hervet C, Quillet E. A synthetic rainbow trout linkage map provides new insights into the salmonid whole genome duplication and the conservation of synteny among teleosts. BMC Genet. 2012;13:15.

175. Jiao W, Fu X, Dou J, Li H, Su H, Mao J, Yu Q, Zhang L, Hu X, Huang X et al. High-resolution linkage and quantitative trait locus mapping aided by genome survey sequencing: building up an integrative genomic framework for a bivalve mollusc. DNA Res. 2013. doi:10.1093/dnares/dst043.

176. Tsigenopoulos CS, Louro B, Chatziplis D, Lagnel J, Vogiatzi E, Loukovitis D, Franch R, Sarropoulou E, Power DM, Patarnello T, et al. Second generation genetic linkage map for the gilthead sea bream *Sparus aurata L.* Mar Genomics. 2014;18 Pt A:77–82.

177. Baranski M, Gopikrishna G, Robinson NA, Katneni VK, Shekhar MS, Shanmugakarthik J, Jothivel S, Gopal C, Ravichandran P, Kent M, et al. The Development of a High Density Linkage Map for Black Tiger Shrimp (*Penaeus monodon*) Based on cSNPs. PLoS One. 2014;9(1):e85413.

178. Andriantahina F, Liu X, Huang H. Genetic map construction and quantitative trait locus (QTL) detection of growth-related traits in *Litopenaeus vannamei* for selective breeding applications. PLoS One. 2013;8(9):e75206.

179. Yu Y, Zhang X, Yuan J, Li F, Chen X, Zhao Y, Huang L, Zheng H, Xiang J. Genome survey and high-density genetic map construction provide genomic and genetic resources for the Pacific White Shrimp *Litopenaeus vannamei*. Sci Rep. 2015;5:15612.

180. Lee BY, Lee WJ, Streelman JT, Carleton KL, Howe AE, Hulata

G, Slettan A, Stern JE, Terai Y, Kocher TD. A second-generation genetic linkage map of tilapia (*Oreochromis*spp.). Genetics. 2005;170(1):237–44.

181. Liu F, Sun F, Li J, Xia JH, Lin G, Tu RJ, Yue GH. A microsatellite-based linkage map of salt tolerant tilapia (*Oreochromis mossambicus* x *Oreochromis spp.*) and mapping of sex-determining loci. BMC Genomics. 2013;14:58.
182. Ohara E, Nishimura T, Nagakura Y, Sakamoto T, Mushiake K, Okamoto N. Genetic linkage maps of two yellowtails (*Seriola quinqueradiata* and *Seriola lalandi*). Aquaculture. 2005;244(1–4):41–8.
183. Aoki J-y, Kai W, Kawabata Y, Ozaki A, Yoshida K, Koyama T, Sakamoto T, Araki K. Second generation physical and linkage maps of yellowtail (*Seriola quinqueradiata*) and comparison of synteny with four model fish. BMC Genomics. 2015;16:406.
184. Ozaki A, Aoki J-y, Usuki H, Yoshida K, Noda T, Hotta T, Mizuochi H, Chujo T, Shima Y, Araki K: A high density genetic linkage map for yellowtail (Seriola quinqueradiata) containing EST-based SNPs. In: Plant and Animal Genome XXIV Conference: Plant and Animal Genome. San Diego; 2016
185. Küttner E, Moghadam HK, Skúlason S, Danzmann RG, Ferguson MM. Genetic architecture of body weight, condition factor and age of sexual maturation in Icelandic Arctic charr (*Salvelinus alpinus*). Mol Gen Genomics. 2011;286(1):67–79.
186. Norman JD, Danzmann RG, Glebe B, Ferguson MM. The genetic basis of salinity tolerance traits in Arctic charr (*Salvelinus alpinus*). BMC Genet. 2011;12:81.
187. Liu P, Wang L, Wan ZY, Ye BQ, Huang S, Wong S-M, Yue GH. Mapping QTL for resistance against viral nervous necrosis disease in Asian seabass. Mar Biotechnol. 2016;18(1):107–16.
188. Wang CM, Lo LC, Zhu ZY, Yue GH. A genome scan for quantitative trait loci affecting growth-related traits in an F1 family of Asian seabass (*Lates calcarifer*). BMC Genomics. 2006;7:274.
189. Xia JH, Lin G, He X, Yunping B, Liu P, Liu F, Sun F, Tu R, Yue GH. Mapping quantitative trait loci for omega-3 fatty acids in Asian seabass. Mar Biotechnol. 2014;16(1):1–9.

190. Baranski M, Moen T, Våge DI. Mapping of quantitative trait loci for flesh colour and growth traits in Atlantic salmon (*Salmo salar*). Genet Sel Evol. 2010;42:17.

191. Tsai H, Hamilton A, Guy D, Houston R. Single nucleotide polymorphisms in the insulin-like growth factor 1 (IGF1) gene are associated with growth-related traits in farmed Atlantic salmon. Anim Genet. 2014;45(5):709–15.

192. Tsai HY, Hamilton A, Guy DR, Tinch AE, Bishop SC, Houston RD. The genetic architecture of growth and fillet traits in farmed Atlantic salmon (*Salmo salar*). BMC Genet. 2015;16:51.

193. Gutierrez AP, Lubieniecki KP, Fukui S, Withler RE, Swift B, Davidson WS. Detection of quantitative trait loci (QTL) related to grilsing and late sexual maturation in Atlantic salmon (*Salmo salar*). Mar Biotechnol. 2014;16(1):103–10.

194. Gonen S, Baranski M, Thorland I, Norris A, Grove H, Arnesen P, Bakke H, Lien S, Bishop SC, Houston R. Mapping and validation of a major QTL affecting resistance to pancreas disease (salmonid alphavirus) in Atlantic salmon (*Salmo salar*). Heredity. 2015;115(5):405–14.

195. Zhang Y, Xu P, Lu C, Kuang Y, Zhang X, Cao D, Li C, Chang Y, Hou N, Li H. Genetic linkage mapping and analysis of muscle fiber-related QTLs in common carp (*Cyprinus carpio L.*). Mar Biotechnol. 2011;13(3):376–92.

196. Boulton K, Massault C, Houston RD, de Koning DJ, Haley CS, Bovenhuis H, Batargias C, Canario AV, Kotoulas G, Tsigenopoulos CS. QTL affecting morphometric traits and stress response in the gilthead seabream (*Sparus aurata*). Aquaculture. 2011;319(1):58–66.

197. Laghari MY, Lashari P, Zhang X, Xu P, Narejo NT, Liu Y, Mehboob S, Al-Ghanim K, Zhang Y, Sun X. Mapping QTLs for swimming ability related traits in *Cyprinus carpio L.* Mar Biotechnol. 2014;16(6):629–37.

198. Louro B, Kuhl H, Tine M, de Koning D-J, Batargias C, Volckaert FA, Reinhardt R, Canario AV, Power DM. Characterization and refinement of growth related quantitative trait loci in European sea bass (*Dicentrarchus labrax*) using a comparative approach.

Aquaculture. 2016;455:8–21.

199. Massault C, Hellemans B, Louro B, Batargias C, Van Houdt J, Canario A, Volckaert F, Bovenhuis H, Haley C, De Koning D. QTL for body weight, morphometric traits and stress response in European sea bass *Dicentrarchus labrax*. Anim Genet. 2010;41(4):337–45.
200. Robinson NA, Gopikrishna G, Baranski M, Katneni VK, Shekhar MS, Shanmugakarthik J, Jothivel S, Gopal C, Ravichandran P, Gitterle T. QTL for white spot syndrome virus resistance and the sex-determining locus in the Indian black tiger shrimp (*Penaeus monodon*). BMC Genomics. 2014;15:731.
201. Wang L, Fan C, Liu Y, Zhang Y, Liu S, Sun D, Deng H, Xu Y, Tian Y, Liao X. A genome scan for quantitative trait loci associated with Vibrio anguillarum infection resistance in Japanese flounder (*Paralichthys olivaceus*) by bulked segregant analysis. Mar Biotechnol. 2014;16(5):513–21.
202. Sauvage C, Boudry P, De Koning DJ, Haley CS, Heurtebise S, Lapègue S. QTL for resistance to summer mortality and OsHV-1 load in the Pacific oyster (*Crassostrea gigas*). Anim Genet. 2010;41(4):390–9.
203. Negrín-Báez D, Navarro A, Afonso JM, Ginés R, Zamorano MJ. Detection of QTL associated with three skeletal deformities in gilthead seabream (*Sparus aurata L.*): Lordosis, vertebral fusion and jaw abnormality. Aquaculture. 2015;448:123–7.
204. Loukovitis D, Sarropoulou E, Tsigenopoulos CS, Batargias C, Magoulas A, Apostolidis AP, Chatziplis D, Kotoulas G. Quantitative trait loci involved in sex determination and body growth in the gilthead sea bream (*Sparus aurata L.*) through targeted genome scan. PLoS One. 2011;6(1):e16599.
205. Massault C, Franch R, Haley C, De Koning D, Bovenhuis H, Pellizzari C, Patarnello T, Bargelloni L. Quantitative trait loci for resistance to fish pasteurellosis in gilthead sea bream (*Sparus aurata*). Anim Genet. 2011;42(2):191–203.
206. Kocmarek AL, Ferguson MM, Danzmann RG. Co-localization of growth QTL with differentially expressed candidate genes in rainbow trout. Genome. 2015;58(9):393–403.
207. Wringe BF, Devlin RH, Ferguson MM, Moghadam HK, Sakhrani

D, Danzmann RG. Growth-related quantitative trait loci in domestic and wild rainbow trout (*Oncorhynchus mykiss*). BMC Genet. 2010;11:63.

208. Leder E, Danzmann R, Ferguson M. The candidate gene, clock, localizes to a strong spawning time quantitative trait locus region in rainbow trout. J Hered. 2006;97(1):74–80.
209. Easton A, Moghadam H, Danzmann R, Ferguson M. The genetic architecture of embryonic developmental rate and genetic covariation with age at maturation in rainbow trout *Oncorhynchus mykiss*. J Fish Biol. 2011;78(2):602–23.
210. Miller MR, Brunelli JP, Wheeler PA, Liu S, REXROAD CE, Palti Y, Doe CQ, Thorgaard GH. A conserved haplotype controls parallel adaptation in geographically distant salmonid populations. Mol Ecol. 2012;21(2):237–49.
211. Perry GM, Martyniuk CM, Ferguson MM, Danzmann RG. Genetic parameters for upper thermal tolerance and growth-related traits in rainbow trout (*Oncorhynchus mykiss*). Aquaculture. 2005;250(1):120–8.
212. Baerwald M, Petersen J, Hedrick R, Schisler G, May B. A major effect quantitative trait locus for whirling disease resistance identified in rainbow trout (*Oncorhynchus mykiss*). Heredity. 2011;106(6):920–6.
213. Campbell NR, LaPatra SE, Overturf K, Towner R, Narum SR. Association mapping of disease resistance traits in rainbow trout using restriction site associated DNA sequencing. G3: Genes/Genomes/Genetics. 2014;4(12):2473–81.
214. Rodriguez MF, LaPatra S, Williams S, Famula T, May B. Genetic markers associated with resistance to infectious hematopoietic necrosis in rainbow and steelhead trout (*Oncorhynchus mykiss*) backcrosses. Aquaculture. 2004;241(1):93–115.
215. Le Bras Y, Dechamp N, Krieg F, Filangi O, Guyomard R, Boussaha M, Bovenhuis H, Pottinger TG, Prunet P, Le Roy P. Detection of QTL with effects on osmoregulation capacities in the rainbow trout (*Oncorhynchus mykiss*). BMC Genet. 2011;12(1):46.
216. Rexroad CE, Vallejo RL, Liu S, Palti Y, Weber GM. Quantitative trait loci affecting response to crowding stress in an F2 generation

of rainbow trout produced through phenotypic selection. Mar Biotechnol. 2013;15(5):613–27.

217. Liu S, Vallejo RL, Gao G, Palti Y, Weber GM, Hernandez A, Rexroad III CE. Identification of single-nucleotide polymorphism markers associated with cortisol response to crowding in rainbow trout. Mar Biotechnol. 2015;17(3):328–37.

218. Sánchez-Molano E, Cerna A, Toro MA, Bouza C, Hermida M, Pardo BG, Cabaleiro S, Fernández J, Martínez P. Detection of growth-related QTL in turbot (*Scophthalmus maximus*). BMC Genomics. 2011;12:473.

219. Rodríguez-Ramilo ST, Toro MA, Bouza C, Hermida M, Pardo BG, Cabaleiro S, Martínez P, Fernández J. QTL detection for Aeromonas salmonicida resistance related traits in turbot (*Scophthalmus maximus*). BMC Genomics. 2011;12:541.

220. Rodríguez-Ramilo S, Fernández J, Toro M, Bouza C, Hermida M, Fernández C, Pardo B, Cabaleiro S, Martínez P. Uncovering QTL for resistance and survival time to Philasterides dicentrarchi in turbot (*Scophthalmus maximus*). Anim Genet. 2013;44(2):149–57.

221. Rodríguez-Ramilo ST, De La Herrán R, Ruiz-Rejón C, Hermida M, Fernández C, Pereiro P, Figueras A, Bouza C, Toro MA, Martínez P. Identification of quantitative trait loci associated with resistance to viral haemorrhagic septicaemia (VHS) in turbot (*Scophthalmus maximus*): a comparison between bacterium, parasite and virus diseases. Mar Biotechnol. 2014;16(3):265–76.

222. Liu F, Sun F, Xia JH, Li J, Fu GH, Lin G, Tu RJ, Wan ZY, Quek D, Yue GH. A genome scan revealed significant associations of growth traits with a major QTL and GHR2 in tilapia. Sci Rep. 2014;4:7256.

223. Wang R, Li L, Huang Y, Luo F, Liang W, Gan X, Huang T, Lei A, Chen M, Chen L. Comparative genome analysis identifies two large deletions in the genome of highly-passaged attenuated Streptococcus agalactiae strain YM001 compared to the parental pathogenic strain HN016. BMC Genomics. 2015;16:897.

224. Palaiokostas C, Bekaert M, Khan MG, Taggart JB, Gharbi K, McAndrew BJ, Penman DJ. A novel sex-determining QTL in Nile tilapia (*Oreochromis niloticus*). BMC Genomics. 2015;16:171.

CHAPTER

6

OPTIMIZING THE CREATION OF BASE POPULATIONS FOR AQUACULTURE BREEDING PROGRAMS USING PHENOTYPIC AND GENOMIC DATA AND ITS CONSEQUENCES ON GENETIC PROGRESS

Jesús Fernández [1], Miguel Á. Toro[2], Anna K. Sonesson[3] and Beatriz Villanueva[1]

[1]Departamento de Mejora Genética Animal, INIA, Madrid, Spain

[2]Departamento de Producción Animal, ETSI Agrónomos, Universidad Politécnica de Madrid, Madrid, Spain

[3]Nofima, Ås, Norway

ABSTRACT

The success of an aquaculture breeding program critically depends on the way in which the base population of breeders is constructed since all the genetic variability for the traits included originally in the breeding goal as well as those to be included in the future is contained in the initial founders. Traditionally, base populations were created from a number of wild strains by sampling equal numbers from each strain. However, for some aquaculture species improved strains are already available and, therefore, mean phenotypic values for economically important traits can be used as a criterion to optimize the sampling when creating base populations. Also, the increasing availability of genome-wide genotype information in aquaculture species could help to refine the estimation of relationships within and between candidate strains and, thus, to optimize the percentage of individuals to be sampled from each strain. This study explores the advantages of using phenotypic and genome-wide information when constructing base populations for aquaculture breeding programs in terms of initial and subsequent trait performance and genetic diversity level. Results show that a compromise solution between diversity and performance can be found when creating base populations. Up to 6% higher levels of phenotypic performance can be achieved at the same level of global diversity in the base population by optimizing the selection of breeders instead of sampling equal numbers from each strain. The higher performance observed in the base population persisted during 10 generations of phenotypic selection applied in the subsequent breeding program.

Keywords: base populations, aquaculture breeding programs, SNP markers, optimal contributions, fish genomics

INTRODUCTION

The success of an aquaculture breeding program critically depends on the way in which the base population of breeders is constructed (Hayes et al., 2006; Holtsmark et al., 2006, 2008a,b). In particular, all the genetic variability that can be used for selection on the traits initially included in the breeding objective is that found in the original breeders. Also, the decisions taken when creating the base population will have consequences on the genetic progress for any other additional trait that may be part of future breeding goals, whatever production or fitness related traits.

Traditionally, base populations in aquaculture breeding programs were created from wild or from domesticated strains not subjected to any formal selection program (see Gjedrem et al., 1991, for details on the formation of the base population in the Norwegian breeding program for Atlantic salmon and Eknath et al., 1993, 1998, for details on the formation of the base population in the GIFT breeding program for tilapia). In this context, Holtsmark et al. (2006, 2008a) investigated through computer simulation the effects of the number of strains contributing to the base population and the mating strategy (within and across strains) on the genetic gain achieved in the subsequent selection program. These studies assumed wild populations and no knowledge of the genetic structure (i.e., within and between strain diversity) or phenotypic levels for the trait of interest when setting up the base population. Consequently, the simulated strategy was randomly sampling the same number of individuals from each strain to form the base population. However, the solution leading to the highest level of diversity and acceptable levels of phenotypic performance when starting the breeding program surely will depart from equal proportions.

In the absence of genealogies (which could be the case even in improved commercial strains), molecular markers can be used to estimate relationships between and within populations. Hayes et al. (2006)compared random with molecular-based optimized selection of breeders in terms of the genetic variance captured for growth and for two disease traits in Atlantic salmon. However, no considerations about the phenotypic level of candidates were made when sampling the breeders and the number of markers they used was scarce (237 AFLP).

The present availability of large panels of SNPs makes marker diversity more informative on the global genetic variability in the genome than diversity computed from genealogical data (Gómez-Romano et al., 2013). Nowadays, dense SNP chips have been already developed for Atlantic salmon (130K, Houston et al., 2014), rainbow trout (57K, Palti et al., 2014), catfish (250K, Liu et al., 2014) and carp (250K, Xu et al., 2014). This development of an increasing number of markers for aquaculture species will make it possible in a near future to have accurate measures of genetic relationships between and within strains. In this new scenario, the optimal number of individuals to be sampled from each strain could be calculated in a similar way as when optimizing the construction of mixed populations from different origins in conservation programs aimed at capturing the highest levels of genetic diversity (Eding and Meuwissen, 2001; Caballero and Toro, 2002; Eding et al., 2002).

Nowadays, for some aquaculture species already genetically improved strains are available. The use of improved rather than wild strains when creating base populations would allow the new breeding program to begin from higher phenotypic levels for the trait of interest, making it competitive from the start. The necessity of taking into account the phenotypic value of each candidate strain is even clearer when searching for new breeders to be included in an already established breeding program whose genetic variability has been greatly reduced. The increase in genetic variability should be achieved without compromising the gain in performance previously obtained through artificial selection.

The objective of this paper was to study, using computer simulations, the consequences of using genome-wide molecular information to compute genetic relationships and phenotypic records to optimize the sampling of individuals from different strains when creating base populations in aquaculture breeding programs. Different scenarios varying in the type of strains (wild or commercial), the degree of relationships within and between strains and the level of information (individual or strain information) were considered for generic aquaculture population designs. Results were compared in terms of phenotypic level for the trait of interest and the diversity achieved in the base population and in subsequent generations of selection.

MATERIALS AND METHODS

Genome Structure

Diploid individuals were simulated with a genome comprising 20 chromosomes of 1 Morgan each. This genome length could be representative of the genomic architecture of the most cultivated fish species in aquaculture (which have genomes ranging from 10 to 35 Morgans). Each chromosome carried 25,000 neutral biallelic loci that will be referred to as "non-marker loci" thereafter. Additionally, 5–5000 biallelic markers per chromosome (equivalent to SNPs) were simulated interspersed with the neutral loci. Therefore, the total number of available markers ranged from 100 to 100,000 and the density of markers ranged from 5 to 5000 markers/M. All loci were evenly spaced within each chromosome. Markers were used to optimize the contributions of the different strains to the base population while non-marker loci (i.e., the 25,000 neutral loci per chromosome) were used to monitor the effect on genetic diversity of the different strategies evaluated (see below). With dense marker panels we expect that most non-marker loci in the genome will be in linkage disequilibrium (LD) with at least some of the markers and, thus, managing diversity using marker genotypes will lead to the maintenance of diversity in the rest of the genome.

Generation of Candidates

The creation of the candidates to contribute to the base population followed a two-step process. First, a large population in mutation-drift equilibrium was generated. Then, individuals were sampled from this population to form different strains which were allowed to diverge for several generations before being available for the selection of the individuals to be included in the base population.

Equilibrium population

First, to obtain a realistic pattern of linkage disequilibrium, a large population (N = 1000 with equal number of males and females) under random selection was simulated for 1000 discrete generations. Each generation, sires and dams were sampled with replacement and population size was kept constant across generations. Initially, genotypes for the individuals were assigned at random independently for each locus (markers and non-markers). Consequently, the initial allelic frequencies were 0.5 for all loci and Hardy-Weinberg and linkage equilibria existed within and between loci, respectively. During this period, mutation was allowed to occur throughout the genome. The mutation rate per locus and generation was $\mu = 2.5 \times 10^{-3}$ for both types of loci (marker and non-marker). The number of new mutations simulated at every generation was sampled from a Poisson distribution with mean $2Nn_c\mu n_l$ where n_c is the number of chromosomes and n_l is the total number of loci (markers and non-markers) per chromosome. Mutations were then randomly distributed across individuals, chromosomes and loci, switching allele 0 to allele 1 and vice versa. When generating the gametes, the number of crossovers per chromosome was drawn from a Poisson distribution with mean equal to 1. Crossovers were randomly distributed without interference. At the end of the process the expected heterozygosity of the population had already reached an equilibrium value.

Creation of strains

The second step of the process consisted in randomly sampling 10 different groups of individuals (mimicking 10 different strains) from the population at equilibrium. At this step, a quantitative trait with phenotypic mean (μ), initial phenotypic variance [$V_{P(0)}$] and heritability [$h^2_{(0)}$] of 100, 30, and 0.4, respectively, was defined. The trait, measured in both sexes, was controlled by 1000 additive loci (thereafter called selective loci). These selective loci were chosen at random from the previously simulated loci (markers and non-markers). The additive effect of locus *i* (a_i) was sampled from a normal distribution with mean 0 and variance $V_A(0)/[2p(1-p)n_{sel}]$, where $V_{A(0)}$ is the initial additive variance ($h^2_{(0)}V_P(0) = 12$), *p* is the average frequency across selective loci and n_{sel} is the number of selective loci (i.e., 1000). Note that, in this way, the expected additive variance summed over all loci equals $V_A(0)$, assuming that covariances between loci generated by LD are negligible. The phenotypic value for a particular individual *j* was obtained as $P_j = \mu + \sum_{i=1}^{nsel} x_i a_i + e_j$, where x_i is an indicator variable that takes values 1, 0 or −1 for homozygous 11, heterozygous or homozygous 00, respectively, and e_j is the individual environmental deviation that was sampled from a normal distribution with mean 0 and variance $V_E = V_A(0)(1-h^2_{(0)})/h^2_{(0)}$. The environmental variance (V_E) was initially calculated for each replicate and strain in order to assure that all started with the same $h^2_{(0)}$ value and was kept constant across generations.

Once the strains were created, they were allowed to diverge during 20 discrete generations (with constant population size) under three different regimes: (i) random selection; (ii) artificial selection with different selection pressures in order to mimic already improved strains; and (iii) stabilizing selection with different optima in order to mimic wild strains with local adaptations in nature. Four different scenarios varying in the type of strains they included were then defined:

- Drift Scenario. All strains were created under random selection but they differed in size (from 10 to 40 individuals, half of each sex) in order to simulate different degrees of genetic drift. Smaller sizes will lead to lower within strain diversity but higher differentiation between strains.

- Selection Scenario. All strains were created under artificial directional selection for the simulated trait. They had all the same size (20 males and 20 females) but differed in the strength of the selection imposed. The highest selection pressure corresponded to selecting the two males and the two females with the highest phenotypic values for the trait, and the lowest pressure corresponded to selecting five males and five females.
- Stabilizing Scenario. All strains were created under stabilizing selection for the simulated trait. The probability of survival of an individual with phenotypic value P was modeled using a Gaussian distribution (Turelli, 1984; Bürger et al., 1989):

$$W(P) = \exp\left(-\frac{(P - P_{opt})^2}{2\omega^2}\right)$$

where P_{opt} is the optimum phenotype in a particular environment and ω^2 is an inverse measure of the strength of stabilizing selection. For some strains P_{opt} was set to the original mean (100), for others the optimum was set to a lower value (90) and for others the optimum was set to a higher value (110). Selection pressure was relatively strong ($\omega^2 = 5$) and population size was equal in all strains (20 males and 20 females).

- Mixed Scenario. The set comprised strains of the three types described above (i.e., randomly selected, under directional selection and under stabilizing selection).

Table 1 summarizes the specific parameters used when creating the 10 strains under each scenario. The sizes of the strains simulated during the 20 generations were rather small in order to force a rapid divergence between them and, thus, real differences in relatedness and phenotypic levels at the end of the differentiation stage. After this period, subpopulations were expanded in order to have a large number of candidates to form the base population. Specifically, the population size of each strain increased to 50 males and 50 females in a single generation of random selection and mating.

Table 1: Parameters used to generate each strain for the four different scenarios simulated

	Drift	Selection		Stabilizing		Mixed		
Strains	Size	Size	Numbers selected	Size	Optimum phenotype	Size	Optimum phenotype	Numbers selected
1	5	20	2	10	90	5	-	-
2	5	20	2	10	90	10	-	-
3	5	20	3	10	90	10	-	-
4	10	20	3	10	100	20	-	-
5	10	20	4	10	100	10	90	-
6	10	20	4	10	100	10	100	-
7	20	20	4	10	100	10	110	-
8	20	20	5	10	110	10	-	5
9	20	20	5	10	110	10	-	3
10	20	20	5	10	110	10	-	1

Foundation of the Base Population

From the 1000 available candidates (50 males and 50 females from each strain) for each particular scenario, base populations were constructed by selecting 100 males and 100 females following different strategies: (i) Taking at random equal numbers of individuals from each strain (strategy E); (ii) Determining optimal strain proportions for maximizing the expected heterozygosity (H_e) calculated from the mean coancestry values within and between strains (strategy MC); (iii) Determining optimal strain proportions for maximizing the mean (strain) phenotypic value with a restriction on coancestry (strategy MP); (iv) as in (ii) but using individual relationships instead of strain means (strategy IC); and (v) as in (iii) but maximizing individual values of the selected individuals instead of strain means (strategy IP). Note that, in these abbreviations, M and I stand for mean and individual information, respectively, C indicates that the objective is just to minimize coancestry (i.e., maintain high levels of H_e) and P indicates that the objective is to create base populations with a high phenotypic level.

Strategy E is equivalent to that used by Holtsmark et al. (2006, 2008a,b) and provides a reference point for comparisons. Strategy MC followed the methodology presented by Eding and Meuwissen (2001) and Caballero and Toro (2002). The particular objective function to minimize was $\sum_{i=1}^{N_s} \sum_{j=1}^{N_s} c_i c_j f_{ij}$, where N_s is the number of strains, c_i is the proportion of individuals to be sampled from strain i and f_{ij} is the mean coancestry coefficient between strains i and j calculated from marker genotypes. Contributions were forced to be in the interval [0,1] and to sum up to 1.

Strategy MP searched for the solution that maximized the objective function $\sum_{i=1}^{N_s} c_i P_i$, where P_i is the mean phenotype of strain i but imposing the restriction $\sum_{i=1}^{N_s} \sum_{j=1}^{N_s} c_i c_j f_{ij} \leq C_E$, where C_E is the mean coancestry of the base population obtained under strategy E. The restrictions imposed in MC were also applied to MP. In the two strategies relying on strain mean values (MC and MP) the actual number of individuals to be sampled from each strain was obtained by multiplying c_i by the total number of individuals to be selected and rounding to the nearest even integer. This procedure was implemented to ensure that half of the individuals from each strain were males and half females.

The objective under strategies IC and IP was to minimize $\sum_{i=1}^{N} \sum_{j=1}^{N} x_i x_j f_{ij}$ and to maximize $\sum_{i=1}^{N} x_i P_i$, respectively. Here N is the total number of candidates (i.e., all individuals from every strain), f_{ij} is the coancestry between individuals i and j, P_i is the phenotype of individual i and x_i is an indicator variable that takes a value of 1 if individual i is to be selected and 0 otherwise. The sum of x's for males and females was forced to be equal to the number of individuals to be selected for creating the base population (i.e., 100 of each sex). Strategy IP also included a restriction to guarantee that solutions had a global coancestry lower or equal than that obtained under E strategy. All the optimizations were performed using "simulated annealing" algorithms (Kirkpatrick et al., 1983).

It must be emphasized that in strategies MC and MP the output of the optimizations is the proportion of individuals to be taken from each strain but the specific individuals are sampled at random. This could be the situation when prior knowledge of the mean genetic relationship or performance of the strains is available but individual information for the candidates is not. In strategies IC and IP the outputs are the particular individuals to be selected as we assume that their genotypes and phenotypic values are known.

As described above, restrictions imposed under strategies MP and IP were set to the level of coancestry (calculated on the markers) obtained under strategy E. Thus, these three strategies could be compared in terms of phenotypic level at the same diversity level.

Artificial Selection

To explore the consequences of each strategy for creating the base population on genetic gain and diversity in the subsequent breeding programme, 10 generations of artificial selection were simulated for each combination of marker density, type of available strains and strategy. At generation $t = 0$ (i.e., base population), founders were mated at random to form 100 families and 10 offspring were obtained from each couple. Consequently, 1000 individuals were available as candidates for selection. The 100 males and 100 females with the highest phenotypic value for the simulated trait were selected to produce generation $t+1$ (i.e., phenotypic truncation selection was conducted). Selected individuals were mated at random and, again, 10 offspring were generated from each couple. Therefore, the proportion of selected individuals was 20% that corresponds to a selection intensity of 1.4.

Variables for Comparison

In the base population, comparisons between strategies were made in terms of the mean phenotypic value ($\overline{\mathrm{P}}$) and H_e of the group of selected individuals. Note that H_e, calculated on the non-marker loci, is a measure of the genetic variation of the population and its ability to adapt to new environments. The contributions of strains to the base population, measured as the proportion of breeders selected from each of the strains, was also considered in the comparisons.

In the artificial selection step, the strategies were also compared in terms of mean breeding value (BV) and additive variance (V_A) for the target trait, genealogical inbreeding (F) and coancestry (f) coefficients and rates of gain, inbreeding (ΔF) and coancestry (Δf). For the computation of inbreeding and coancestry, founders in the base population were assumed to be unrelated and non-inbred. In all scenarios, values presented are averages of 100 replicates.

RESULTS

Contributions of Strains to the Base Population

The proportional contribution of each available strain to the base population is shown in Table 2 for the most extreme marker densities, the four scenarios and the five strategies simulated. In general, the observed patterns were the same when using high or low density of markers. Particular differences in performance due to marker density are highlighted below.

Table 2: Contributions of each strain to the base population (in percentage) under different strategies to select individuals for the base population for the four scenarios considered and for different number of markers

Markers			Drift				Selection				Stabilizing				Mixed			
		E	MC	MP	IC	IP	MC	MP	IC	IP	MC	MP	IC	IP	MC	MP	IC	IP
100	1	10.0	−3.3	−3.6	−3.6	−2.9	−1.5	−3.8	−1.6	−4.3	0.6	−4.1	−0.1	−7.6	−0.9	−2.9	−1.2	−4.1
	2	10.0	−5.0	−4.0	−4.4	−1.8	−1.8	−4.5	−1.7	−5.0	−0.7	−5.1	−0.6	−7.6	4.0	1.0	3.3	−1.5
	3	10.0	−5.1	−4.4	−4.5	−2.1	0.1	−0.2	−0.3	−0.5	−1.1	−5.4	−0.7	−7.8	2.1	−1.2	2.5	−2.5
	4	10.0	−0.5	0.6	−0.6	1.0	−0.9	−1.5	−0.7	−1.2	1.5	1.7	0.8	0.4	12.4	5.4	9.8	0.1
	5	10.0	−2.6	−1.4	−1.6	0.5	0.5	1.6	0.1	1.3	0.7	0.7	0.9	0.2	−2.3	−7.7	−1.7	−8.9
	6	10.0	−2.2	−3.1	−1.5	−1.9	−0.4	0.1	0.0	0.8	0.3	−0.1	0.6	−0.6	−1.9	−3.8	−1.1	−4.2
	7	10.0	4.6	4.3	3.9	2.0	−0.3	0.7	0.1	1.3	−0.5	−0.7	−0.1	−0.7	−2.7	−0.4	−2.0	1.0
	8	10.0	4.2	3.8	3.9	2.1	2.0	3.2	1.3	2.7	−0.3	4.3	−0.7	7.7	−1.7	7.0	−1.0	10.9
	9	10.0	5.8	5.3	5.1	1.9	0.0	1.0	0.5	1.6	−0.9	3.8	−0.5	7.6	−4.1	3.8	−3.5	8.3
	10	10.0	4.1	2.3	3.5	1.1	2.3	3.3	2.3	3.4	0.4	5.0	0.4	8.5	−4.9	−1.2	−5.1	0.8
		0.0	*16.2*	*12.7*	*12.6*	*3.4*	*1.6*	*6.2*	*1.3*	*7.1*	*0.6*	*13.4*	*0.4*	*36.7*	*23.6*	*15.0*	*16.2*	*31.1*
100,000	1	10.0	−4.0	−3.8	−4.8	−3.3	−2.0	−3.5	−2.0	−3.0	0.4	−1.3	−0.4	−2.0	−1.2	−2.8	−2.0	−3.2
	2	10.0	−4.9	−4.7	−4.8	−3.5	−2.1	−3.6	−1.9	−3.0	−0.6	−2.3	−0.5	−2.2	3.3	0.9	2.7	0.4
	3	10.0	−5.0	−4.8	−4.7	−3.5	−0.9	−0.7	−1.0	−0.9	−0.7	−2.4	−0.4	−2.1	2.5	0.3	2.8	0.6
	4	10.0	−0.9	−0.5	−1.5	−0.3	−1.2	−1.1	−1.0	−0.8	1.3	1.3	0.7	0.6	12.4	8.3	12.2	6.8
	5	10.0	−1.8	−1.3	−1.6	−0.2	1.0	1.8	0.4	0.9	0.4	0.4	0.7	0.6	−2.4	−7.4	−2.2	−6.5
	6	10.0	−1.8	−1.2	−1.6	−0.3	0.0	0.6	0.3	0.8	0.5	0.4	0.7	0.7	−1.6	−3.2	−1.3	−2.6
	7	10.0	4.3	2.5	4.6	1.7	−0.1	0.2	0.2	0.4	0.2	0.3	0.6	0.5	−2.6	−0.6	−2.4	−0.3
	8	10.0	5.2	6.0	4.9	4.0	2.4	2.8	1.7	1.9	−0.2	1.6	−0.4	1.3	−1.4	5.4	−1.2	5.3
	9	10.0	4.5	3.8	4.8	2.7	1.4	1.7	1.7	1.8	−0.6	1.0	−0.4	1.3	−3.9	1.6	−3.7	1.8
	10	10.0	4.5	3.9	4.8	2.8	1.4	1.8	1.6	2.0	−0.7	0.9	−0.5	1.2	−5.1	−2.4	−5.1	−2.2
		0.0	*15.9*	*13.6*	*16.7*	*7.0*	*2.1*	*4.4*	*1.8*	*3.2*	*0.4*	*2.0*	*0.3*	*1.9*	*23.0*	*18.0*	*22.1*	*14.1*

Contributions for strategies MC, MP, IC, and IP are given as deviations from those of strategy, E. The variance of contributions across strains is given in italics. E, equal numbers from each strain; MC, minimization of coancestry based on mean strain values; MP, maximization of phenotype with restriction on coancestry based on mean strain values; IC, minimization of coancestry based on individual values; IP, maximization of phenotype with restriction on coancestry based on individual values.

In the Drift scenario no meaningful differences in contributions were observed between strategies MC, MP, and IC for a particular strain. This was due to the fact that no selection on the quantitative trait was exerted during the generation of the strains and, therefore, the phenotypic mean was equal to the initial value before the divergence period (i.e., 100) for all strains. However, differences arose between strains. The higher the population size of a particular strain the higher was its contribution. Strains with a small size (strains 1, 2, and 3) had the lowest contributions due to the large loss of genetic diversity during the divergence period. Therefore, these strains were less useful for increasing the amount of diversity stored in the synthetic base population. The opposite happened with large size strains (7, 8, 9, and 10) that contributed more than 14% each to the base population (see Drift scenario in Table 2). The lowest variance of contributions between strains (beyond strategy E) was found under strategy IP. This could be explained by the fact that strong drift in small populations could result in the existence of individuals with extreme high phenotype. Thus, as the main objective of strategy IP is achieving a high phenotypic level in the base population, it would be worthy to keep individuals not only from large but also from small strains. In this situation a high phenotypic level can be obtaining without reducing too much the diversity maintained.

In the Selection scenario, the variance of contributions across strains was much lower than in the Drift scenario (1.3–6.2 vs. 12.6–16.7; Table 2) except for strategy IP with low marker density. This is a consequence of the greater uniformity between strains, at least for the genetic variability of the trait. In the Selection scenario all strains had been under directional selection for the same trait and all had the same size. The only difference between strains was the strength of selection. Those under the weakest selection pressure (8, 9, and 10) had a higher effective population size (N_e), maintained higher levels of genetic diversity and, thus, in average contribute more to the base population. Strains subjected to a strong selection pressure were those with the lowest contributions, even under the strategies directed to keep high levels of trait performance (see strains 1 and 2, Selection scenario in Table 2). This is due to the relative long time period since the separation of the strains (20 generations). The small N_e induced by the selection pressure erodes rapidly not only neutral variability but also the genetic variance of the trait. Therefore, strains

1 and 2 reached a selection limit before the rest and presented lower phenotypic mean values at the time the base population was created (data not shown). Consequently, they cannot contribute much to diversity nor to trait value either.

In the Stabilizing scenario, when the criteria for choosing individuals to constitute the base population included considerations on the phenotypic performance (i.e., strategies MP and IP) the contribution of a strain was proportional to its phenotypic mean (Table 2). Contrarily, contributions were almost equalized when the only concern was to keep the highest levels of diversity (strategies MC and IC), given that all strains had an identical population size and a similar selection pressure in the divergence period. An interesting observation was that the variance of contributions across strains greatly decreased for strategies MP and IP when a large panel of SNPs (100,000) was used (see lower section of Table 2). With dense genotyping, diversity at selective loci is tightly linked to neutral diversity and, thus, groups of individuals with high phenotype will also have low diversity at the markers. Therefore, optimal solutions include the selection of fewer individuals from the same high performance strain to cope with the restriction on genetic diversity. The lower diversity of high performance groups of individuals is not detected with sparse marker coverage as diversity at selective loci is loosely linked to neutral diversity.

The Mixed scenario included all kind of strains and, thus, the performance was somehow more complex. Notwithstanding, the general patterns highlighted before (i.e., dependence of contributions on the historical size, the intensity of selection and the optimal phenotypic value, respectively) are still observable (see right part of Table 2). A particular observation was that the contribution of strains adapted to a low phenotypic value (i.e., strains created under stabilizing selection with a low optimum) was no longer required even with massive genotyping. Also, it was observed that under strategies MP and IP the low diversity of high performance individuals may be compensated for the high diversity of the "drift" strains with reasonable phenotypic levels.

Diversity and Phenotypic Level in the Base Population

The mean phenotypic value for the trait of interest and the genetic diversity captured in the group of individuals conforming the base population are presented in Table 3. It must be pointed out that all strategies except E are explicitly concerned with the maintenance of diversity. However, only MP and IP strategies included considerations about the phenotypic level in the objective function to optimize.

Table 3: Average phenotypic value and expected heterozygosity (in percentage) under different strategies to select individuals for the base population for the four scenarios considered and for different number of markers (n_m)

Scenario	n_m	Phenotypic value					Expected heterozygosity				
		E	MC	MP	IC	IP	E	MC	MP	IC	IP
Drift	100	99.93	99.88	101.44	99.97	106.14	46.21	46.23	45.63	46.32	45.77
	1000	99.81	99.73	100.98	99.76	105.78	46.22	46.49	46.11	46.51	46.14
	100,000	100.06	100.00	101.29	99.98	105.97	46.18	46.48	46.12	46.60	46.19
Selection	100	129.02	129.12	130.45	129.09	134.41	44.54	44.17	43.69	44.26	43.64
	1000	127.96	128.11	128.82	128.07	132.81	44.55	44.56	44.41	44.58	44.36
	100,000	128.26	128.37	129.01	128.04	132.87	44.51	44.57	44.47	44.67	44.51
Stabilizing	100	100.00	99.95	102.70	100.01	107.94	45.14	44.78	44.37	44.87	43.81
	1000	99.99	100.03	102.65	100.08	107.91	45.13	44.75	44.33	44.84	43.80
	100,000	99.02	98.95	99.97	98.99	104.15	44.66	45.00	44.62	45.00	44.65
Mixed	100	106.27	104.07	109.49	104.32	115.50	45.24	45.42	44.70	45.49	43.93
	1000	105.95	103.94	108.15	104.13	113.05	45.34	45.71	45.27	45.74	45.08
	100,000	106.08	103.97	108.12	103.97	112.58	45.33	45.72	45.31	45.84	45.33

Using strategy IP with a large number of markers in the Mixed scenario led to a base population with mean phenotypic value for the trait 7% higher than under strategy E at the same diversity. The advantage of strategy IP over E in other scenarios ranged from about 4% (Selection scenario) to 6% (Drift scenario), being proportional to the degree of differentiation between strains. This is a logical result if we realize that the margin for improvement with unequal contributions was lower when strains were more similar (for example in the Selection scenario). When relying on the information of few markers, IP led to higher trait means than with dense genotyping (left part of Table 3) because diversity at selective loci and at the markers was more loosely linked. Then, groups of individuals with high phenotypic mean for the trait can be found also showing high levels of diversity at the markers and, thus, coping with the restrictions in the optimization. However, the global diversity at non-marker loci will be low (as stated before) yielding solutions that do not achieve the intended balance between phenotype and diversity (right part of Table 3).

In general, the higher the number of markers used to estimate relationships the higher the H_e retained in the base population. However, the differences in H_e with different marker densities were small. The largest difference occurred in the Mixed scenario under strategy IP (3% higher H_e when using 100,000 markers instead of 10 markers). Improvements in the level of H_e maintained under IC, which should be the most efficient strategy in terms of diversity captured were never larger than 1% (Table 3).

The genetic diversity maintained in the base population under different strategies was very similar across scenarios. Except for some cases with low number of markers, the strategy capturing the highest levels of neutral H_e was IC, because no other factor but diversity was included in the objective and decisions were taken on the genotypes of the individual candidates and not on the mean strain values. The advantage of this strategy compared to sampling equal number of individuals from each strain (strategy E) ranged from 0.4% (in the Selection scenario) to 1.1% (in the Mixed scenario). This result was obtained because the Selection scenario and Mixed scenario present the highest and the lowest degree of homogeneity between the available strains, respectively.

When using the average strain values (i.e., all individuals from the same strain assumed equivalent) levels of diversity obtained under MC were always lower than with IC although, as stated before, differences were small. With increasing number of markers differences diminished and, eventually, disappeared (see Table 3, Stabilizing scenario).

Strategies MP and IP were intended to select individuals with high phenotypic performance but keeping the same level of diversity than E. When using a low number of markers these strategies did not fulfill the restriction (i.e., H_e was lower under MP or IP than under E; see Table 3) as this was introduced in the formulation through the molecular coancestry calculated from the markers but diversity results were obtained from the non-markers genotypes. With the largest panel (i.e., 100,000 SNPs) IP did maintain the same diversity level than E but MP maintained slightly lower values because of the random sampling of individuals within strains.

Genetic Gain and Inbreeding from the Breeding Program

The capability to respond to artificial selection depends on the amount of additive genetic variance (V_A) present for the trait. Figure 1 shows V_A along the 10 generations of phenotypic truncation selection for all scenarios and strategies used to construct the base population. The highest initial V_A corresponded to the Mixed scenario as this was the most heterogeneous scenario in terms of types of available strains. The order for the rest of scenarios was Stabilizing, Drift and Selection, following thus the same pattern as that observed for general variability described in the previous section. Within each scenario, patterns of V_A for each strategy were very similar. The highest values were observed for strategy E and the lowest for strategy IP. A large decrease in V_A was observed in early generations in all scenarios due to the Bulmer effect (Falconer and Mackay, 1996).

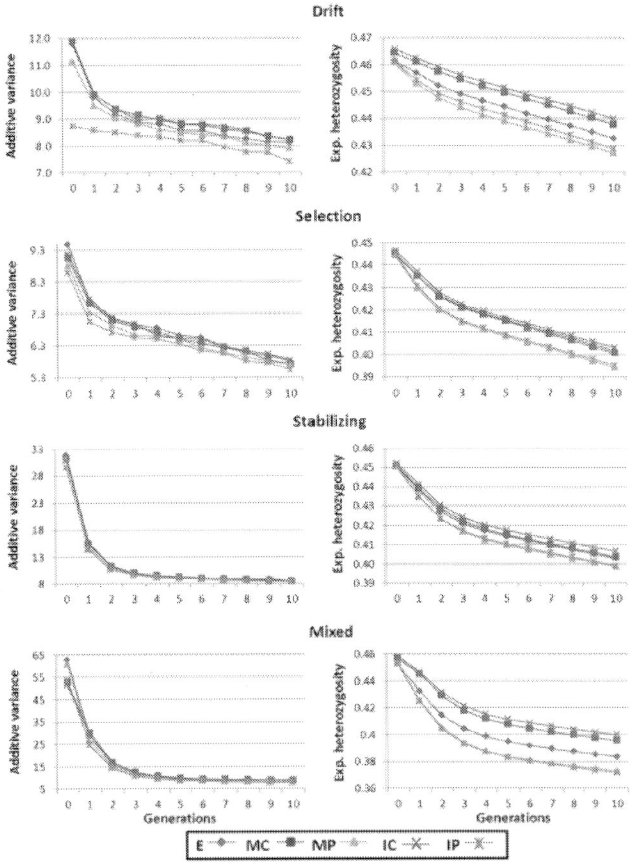

Figure 1: Additive genetic variance for the selected trait and expected heterozygosity for the non-marker loci along the generations of selection. Results shown correspond to base populations obtained using 100,000 markers. E,

equal numbers from each strain; MC, minimize mean strain coancestry values; MP, maximize mean strain phenotypic value with a restriction on coancestry; IC, minimize individual coancestry; and IP, maximize individual phenotypic value with a restriction on coancestry.

Higher V_A values in the founders of the breeding program turned into higher initial responses to selection (right panels in Figure 2). After the initial generations of selection, gain increased at a lower rate and, after the 10 generations observed values were, in general, inversely related to the initial V_A (i.e., scenarios with higher initial V_A maintained lower BV gain for the whole period of selection).

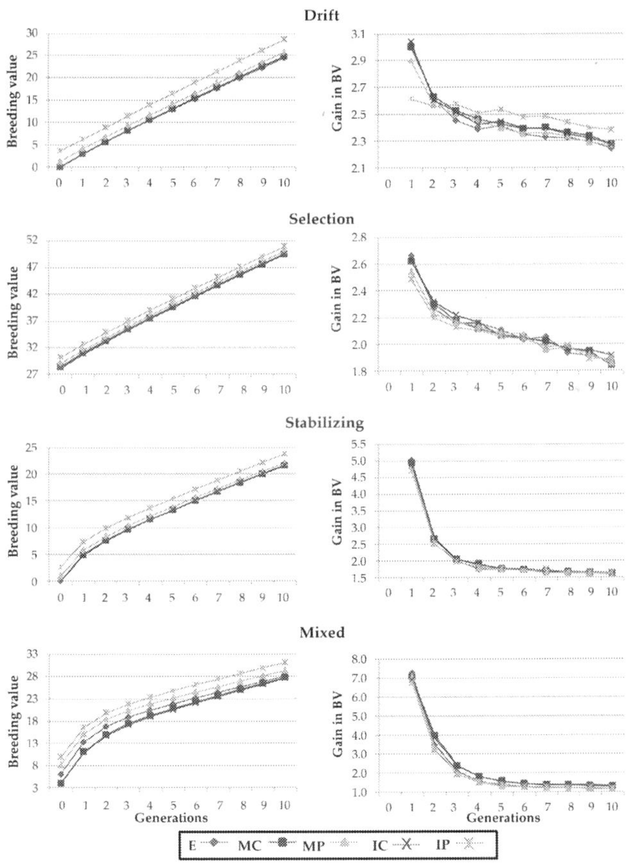

Figure 2: Average breeding value and gain in breeding value along the generations of selection. Results shown correspond to base populations obtained using 100,000 markers. E, equal numbers from each strain; MC, minimize mean

strain coancestry values; MP, maximize mean strain phenotypic value with a restriction on coancestry; IC, minimize individual coancestry; and IP, maximize individual phenotypic value with a restriction on coancestry.

The mean breeding value (BV) of the base population was higher for those scenarios including already selected strains (i.e., Selection and Mixed) and close to zero for the other two scenarios (left panels in Figure 2). Irrespective of the differences in the rate of gain for each scenario pointed out before, the rank of scenarios in terms of BV remained the same for the 10 generations of selection.

For a particular scenario, strategy IP always provided the highest BVs (Figure 2), even although generally started from the lowest V_A (Figure 1). When the number of markers used in the creation of the base population was small the advantage (in terms of mean BV) of strategy IP was greater than that presented in Figure 2 (data not shown) due to the higher initial differences with the rest of strategies in trait performance already shown when constructing the base population. But, as discussed before, strategy IP also resulted in lower H_e values (Table 3). In all scenarios, strategies MC and IC performed almost identically for V_A, BV and gain, especially when using a large panel of markers (left panel in Figure 1 and both in Figure 2). Strategy E yielded similar results to those from MC and IC except for the Mixed scenario at early generations where the average BV was higher for E (left panels in Figure 2). However, at the end of the 10 generations of selection average BV for E, MC, and IC equalized.

It must be highlighted that the trait under selection was simulated with an additive gene action within and across loci. This is the reason for a continuous decay of V_A for the trait and the corresponding decrease of genetic gain between consecutive generations (Figures 1, 2, respectively). In traits with an important non-additive component the selection process may generate new additive variance which could lead to the maintenance of levels of response to selection larger than expected under a pure additive model.

The Drift scenario started from the highest H_e levels and also showed the lowest rate of loss of diversity along the breeding program (right panels in Figure 1). On the other hand, the fastest decrease in diversity was observed in the Mixed scenario and this was related to the large initial responses obtained under this scenario.

For all scenarios, populations arising from strategies accounting for the phenotypic level of the founders (i.e., MP and IP) lost more H_e during the 10 generation of selection than strategies aiming just at keeping diversity (right panels in Figure 1). This could be due to the fact that in groups of selected individuals the genetic variance for the trait would be more correlated to the global genetic diversity under MP and IP strategies than in the other strategies. Consequently, during the breeding programme the reduction in V_A inherent to the selection process also imply greater reductions in H_e across generations. When initial breeders were chosen based on the genotypes for few markers, populations obtained following IC strategy maintained higher levels of diversity along the generations of selection than when using MC (data not shown) but when a large panel of SNPs was used the performance of both strategies was similar (right panels in Figure 1).

As mating was at random throughout the selection process, average inbreeding (F) and coancestry (f) coefficients run in parallel, with the expected lag for F. Therefore, only results for f are shown. Especially for the Mixed scenario (and to a lesser extent for the Stabilizing scenario) f was higher for strategies MC and IC than for the rest (left panels in Figure 3). The reason is that strategies MC and IC keep higher numbers of "low performance" individuals whose descendants will not be selected, leading to higher Δf and, thus, to lower N_e than expected. This effect was not detectable in the Selection and Drift scenarios due to the higher homogeneity between strains and individuals for the phenotypic level of the trait.

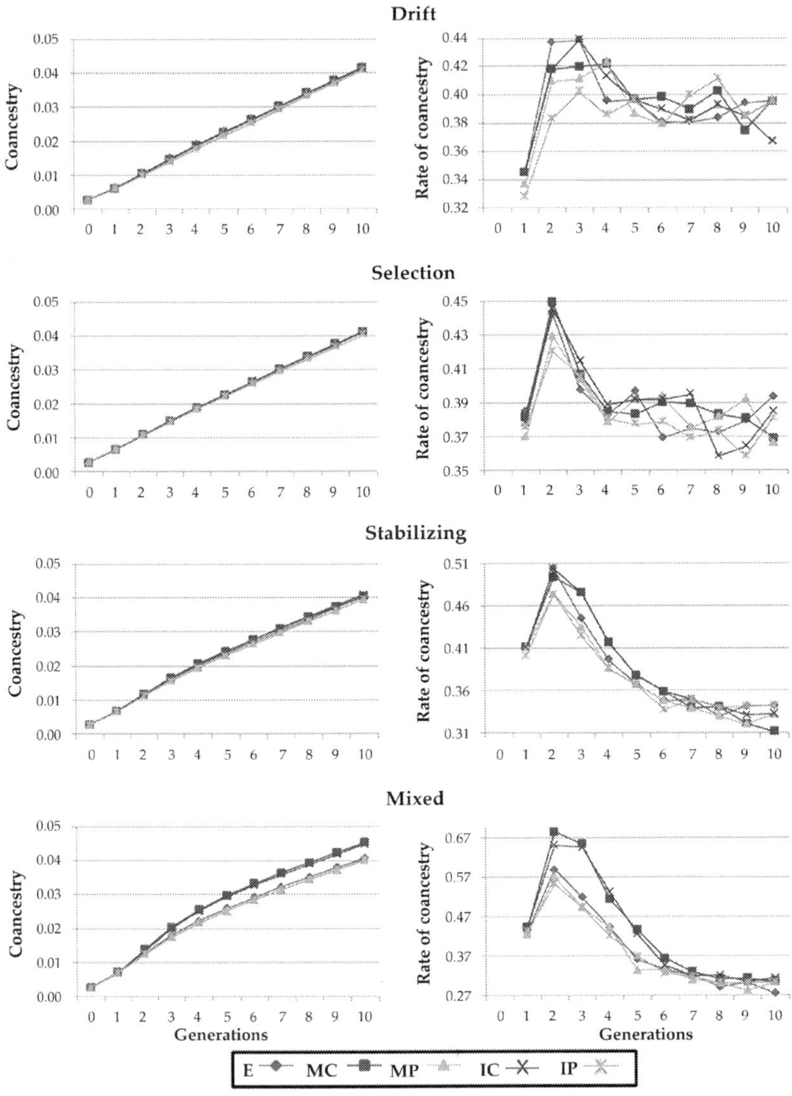

Figure 3: Average genealogical coancestry coefficient and rate of coancestry along the generations of selection. Results shown correspond to base populations obtained using 100,000 markers. E, equal numbers from each strain; MC, minimize mean strain coancestry values; MP, maximize mean strain phenotypic value with a restriction on coancestry; IC, minimize individual coancestry; and IP, maximize individual phenotypic value with a restriction on coancestry.

Irrespective of the scenario, at the beginning of the breeding program there was an increase in Δf (right panels in Figure 3) that was due to the removal of individuals with low genetic BV for the trait. Afterwards, Δf stabilized in the Drift and Selection scenarios around 0.4%. This figure is higher than the expected rate (Δf = 0.25%) for a random selection population of size 200 (i.e., the number of selected individuals each generation; Woolliams and Bijma, 2000) because between-family selection occurs. In the Stabilizing and Mixed scenarios, Δf monotonically decreased reaching levels closer to 0.25%.

DISCUSSION

The present study has shown that the use of phenotypic information of the candidate strains and the use of genome-wide marker information to infer relationships within and between strains can help to optimize the proportion of individuals to be sampled from each strain when creating base populations for breeding programs in aquaculture species. The advantage of using this information is reflected in the phenotypic and breeding values obtained at the beginning of the program and at the genetic diversity captured by the base population. The advantage remains during the subsequent generations of selection, making the breeding program more profitable. The study has not been designed for a particular species. Instead, we have considered a general genome architecture and a population structure that fit most aquaculture species.

Traditionally, when creating base populations for the establishment of a breeding program, no information was available about the genetic relationships between candidate strains or between individuals within strains. Consequently, the usual strategy was to collect equal number of individuals from as many strains as possible (Holtsmark et al., 2006, 2008a,b). However, the increasing amount of molecular markers developed for aquaculture species provide us with the opportunity of estimating genetic relationships within and between strains and to optimize the contribution of each strain. In this study it has been shown that strains harboring low levels of genetic diversity should contribute less individuals in order to maximize the global diversity of the base population.

In the creation of base populations, one important point to determine beforehand is if the objective is to maximize the genetic variance for a particular trait (i.e., the trait in the breeding goal) or to maintain the highest global diversity. For the former objective, Bennewitz and Meuwissen (2005) showed that the optimal strategy is what they called maximum variance total (MVT), which gives more weight to the variance between strains than within strains. However, although the profitability of the breeding program depends on the performance for the target trait, diversity must be also maintained for other traits that are likely to be included in future breeding objectives and for fitness related traits. Following this logic the methodology should be minimizing the global coancestry which poses the same weight to within and between strains diversity. The latter method is similar to minimizing the long-term inbreeding of the population as demonstrated by Eding and Meuwissen (2001) and was the chosen strategy for the present study. Accordingly, in our results for three of the simulated scenarios, strategy E yielded the highest V_A for the target trait in the base population. The lower levels of V_A observed under strategies MC and IC were due to the fact that the objective was to maximize the global genetic diversity measured as H_e across all the genome. Hayes et al. (2006) compared random with marker-based optimized selection of breeders from a single population of Atlantic salmon in terms of the genetic variance captured for three different traits (growth and two disease traits). They followed an equivalent methodology to strategy IC presented in this study and found higher additive variances in the breeders for the disease resistance traits but a lower variance for growth when optimizing the selection than when breeders were chosen at random. The explanation for these contrasting results was the different genetic architecture of the traits. It must be noticed that the simulated trait in the present study was controlled by a large number of additive loci and had an intermediate heritability typical for growth. This is the reason for similar performance (i.e., highest levels under E than IC strategy) observed in Hayes et al. (2006) for growth and in the present study (at least for three of the simulated scenarios). Another problem for the interpretation of the results in Hayes et al. (2006) is that they employed 237 AFLPs and this number may be not enough for obtaining a high correlation between diversity at markers and at loci controlling growth.

In concordance with the previous considerations, in this study the highest levels of global diversity (H_e measured at the non-marker loci) were captured when optimizing the creation of the base population using individual genotypes (strategy IC) with a large number of SNPs, although differences with the strategy equalizing proportions (strategy E) were small. Surprisingly, scenarios with a limited number of markers (i.e., 100) implied only a loss in H_e of 1% when comparing with results from using large numbers. In any case, when using few markers the H_e maintained under IC was sometimes lower than under the E strategy because of the lack of correlation between diversity at markers and diversity in the rest of the genome.

When individual information was absent (i.e., strategy MC) there was a reduction in the ability to capture diversity respect IC strategy whatever the number of markers used. Notwithstanding, values of H_e obtained when relying on strain averages were less than 2% lower than those observed when individual genotypes of candidates were available. This is an appealing result for cases where the budget is low and no all candidates can be genotyped.

Beyond all considerations about the genetic diversity, we must remember that the short-term profitability of a breeding program depends on actual mean levels of the phenotypic value for the trait of interest (as long as the breeding goal does not change and no fitness troubles arise in the population). The present study has shown that the mean phenotype of the selected individuals should be also accounted for when constructing the base population if that information is available. The loss of profit resulting from including low performance individuals in the base population may be not economically compensated in a reasonable period of time even if the response to selection is high due to a wider genetic variance for the trait. In fact our results showed that superiority of individuals selected under strategy IP last for the 10 generations of selection.

The presented results showed that, when reliable information is available for a fixed set of strains, a compromise solution between diversity and performance can be found when creating base populations. Having as a reference point the strategy randomly sampling the same number of individuals from each strain (strategy E, equivalent to the methodology used in Holtsmark et al., 2006, 2008a,b) up to 7% higher levels of phenotypic performance can be achieved under strategy IP at the same level of global diversity (H_e measured at the non-marker loci) when using a large panel of SNPs to genotype all candidates. Depending on the market value for the increase of one unit of the target trait this could translate into a large economic gain. Moreover, results showed that phenotypic values remain higher during all generations of artificial selection that were simulated from the base population under the IP strategy. Therefore, there was a clear superiority of fishes obtained using this strategy.

It must be realized that giving a large weight to the phenotypic value for a particular trait will have consequences on other correlated traits. Special attention should be paid for traits negatively correlated with the target trait which could be of potential interest. In such situations a useful strategy would be to use an index that includes several traits in the objective function to be maximized or to include an additional restriction in the optimization to ensure a minimum acceptable phenotypic level for the secondary traits.

It must be stressed that genotyping for a limited number of markers may give undesirable results because diversity at those markers will be loosely related to diversity at non-marker loci and to diversity at loci controlling the trait. Consequently, individuals with a high phenotypic performance may actually maintain little global diversity but still exhibit by chance high levels of diversity at the markers. Our results show that, when relaying in few markers, lower H_e levels were found when optimizing contributions using the IP strategy than using the E strategy even when a restriction was imposed to keep the same level of diversity (Table 3). However, results from our simulations suggest that, although it will depend on the particular characteristics of the species under management and the genetic architecture of the available strains, about 1000 SNPs could be enough to efficiently create base populations in aquaculture as no relevant improvements are obtained by increasing further the number of markers.

For all scenarios, strategies accounting for the phenotypic level of the founders (i.e., MP and IP) started the selection program from lower values of V_A for the target trait. This fact did not preclude these strategies to maintain the initial advantage in performance during the 10 generations of selection.

Results of this study suggest that if "healthy" commercial strains (i.e., where diversity is not exhausted and no problems of inbreeding exist) are available they should be used to form the base population because they provide a higher starting performance level. This is even clearer when the objective is to complement the breeders› population in an ongoing selection program. Contrarily, the use of wild adapted strains with low performance would be only recommended if we suspect that unique information for other traits of interest is present in them. This could be the situation for strains naturally resistant to a particular disease. Otherwise, the general diversity that they could provide will not compensate for the lower trait phenotypic mean.

The small differences in H_e observed under different strategies in our simulations could be due to the large number of individuals selected to form the base population (200) making strategy E to perform so well that the other strategies have difficulties in improving H_e. In an extra scenario run with a smaller set of candidates (only four strains with 20 individuals each), harboring lower levels of diversity (mutation-drift equilibrium reached for a population of 100 individuals) and selecting a lower number of breeders (24), strategies implying optimized proportions showed still only slightly greater advantages over strategy E (2% increase). In any case, the differences observed in phenotypic values make worthy to optimize the construction of base populations in aquaculture and levels of diversity should be also accounted for in that task to get an appropriate balance.

Another advantage of using molecular information, beyond balancing phenotypic values and diversity when optimizing the construction of a base population, is that it provides us with the possibility of estimating the actual relationships between breeders in the base population itself. These relationships can be used for calculating EBVs through BLUP methodology and also for controlling the rate of inbreeding through Optimal Contribution strategies. Holtsmark et al. (2008b) studied the effects on the performance of the breeding program across generations of assuming unrelated and non-inbred founders when they are not. They concluded that an incorrect estimation of the relationships between and within strains and individuals leads to sub-optimal use of subpopulations with an increased risk of loss of alleles of direct and strategic relevance to the breeding program.

If genotypes for dense panel of markers and phenotypes are available for the same individuals (the candidates to be part of the base population or related individuals), the additive effect of each SNP can be calculated in the same way as in the Genomic Selection methodology (Meuwissen et al., 2001). Thus, genomic value of candidates can be calculated and used to take decisions instead of their phenotypic value.

In the present study the selected trait was simulated with an additive gene action both within and across loci. However, traits with commercial interest may have an important dominant component. Dominant effects can be also estimated and then used to design the mating scheme between breeders, at least to form the families from which the selection program will start. This way, the effects of heterosis can be accounted for and extra responses can be obtained in the first round of selection (Toro and Varona, 2010).

In short life species, where individuals may be not reproductively active by the time their phenotypic records and genotypes are available, it would be difficult to implement strategies based on individual information (i.e., IC and IP). However, even in such situations strain information can be used to optimize the creation of the base population, as proved in the present study. The impossibility of controlling the specific matings with mass spawning species does not interfere with the optimization of the base population either. Phenotypic levels and diversity can be optimized at the start of the program although responses in subsequent generations of selection may differ to those shown here given that selected breeders would need to be mated in groups.

In livestock terrestrial species, breeding programs have been running for many years and, thus, it is not very likely that there is a need for creating new base populations. Notwithstanding, our conclusions go beyond the scope of creating base populations. For instance, they may help to take decisions when creating a gene bank for any species. When the aim of such a bank is to store the genetic diversity from the available strains, the same strategies can be applied as when creating the base population of breeders. When, in the future, the stored material will be used for creating a live population or for complementing a breeding program the same results and consequences than those presented here are expected. If the phenotypic values of the candidates are not taken into consideration when determining the sampling scheme the starting population will show low levels for the trait of interest. Finally, the methodology presented in this study is also useful when the objective is to create a “core” live population within an ex-situ conservation program that aim at collecting the genetic diversity that exist in all available populations. This scenario is common for local breeds of livestock species.

CONFLICT OF INTEREST STATEMENT

The authors declare that the research was conducted in the absence of any commercial or financial relationships that could be construed as a potential conflict of interest.

ACKNOWLEDGMENTS

The research leading to these results has received funding from the European Union's Seventh Framework Programme (KBBE.2013.1.2-10) under grant agreement n° 613611.

REFERENCES

1. Bennewitz, J., and Meuwissen, T. H. E. (2005). A novel method for the estimation of the relative importance of breeds in order to conserve the total genetic variance. *Genet. Sel. Evol.* 37, 315–337. doi: 10.1186/1297-9686-37-4-315
2. Bürger, R., Wagner, G. P., and Stettinger, F. (1989). How much heritable variation can be maintained in finite populations by mutation-selection balance? *Evolution* 43, 1748–1766. doi: 10.2307/2409390
3. Caballero, A., and Toro, M. A. (2002). Analysis of genetic diversity for the management of conserved subdivided populations. *Conserv. Genet.* 3, 289–299. doi: 10.1023/A:1019956205473
4. Eding, H., and Meuwissen, T. H. E. (2001). Marker based estimates of between and within population kinships for the conservation of genetic diversity. *J. Anim. Breed. Genet.* 118, 141–159. doi: 10.1046/j.1439-0388.2001.00290.x
5. Eding, J. H., Crooijmans, R. P. M. A., Groenen, M. A. M., and Meuwissen, T. H. E. (2002). Assessing the contribution of breeds to genetic diversity in conservation schemes. *Genet. Sel. Evol.* 34, 613–633. doi: 10.1186/1297-9686-34-5-613
6. Eknath, A. E., Dey, M. M., Rye, M., Gjerde, B., Abella, T. A.,

Sevilleja, R., et al. (1998). "Selective breeding for Nile tilapia for Asia," in *Proceedings of the 6th World Congress on Genetics Applied in Livestock Production*, Vol. 27 (Guelph, ON), 89–96.

7. Eknath, A. E., Tayamen, M. M., Palada-de Vera, M. S., Danting, J. C., Reyes, R. A., Dionisio, E. E., et al. (1993). Genetic improvement of farmed tilapia: the growth performance of eight strains of *Oreochromis niloticus* tested in different environments. *Aquaculture* 111, 171–188. doi: 10.1016/0044-8486(93)90035-W
8. Falconer, D. S., and Mackay, T. F. C. (1996). *Introduction to Quantitative Genetics, 4th Edn*. Essex: Longman.
9. Gjedrem, T., Gjøen, H. M., and Gjerde, B. (1991). Genetic origin of Norwegian farmed Atlantic salmon. *Aquaculture* 98, 41–50. doi: 10.1016/0044-8486(91)90369-I
10. Gómez-Romano, F., Villanueva, B., de Cara, M. A. R., and Fernández, J. (2013). The benefit of using molecular coancestry in the management of populations under conservation and its dependency on effective population size and marker density. *Genet. Sel. Evol.* 45:38. doi: 10.1186/1297-9686-45-38
11. Hayes, B., He, J., Moen, T., and Bennewitz, J. (2006). Use of molecular markers to maximise diversity of founder populations for aquaculture breeding programs *Aquaculture* 255, 573–578. doi: 10.1016/j.aquaculture.2005.11.038
12. Holtsmark, M., Klemetsdal, G., Sonesson, A. K., and Woolliams, J. A. (2008a). Establishing a base population for a breeding program in aquaculture, from multiple subpopulations, differentiated by genetic drift: I. Effects of the number of subpopulations, heritability and mating strategies using optimum contribution selection. *Aquaculture* 274, 232–240. doi: 10.1016/j.aquaculture.2007.11.042
13. Holtsmark, M., Klemetsdal, G., Sonesson, A. K., and Woolliams, J. A. (2008b). Establishing a base population for a breeding program in aquaculture, from multiple subpopulations, differentiated by genetic drift: II. Sensitivity to assumptions on the additive genetic relationships of base animals. *Aquaculture* 274, 241–246. doi: 10.1016/j.aquaculture.2007.11.043
14. Holtsmark, M., Sonesson, A. K., Gjerde, B., and Klemetsdal, G. (2006). Number of contributing subpopulations and mating design

in the base population when establishing a selective breeding program for fish. *Aquaculture* 258, 241–249. doi: 10.1016/j.aquaculture.2006.03.039

15. Houston, R. D., Taggart, J. B., Cézard, T., Bekaert, M., Lowe, N. R., Downing, A., et al. (2014). Development and validation of a high density SNP genotyping array for Atlantic salmon (*Salmo salar*). *BMC Genomics* 15:90. doi: 10.1186/1471-2164-15-90
16. Kirkpatrick, S., Gelatt, C. D., and Vecchi, M. P. (1983). Optimization by simulated annealing. *Science* 220, 671–680. doi: 10.1126/science.220.4598.671
17. Liu, S., Sun, L., Li, Y., Sun, F., Jiang, Y., Zhang, Y., et al. (2014). Development of the catfish 250K SNP array for genome-wide association studies. *BMC Res. Notes* 7:135. doi: 10.1186/1756-0500-7-135
18. Meuwissen, T. H. E., Hayes, B. J., and Goddard, M. E. (2001). Prediction of total genetic value using genome wide dense marker maps. *Genetics* 157, 1819–1829.
19. Palti, Y., Gao, G., Liu, S., Kent, M. P., Lien, S., Miller, M. R., et al. (2014). The development and characterization of a 57K single nucleotide polymorphism array for rainbow trout. *Mol. Ecol. Res.* doi: 10.1111/1755-0998.12337. [Epub ahead of print].
20. Toro, M. A., and Varona, L. (2010). A note on mate allocation for dominance handling in genomic selection. *Genet. Sel. Evol.* 42:33. doi: 10.1186/1297-9686-42-33
21. Turelli, M. (1984). Heritable genetic variation via mutation-selection balance: Lerch's zeta meets the abdominal bristle. *Theor. Popul. Biol.* 25, 138–193. doi: 10.1016/0040-5809(84)90017-0
22. Woolliams, J. A., and Bijma, P. (2000). Predicting rates of inbreeding in populations undergoing selection. *Genetics* 154, 1851–1864.
23. Xu, J., Zhao, Z., Zhang, X., Zheng, X., Li, J., Jiang, Y., et al. (2014). Development and evaluation of the first high-throughput SNP array for common carp (*Cyprinus carpio*). *BMC Genomics* 15:307. doi: 10.1186/1471-2164-15-307

CHAPTER

7

GENETICALLY-IMPROVED TILAPIA STRAINS IN AFRICA: POTENTIAL BENEFITS AND NEGATIVE IMPACTS

Yaw B. Ansah [1,2], Emmanuel A. Frimpong [1] and Eric M. Hallerman [1]

[1]Department of Fish and Wildlife Conservation, Virginia Polytechnic Institute and State University, 100 Cheatham Hall, Blacksburg, VA 24061, USA

[2]Department of Agricultural and Applied Economics, Virginia Polytechnic Institute and State University, 208 Hutcheson Hall, Blacksburg, VA 24061, USA

ABSTRACT

Two genetically improved tilapia strains (GIFT and Akosombo) have been created with *Oreochromis niloticus*(Nile tilapia), which is native to Africa. In particular, GIFT has been shown to be significantly superior to local African tilapia strains in terms of growth rate. While development economists see the potential for food security and poverty reduction in Africa from culture of these new strains of tilapia, conservationists are wary of potential ecological and genetic impacts on receiving ecosystems and native stocks of tilapia. This study reviews the history of the GIFT technology, and identifies potential environmental and genetic risks of improved and farmed strains and tilapia in general. We also estimate the potential economic gains from the introduction of genetically improved strains in Africa, using Ghana as a case country. Employing a combination of the Economic-Surplus model and Monte Carlo simulation, we found the mean net present value (NPV) of the introduction of the GIFT strain in Ghana to be approximately 1% of the country's gross domestic product. Sensitivity analysis indicated that the difference in growth or yield between the GIFT and locally-available strains has the largest effect on mean NPV. We conclude that improvements in management practices and infrastructure could increase the yield and profitability of the local strains even if genetically-improved strains are not introduced. These improvements also will ensure the realization of the full potential of introduced strains.

Keywords: sustainable aquaculture; native species; genetic improvement; Monte Carlo simulation; economic surplus; risk analysis; Ghana

THE IMPORTANCE OF AQUACULTURE TO AFRICA'S DEVELOPMENT

Capture fisheries production has levelled off and is no longer considered capable of sustaining the supply of fisheries products needed to meet growing global demand [1]. Aquaculture, especially of tilapias, has the potential to play a leading role in the fight against food insecurity, malnutrition, and poverty in Africa [2]. The continent has an immense biological diversity of native fish resources. However, due to poor management and genetic erosion, most aquaculture stocks in current use on the continent are genetically inferior to wild, undomesticated stocks [3,4]. It is widely accepted that successful aquaculture development in Africa requires improvements in feed quality and availability, business and marketing models, and local technical capacity. Another important factor that should be considered is the effective utilization and management of fish genetic resources [4,5]. Specifically, improved strains that are faster growing, resistant to disease, and suited for culture in a variety of fish farming conditions could go a long way to meet the demand for fish protein [6].

TILAPIA CHARACTERISTICS AND PRODUCTION

Tilapias (Family: Cichlidae) are suitable for various aquaculture systems due to their ease of propagation, tolerance to handling, fast growth on both natural and manufactured feeds, tolerance of a wide range of environmental conditions, and high palatability, marketability and nutrient content [7]. They are especially well-suited for culture in developing countries due to their fast growth and short generation time, tolerance to a wide range of environmental conditions, resistance to stress and disease, ability to reproduce in captivity, and their acceptance of artificial feeds right after yolk-sac absorption [8].

Global aquaculture production of tilapias increased from 28,000 tonnes to over 3 million tonnes from 1970 to 2010 [9]. Globally, the tilapias were the dominant species group caught in inland fisheries between 2000 and 2005 (the tilapias were surpassed in 2005 by the cyprinids [10]). In terms of aquaculture production, the tilapias comprise approximately 5 percent of total global fish farming, second to the carps, which account for more than 70 percent [11]. However, aquaculture of tilapia in Africa constitutes only approximately 19% of the world's tilapia production [12].

SOCIAL BENEFITS OF TILAPIA

Historically and from a social standpoint, the most important use of tilapias has been production for home consumption, with millions of small-scale fish farmers in more than 100 countries supplementing their diets with tilapia [13]. There also has been a steady increase in the number of family-owned tilapia marketing microenterprises in many countries. The fish often are retrieved from nearby ponds or tanks, cleaned, fried, and offered for sale. This fried tilapia provides a considerable proportion of dietary protein and calories in these developing countries [13]. More commonly, fresh tilapia also is provided for sale either at the farm gate or in local markets. However, tilapias have grown in importance from being just a low-cost, high-protein food fish ("aquatic chicken" [14]) employed by development agencies to feed the poor in the world's rural areas, to a highly-domesticated "livestock" with annual sales amounting to over $2 billion globally [10,13]. In terms of economic importance, tilapia surpassed the salmonids in 2004, and they are expected to eventually equal the carps [13]. The tilapias have been referred to as the "most important global whitefish commodity" [15].

DISTRIBUTION OF TILAPIAS

The natural distribution of tilapias is restricted to Africa, Jordan, and Israel, where 112 species and subspecies of the genera *Oreochromis*, *Sarotherodon*, and *Tilapia* have been identified [8,16,17,18,19]. However, only a few of these species are commercially important, and fewer still are of aquacultural importance [11]. *Oreochromis niloticus*, *O. aureus*, and various hybrids of these with *O. mossambicus* are regarded as the most important aquaculture species [11]. For example, in China, of the reported 1.1 million tonnes of *O. niloticus* produced in 2008, approximately one-quarter was a hybrid between Nile tilapia (*O. niloticus*) and blue tilapia (*O. aureus*) [10]. All of the important aquaculture species have been introduced extensively outside of their native range. In the 20th century alone, tilapias were introduced into 90 countries for aquaculture, fisheries, the aquarium trade, or inadvertently [20,21,22]. However, their intolerance to low temperatures (below 20 °C) restricts culture to warmer areas [11].

BACKGROUND TO GENETIC IMPROVEMENT OF TILAPIAS

Due to the increasing importance of the tilapias in global fish farming, the intensity and diversity of efforts to improve the genetic baseline of these species have intensified over the last few decades. Ponzoni *et al.* [23] showed that genetic improvement is one of the most powerful and least expensive means of increasing the efficiency of aquaculture. Both traditional animal breeding and science-based quantitative genetic approaches have been used to improve tilapia phenotypes [24]. Other important genetic improvement methods include innovative chromosomal manipulations, physiological alteration of sex determination, gene transfer, and genetic marker-assisted breeding [24,25]. Traditional animal breeding approaches are still the most practical means of improving tilapia stocks for low-tech producers in most countries, and these approaches basically exploit additive and non-additive gene effects [24].

One common traditional approach to genetic improvement is the practice of individual selection, or selective breeding, which is based on the underlying principle that some significant portion of the variation in observable performance is due to individual genotypes, and that a component of these genotypic influences is directly heritable from parent to offspring [24]. Even in cases of high heritability, a measurable amount of phenotypic variation is needed to enhance growth rate through selection. Random genetic drift and excessive inbreeding result in lower heritabilities due to reduced genetic variation, posing a frequent problem in tilapia culture since a large number of broodstock on any particular farm most likely originated from a few individuals [24]. Hence, selection is usually a viable approach for tilapia genetic improvement where sufficient genetic variation exists [24]. According to Ponzoni *et al.* [23], selective breeding has a number of advantages over other genetic approaches: continuous genetic gain is possible, genetic gains can be handed down from one generation to the next, and gains in a nucleus can be multiplied and expressed in millions of individuals in the production sector.

The Genetic Improvement of Farmed Tilapia (GIFT) project, one of the most significant recent innovations in tilapia culture, succeeded in part because it was based upon using selective breeding of a highly diverse synthetic base population [8]. The GIFT technology has been applied in a number of countries to improve local strains of tilapia. One example of applying the GIFT methodology is the Akosombo strain, which was developed in Ghana by the Aquaculture Research and Development Center (ARDEC) in collaboration with the World Fish Center in 2003 [26,27,28].

THE GENETIC IMPROVEMENT OF FARMED TILAPIA (GIFT) PROJECT

The GIFT project was a collaborative research effort involving five separate research institutions that was implemented in the Philippines from 1988 to 1997 [29,30]. This project had the overarching goal of "increasing the quantity and quality of protein consumed in low income rural and urban populations in tropical developing countries, and in all regions of the world, leading to an increase in the income of low-income producers" [31]. The collaborating institutions were the International

Center for Living Aquatic Resources Management (ICLARM, now the World Fish Center), the Philippines Bureau of Fisheries and Aquatic Resources (BFAR), the Freshwater Aquaculture Center of the Central Luzon State University (FAC-CLSU), the Marine Science Institute of the University of the Philippines (UPMSI), and the Institute of Aquaculture Research, Ltd., in Norway [30]. Funding for this project was provided by the Asian Development Bank, the United Nations Development Program, and ICLARM [30].

The GIFT project had three specific objectives: (1) to develop improved breeds of Nile tilapia (*Oreochromis niloticus*) and provide those fish breeds to national testing programs and then to the fish farmers; (2) to strengthen national institutions in aquaculture genetics research; and (3) to establish a mechanism for international exchange and evaluation of improved breeds and research methods [31]. Nile tilapia was chosen as the focal species for a number of reasons. It has a short generation time (approximately 8 months), which made it the perfect species for a breeding program [32]. This species also was growing rapidly in importance in aquaculture [14,33]. Additionally, the omnivorous diet of the Nile tilapia makes it an excellent fit for low-cost aquaculture, in contrast to carnivorous species that rely heavily on fishmeal or other expensive animal protein [31]. Also, numerous potentially useful Nile tilapia resource stocks existed in several countries in Africa and the Middle East.

Before the commencement of field trials, a number of consultations were held with fish farmers and experts from various disciplines, and the relative importance of the improved stocks was considered for the various tilapia farming systems. Some of the farming systems considered were cage culture, backyard fish pond, rice-fish integrated culture, and more intensive systems [31]. Fingerling size was set at 3–7 g, and the grow-out period was 90 days, with a harvest weight of about 120 g [31].

Even though the natural tilapia resources are restricted mostly to Africa, the world's main tilapia aquaculture industries are located in Asia [10,32]. Due to the generality that established farmed tilapia stocks in Asia were derived from few founder individuals, there was the suspicion that loss of variation through random genetic drift, inbreeding and introgression of genes from other less-desirable feral tilapia stocks had occurred. Hence, the project collected wild Nile tilapia germplasm from Egypt (May 1988; August 1989; Nile Delta system), Ghana (October 1988; Upper Volta system), Senegal (October 1988; extreme west of distribution) and Kenya (August 1989; Lake Turkana) [34,35]. These collections represented the first-ever direct transfers of *O. niloticus* from Africa to Southeast Asia, with all samples belonging to the sub-species *Oreochromis niloticus niloticus*, with the exception of the Kenyan samples, which belonged to the sub-species *Oreochromis niloticus vulcani*. Four commercial Nile tilapia strains were collected from the Philippines (three strains originated from Ghana, and one stock originated from Egypt [34]).

The most important criterion determining the number of strains to be developed by the GIFT project was relative performance (growth, maturation and fecundity, and hardiness) in different target environments, or the genotype x environment interaction (GxE) [31]. An insignificant GxE effect (in terms of farming relevance) implies that the best strain in one environment will be the best in most or all environments. A high GxE effect, on the other hand, implies that special strains would have to be developed for specific environments. The GxE interaction effect was found to be low, *i.e.*, overall growth performance was found not to differ significantly with environment, in terms of farming relevance. Hence, it would not be necessary to develop different strains for the different farming systems. The principal breeding objective of the GIFT program was growth rate, while monitoring other traits, such as survival, occurrence of disease and maturation rate [31].

Results indicated that, with the exception of the Ghana strain, the strains from Africa performed as well as or better than the then existing commercial or "domesticated" strains in Asia. The Egyptian strain was the best performer in the first generation, while the Kenyan strain was the best in the second. The Ghana strain performed the worst in both generations [34]. One of the widely cultured strains (the Israel strain, originally derived from Ghana) also performed poorly [34]. Crossbreeding (hybridization) of the different strains did not result in significant improvements. Therefore, individuals from the best-performing purebred and crossbred groups were selected, based on the growth performance of a number of different strain combinations, in order to build a stock with a broad genetic base. This constituted the original GIFT strain [31].

It was no surprise that the GIFT project, after just one generation of selection, had generated considerable interest from the tilapia production industry in Asia. Therefore, a consultation meeting was held in 1992 to discuss strategies and safeguards for fish germplasm transfer and distribution [31]. The consultation meeting involved senior scientists from National Agricultural Research Systems (NARS) of developing countries, international experts on fish genetics and biodiversity, representatives of NGOs, and donor institutions. Following the recommendations from this meeting, the improved strain developed in the Philippines was disseminated (from May 1994 to August 1997) to member countries that formed the Dissemination and Evaluation of Genetically Improved Tilapia in Asia (DEGITA).These countries were Bangladesh, People's Republic of China, Philippines, Thailand, and Vietnam. This dissemination allowed detailed evaluation of the genetic and socioeconomic performance and environmental impacts of this strain prior to more widespread commercial production and dissemination [31]. The dissemination was carried out according to standard quarantine procedures developed as part of the GIFT project [36].

Dey [37] studied the potential economic impacts of culturing the GIFT tilapia strain in five Asian countries: Bangladesh, China, the Philippines, Thailand and Vietnam. He concluded that adoption of the improved tilapia strain would benefit both producers and consumers of fish in each of the countries studied. Further investigations concluded that the GIFT strain resulted in body weight 18%–58% higher than "non-GIFT" strains on "average" farms in Asia. The break-even price above variable cost was found to be 7%–36% lower for the GIFT strain than for other *O. niloticus* strains then being farmed [35].

THE GIFT/GENOMAR SUPREME (GST) STRAIN

In 1999, the GIFT Foundation International, which was founded as part of the GIFT project, signed an exclusive agreement with GenoMar, a Norwegian company based in Oslo, for the long-term continuation of the GIFT breeding program [15]. According to GenoMar, and beginning with the 10th GIFT generation DNA marker-assisted selection has been applied in order to increase the selection differential. According to Gjoen [15], the selection differential from this approach, compared to that of the traditional selection approach, was expected to be 40% higher than for the 9th-generation GIFT tilapia in 2000. Genetic maps were generated for tilapia, and experiments were conducted in order to reveal genes that influence traits that are economically important [15,25]. According to Gjoen [15], color, growth, body shape, salt tolerance, and sex determination are some of the traits for which GenoMar detected influential chromosomal regions. GenoMar planned to use this information to facilitate the acceleration of genetic gains, especially for traits such as disease resistance and feed conversion, which are difficult to measure within traditional selective breeding programs [15]. As of 2001, the GIFT/GenoMar strain was officially being distributed in countries in Southeast Asia and Latin America [15]. The GIFT/GenoMar strain is known officially as the GenoMar Supreme Tilapia (GST™) [8].

Ponzoni *et al.* [5] detail the current state of the GIFT strain and also summarize research that the World Fish Center has conducted on the strain since 2000. The World Fish Center took delivery of 63 full-sib groups of 35 fish each (the progeny of single-pair mated parents) at Jitra, Kedah State, Malaysia, towards the end of 2000 and beginning of 2001, from the GIFT Foundation International, Philippines. The aims of this project included, among others, the maintenance and continuous improvement of the GIFT strain, and the distribution to partner countries likely to benefit from its use. According to the authors, the strain has achieved sustained gains of 10%–15% per generation over more than six generations. Importantly, these gains have not been accompanied by any undesirable correlated response to date [5].

DISSEMINATION OF GIFT/GST STRAINS IN AFRICA

The outcome of the GIFT project generated interest from developing countries in Asia, the Pacific, and Africa, both in terms of developing their own aquaculture strains, and also in gaining access to the GIFT germplasm [38]. Africa, the origin of the tilapias, benefits the least from the GIFT strain, even though much of the continent has a high potential for tilapia farming [39]. This situation arose due to the policy of the WorldFish Center not to introduce the GIFT strain into countries where *O. niloticus* is indigenous, concerned that interbreeding of the GIFT strain with locally-adapted native populations might compromise wild aquatic genetic diversity [38]. This decision by the WorldFish Center was given weight by an expert consultation in 2002 in Nairobi, which was sponsored by the WorldFish Center, the Food and Agriculture Organization (FAO), the World Conservation Union (IUCN), the United Nations Environment Program (UNEP), and the Technical Center for Agriculture and Rural Cooperation (CTA) [40]. The expert consultation resulted in the Nairobi Declaration, a set of ten recommendations aimed at realizing the potential of African aquaculture without compromising native ecological and genetic resources [41]. The WorldFish Center, instead, decided to help these countries apply the GIFT methodology to the genetic improvement of indigenous tilapias.

The WorldFish Center established the International Network on Genetics in Aquaculture (INGA) in 1993 to train member-country scientists in quantitative genetics applied to aquaculture, and to coordinate national breeding programs in the 13 member countries (Bangladesh, China, Cote d'Ivoire, Egypt, Fiji, Ghana, India, Indonesia, Malaysia, Malawi, Philippines, Thailand, and Vietnam) using the GIFT methodology to genetically improve their indigenous cultured species [35,38].

The pressure for the dissemination of the actual GIFT germplasm to Africa seemed to reach a head in 2007, when the WorldFish Center approved the Policy on the Transfer of GIFT from Asia to Africa, making the GIFT strain available to any African government that can demonstrate procedures to manage environmental and biodiversity risks, among other conditions [42]. Such a country also must display compliance with the Convention on Biological Diversity (CBD), an international treaty, while at the same time addressing the development objective it intends to achieve with the GIFT introduction [36,42]. The guiding principle for this new policy was that the genetic risks of introducing the actual GIFT strain to Africa was comparable to those associated with the genetic improvement of indigenous *O. niloticus* strains in Africa [42]. Basically, the new policy acknowledged that there was no point in keeping the GIFT strain from Africa if the genetic improvement of tilapia was going to occur on the continent anyway.

The Nogoya Protocol on Access to Genetic Resources and the Fair and Equitable Sharing of Benefits Arising from their Utilization to the CBD, a multi-lateral agreement, was adopted in 2010. This protocol aims at:

Sharing the benefits arising from the utilization of genetic resources in a fair and equitable way, including by appropriate access to the genetic resources and by appropriate transfer of relevant technologies, taking into account all rights over those resources and to technologies, and by appropriate funding, thereby contributing to the conservation of biological diversity and the sustainable use of its components.[43]

It can be argued that the World Fish Center has equitably shared with Africa the benefits arising from the utilization of the genetic resources (*O. niloticus* germplasm) collected from Africa to develop the GIFT strain ([6]; p. 139). The center has been training African scientists in application of the GIFT technology to their own national improvement programs, and then eventually allowing African governments' access (at no cost) to the GIFT strain.

Although the decision to allow the GIFT strain into Africa seems to have been reached through a logical scientific process, the policy evoked mixed reactions in Africa. The Sustainable Aquaculture Research Networks in Sub-Saharan Africa (SARNISSA) is a network of scientists, fish farmers, development partners, and other players in the African aquaculture industry. SARNISSA, hosted by University of Stirling, UK, maintains an online forum on which issues that are pertinent to African aquaculture are discussed, and one of the most hotly-debated topics recently discussed was use of the GIFT strain in Africa. For example, on 29 November 2011, SARNISSA member Hiskia Asino posted an enquiry onto the forum asking from which African countries he could obtain GIFT tilapia fingerlings for farming in Namibia. This request set off a heated debate that lasted several months. While most fish farmers and development agents generally seemed to favor the new policy and its potential benefits towards increasing farm profits and reducing poverty, a number of scientists and conservationists spoke quite passionately against it, citing the potential negative genetic and ecological impacts of the introduction. Instead of the GIFT strain, those against it proposed improvements in the existing aquaculture practices and infrastructure to achieve the same economic goals as with the GIFT. A forum contributor summed up the fears of SARNISSA members opposed to the use of the GIFT strain in Africa, "Once local species or stains are lost to competing non-endemic species or strains, that biodiversity is lost forever".

The GIFT technology has been disseminated to Kenya, Cote d'Ivoire, and Egypt, along with other developing countries in Asia and the Pacific (Dr. Raul Ponzoni, formerly of WorldFish, personal communication). GenoMar also has supplied the GST strain to partner hatcheries in Zambia and Angola, and by 2008 it was carrying out a feasibility study on setting up another partner hatchery in Uganda [36]. In Ghana, the Aquaculture Research and Development Center (ARDEC) took delivery of the GIFT strain officially in 2012, with the expressed objective of comparing its growth performance with the locally improved Akosombo strain (Dr. Joseph Padi, ARDEC, Water Research Institute, Ghana, personal communication).

Considering the porous nature of African borders and inadequate capacity to monitor the transfer of genetic material, we expect that the GIFT/GST strain is currently in several other countries in Africa. An example of the easy movement of non-native species on the African continent is the current production in Omilende, Nigeria, of the Asian catfish *Pangasius* sp. fingerlings for sale, as was reported on the SARNISSA forum on 2 March 2013—another non-native species introduction revelation that was followed by a lengthy and contentious discussion on the forum. It is clear, therefore, that once the GIFT/GST strain is legitimately introduced into a country in Africa, it eventually will be found in unapproved parts of that country or in other unauthorized neighboring countries. It is also clear that both benefits and negative impacts are possible with the introduction of the GIFT strains into Africa (Figure 1).

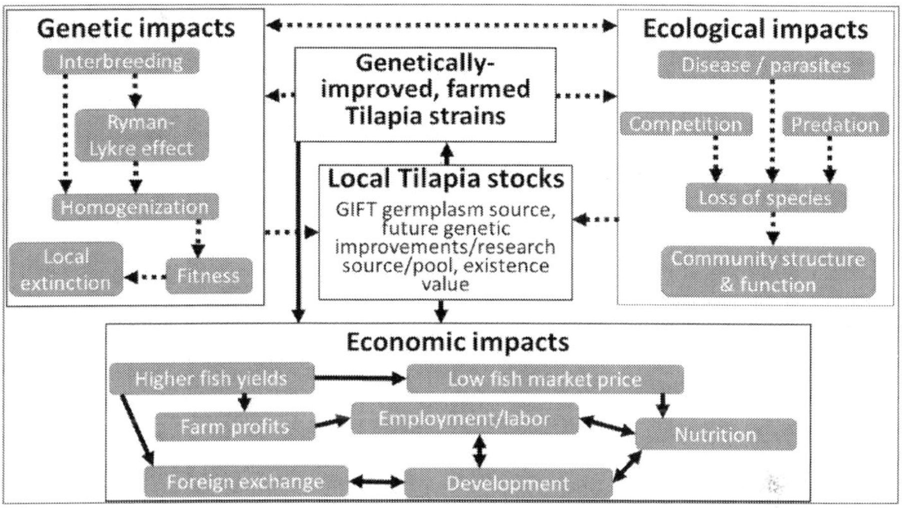

Figure 1: Interdependence of the potential ecological, genetic, and economic impacts with the introduction of genetically-improved tilapia into Africa. Broken arrows indicate negative impacts and solid arrows indicate positive impacts.

POTENTIAL ECOLOGICAL IMPACTS OF TILAPIA INTRODUCTIONS

Tilapia production is well on its way to taking center stage in the global aquaculture industry [13]. However, in the context of development, the success of a species is determined by its social, cultural, economic and environmental impacts, in addition to its contribution to production *per se* [22]. The traditional small-scale, semi-intensive culture of tilapias has proven sustainable in many countries where they are native, with no observable negative ecological impacts on surrounding environments attributable directly to the species (e.g., [8,44]). The recent trend towards intensification, driven by market forces, has been predicted to pose serious environmental and socioeconomic problems [8]. These impacts will arise as a result of the expected increases in use of processed feeds, drugs, hormones [8], and non-indigenous genetic resources in intensive aquaculture. The global transfer of both genetically-altered and unaltered tilapia species outside their native ranges for aquaculture purposes also is expected to result in negative impacts upon natural aquatic ecosystems.

The popularity of the GIFT strain will drive its introduction into African countries outside the natural range of *Oreochromis niloticus*. Characteristics that make tilapias the perfect aquaculture species also make them invasive species, keeping in mind that even though Nile tilapia is native to Africa, it is not native to all parts of the continent. Escape of both altered and unaltered genetic resources from aquaculture installations into natural ecosystems occurs relatively frequently [45,46]. For example, Attipoe *et al.* [27] report the loss of the entire control line of the 2003 spawning season from the ARDEC research facility during the development of the Akosombo strain in Ghana as a result of a violent storm.

Pullin *et al.* [21] studied the establishment success of various tilapine species and found that tilapia species, in general, are moderately to highly invasive, with a 60%–90% probability of becoming established in new open waters. The impacts of any invasive species on an aquatic ecosystem differ significantly, depending on the species, the extent of the introduction (*i.e.*, propagule pressure), and the vulnerability of the invaded ecosystem [18]. Tilapine species invest in significant parental care that facilitates their establishment in novel habitats. The family Cichlidae is grouped into nest-builders (*Tilapia* spp.), and mouth-brooders (*Oreochromis* and *Sarotherodon* spp.) [11]. Mouth-brooders, for example, have no specific substrate requirements for reproduction; they simply carry fertilized eggs and yolk-sac fry in their mouths until yolk sac absorption and dispersal [47,48].

Examples abound in the literature demonstrating the negative impacts of tilapia on natural aquatic ecosystems after their introduction, keeping in mind that Nile tilapia, and tilapia in general, are not native to all parts of Africa. These include predation on eggs and small fish [49], rapid total elimination of submerged and floating aquatic macrophytes that served as essential habitat for native fish species in natural and man-made reservoirs [50,51], and eutrophication (bioturbation and nutrient cycling through ingestion and excretion) stemming from the foraging activity of tilapia on benthic algae [52]. Tilapia introductions into Madagascar, Lake Victoria, and Zimbabwe led to habitat alteration, such as declines in aquatic plants and decreases in the availability of breeding areas for native species [53]. Also, the establishment of *Tilapia graham* in Kenya's Lake Nakuru led to the emergence of a fish-eating bird population [54].

Canonico *et al.* [18] provide an excellent review of impacts of tilapia on receiving aquatic systems using specific case studies from around the world.

Another potential impact of introducing the genetically improved tilapia strain from Asia to Africa is the risk of "hitch-hiker" disease-causing vectors and pathogens from Asia. Two possibilities exist. Firstly, the selectively bred strain might not be able to withstand diseases to which local African stocks have evolved resistance, which could lead to high mortalities on farms producing the selected strains. Secondly, the selectively bred strain, if resistant to certain Asian disease vectors, might carry these vectors and pathogens to Africa and infect local fish stocks. Noga [55] provides detailed descriptions of diseases of tilapia and numerous other fish species. Tilapia diseases include those caused by bacteria of the family Eimeriidae, *Mycobacterium* spp., and *Edwardsiella tarda*.

A number of general procedures exist for conducting risk analysis of non-native fishes (e.g., [56]). There are established procedures for analyzing risk regarding pathogen transfer and for quarantine monitoring [57]. Unfortunately, there is no universally-accepted *a priori* procedure for assessing the potential environmental impacts of non-native tilapias. In fact, scientific and policy decisions on new tilapia introductions frequently are polarized and are largely based on guesswork [21]. As such, a majority of the impacts mentioned above were recognized after introduction and/or establishment.

POTENTIAL GENETIC IMPACTS OF TILAPIA GENETIC IMPROVEMENT ON NATIVE POPULATIONS

There is a paucity of studies on the impacts of aquaculture on locally-adapted gene pools of tilapia in receiving ecosystems. However, potential harms and associated risks stemming from aquaculture escapes have been considered in a more general context, Carvalho and Hauser [58] grouped the genetic impacts of "escapee" genetic resources into direct and indirect effects. Interbreeding of natural fish populations with escaped cultured stocks is arguably the biggest direct effect. When cultured fish escape or are released, the receiving population may experience a reduction in genetically effective population size, N_e, a phenomenon referred to as the Ryman-Laikre effect [59]. In addition, the risk of subsequent inbreeding may be increased if the ratio of "escapees" to the natural population is sufficiently high due to the relatively low N_e of many cultured stocks. Natural selection acts upon alleles at fitness-related loci to ensure adaptation of wild fish populations to their environments. That is, local differences in natural selection over a wide area result in adaptive genetic divergence of populations over time; further, selective forces that operate across adaptively important loci may result in combinations of alleles, or co-adapted gene complexes, which confer fitness upon their carriers [60].

In contrast, selective breeding acts upon alleles at performance-related loci to ensure expression of valued traits within aquaculture systems. Escape of selectively bred individuals into the wild and interbreeding with wild populations may result in offspring that exhibit low fitness, posing the risk of outbreeding depression at a local scale; further, interbreeding of escaped, cultured stocks at multiple sites across a landscape will tend to homogenize among-population variation of wild populations and also could lead to loss of fitness and outbreeding depression. The best-demonstrated case studies involve salmonids, and are reviewed by Ferguson *et al.* [61]. Among case studies, McGinnity *et al.* [62] demonstrated that the interaction of farmed with wild Atlantic salmon resulted in lowered fitness, and that repeated fish escapes caused cumulative fitness depression. Araki *et al.* showed dramatic losses of fitness after just two generations of captive breeding in steelhead salmon [63], as well as reduced fitness of their wild-born descendants [64]. Hindar *et al.* [65] reviewed studies of the effects of escape of cultured fish and introgression with local populations upon genetic differentiation, noting that in a subset of cases genetic swamping of local populations led to loss of indigenous stock structure in several salmonids. Conceptual models have been developed to predict the effects of introductions of maladapted individuals into locally-adapted gene pools [66,67] or to predict the relative likelihoods of beneficial or negative impacts of gene flow on receiving gene pools [68]. However, empirical studies of the effects of interbreeding of escaped cultured fish upon wild populations are lacking for most fish taxa, including tilapias. Hence, we lack the empirical data to parameterize models in order to predict the effects of introgression of selectively bred tilapias into native gene pools. Knowledge of gene flow rates and ecological differences among source and recipient populations [69], carefully controlled, multigenerational experiments, and thoughtful use of data from "experiments" created unintentionally [70] will be needed to advance our understanding.

A number of indirect effects also may be posed by the release or escape of cultured tilapia stocks into the wild. Indirect effects generally highlight the strong relationship between the ecology of tilapia species and their genetics (Figure 1). Released cultured stock may reduce abundance, and hence the effective population sizes (N_e) of critical species in the receiving ecosystem through competition, predation, habitat alteration, or changes in community trophic structure or food webs [63]. This decrease in N_e could cause a loss of genetic variability and adaptive ability to changing selective pressure, possibly leading to an increase in the likelihood of inbreeding and extinction.

The level of impact stemming from establishment of these "escapee" genetic resources in an ecosystem is influenced by three factors: the species' invasiveness, the fitness of the selectively-bred stock, and characteristics (specifically, the invasibility or vulnerability to invasion) of the receiving community [71]. Invasiveness refers to the ability of a cultured stock to escape, disperse, and become feral in aquatic communities, and tilapias, along with other aquaculture species, exhibit great abilities to disperse and become established in non-native ecosystems.

Selective breeding that increases fitness may increase the likelihood of a cultured species becoming established in a receiving ecosystem (e.g., significant impacts on native soil, vegetation and animals by feral hogs in southeastern USA [72], and feral horses in Australia [73]). However, experience from selective breeding in domestic farm animals for production purposes suggests that this is not usually the case, and is strictly case-dependent. Physiological imbalances or growth demands in natural environments with limited food availability tend to decrease the fitness of selectively-bred stock in the wild. Nevertheless, it is possible for selectively-bred stocks to overcome one fitness component if other components, such as juvenile viability, adult viability, age at sexual maturity, female fecundity, male fertility, and mating success are enhanced through the selective breeding process [74]. Whether selectively bred tilapia show increased or decreased fitness in the wild has yet to be assessed experimentally.

A stable community is one whose structure and function return to their initial conditions after a perturbation [75]. A stable ecosystem will quickly recover from a fish escape event. It has been shown that decreases in native species are more common in low-diversity aquatic ecosystems following introductions of tilapias (e.g., high elevation lakes of Madagascar with few native species), than in high-diversity ecosystems (e.g., coastal lakes with many native species) [76]. Despite widespread introductions into Asian waters, explicit evidence on the ecological and genetic impacts of these non-native tilapias as a whole is scanty [22]. The scanty evidence of the impacts of tilapias in Asia could be due to the absence of conspecifics and other related fish species in communities of that region. This lack of information also could be attributed in part to poor assessment of "before" and "after" states of the receiving ecosystems as part of deliberate species introductions, especially for aquaculture.

Release or escape of the GIFT/GST strain into African waters may have the potential to be highly damaging. These selectively-bred strains could mate with wild stocks to set off the genetic impact scenarios described above. Apart from there being related, wild tilapia stocks with which to interbreed, African countries currently lack the capacity to prevent the escape of selectively-bred fish from aquaculture facilities or to prevent the intentional release of these fish into the wild. Also, there is little data on the baseline condition of the receiving ecosystems before the introduction of selectively-bred tilapia.

While Brummett and Ponzoni [77] and Ponzoni *et al.* [23] do not associate use of genetically improved tilapia strains in Africa with a high level of concern, Hallerman and Hilsdorf [78] noted that considerable molecular and adaptive variation exists within the species *O. niloticus*. The species has an exceptional ability to colonize and adapt to a wide range of habitats, ranging from small forest rivers to large drainages and lakes, as well as alkaline pools with hot springs [79,80]. At a general level, the description of seven subspecies based on eco-morphology [79] reflects adaptive divergence. Multiple putative evolutionarily significant units (ESUs) correspond more strongly to bioregions, however, than to subspecies. Bezault *et al.* [81] discuss the hypothesis that *O. n. filoa* and *O. n. cancellatus* are differentially adapted ecotypes rather than valid subspecies; they may constitute ESUs. Additional ESUs may be detected upon detailed survey; for example, Nyingi *et al.* [82] found a unique genetic resource in a recently discovered population from a warm water spring, a tributary of the Loboi Swamp in Kenya. Sex determination systems of natural populations adapted to three extreme thermal regimes showed thermosensitivity of sex differentiation [81], indicating either genotype-environment interaction or epigenetic effects [83] upon sex determination. Observation of genetic differentiation among *O. niloticus* populations within regions supports the existence of multiple management units (MUs) within certain ESUs, for example, in the Ethiopian and Nilotic regions. In the latter, analysis of microsatellite variation among five Egyptian populations [84] indicated distinct groups respectively inhabiting the deeper lotic Nile River, the shallow less lotic Delta lakes, and the upstream Nile River. The economic importance of *O. niloticus* worldwide makes detailed knowledge of its genetic resources pivotal for sustainable use of the species in aquaculture operations [85]. Hence, detailed consideration of adaptive differentiation is needed to defensibly assess genetic risk from culture of selectively bred *O. niloticus* in Africa.

Several international and national agreements address management of genetic resources. The major one is the Cartagena Protocol on Biosafety developed to implement the Convention on Biological Diversity, which came into effect in 2003, that which governs the movement of living modified organisms (LMOs) resulting from modern biotechnology from one country to another [86]; Convention articles 15, 16, and 22 outline guidelines on risk assessment, risk management and capacity building with regards to LMOs. Kapuscinski *et al.* [87] outline the steps to be taken by developing countries to strengthen scientific and technical capacity to address environmental biosafety issues associated with LMOs; while the risks associated with LMOs and selectively bred stocks may differ, the risk assessment framework developed for LMOs could serve as a useful resource for countries considering the prospect of importing existing selectively bred strains or developing their own strains.

POTENTIAL ECONOMIC BENEFITS OF THE GIFT STRAIN IN AFRICA; CASE STUDY OF GHANA

Model

Against this background, we estimated the potential economic benefits of using selectively bred tilapia strains in Africa. We focused upon the GIFT strain for which production traits and prices are known and approached the benefit assessment by applying the economic surplus model, the most common method for analyzing the economic benefits of agricultural research in a partial equilibrium framework [88,89]. This analysis has been used to study, *ex ante*, the benefits of agricultural research for various innovations and for a number of countries (e.g., marker-assisted rice breeding in Southeastern Asia [90], and cassava breeding in sub-Saharan Africa [91]). Total economic benefits associated with a new technology for a non-traded commodity (Ghana currently exports very little freshwater fish; [92]) can be represented by the formula:

$$\Delta TS = PQK\,(1 + 0.5Zn) \qquad (1)$$

where P and Q are the initial equilibrium price and quantity, respectively; $Z = Ke/(e + n)$ is the relative reduction in price due to the supply shift resulting from the new technology; e is supply elasticity, and n is demand elasticity (absolute value), which reflect how responsive the quantity supplied and quantity demanded are to changes in prices, respectively; and K is the shift in the supply curve as a proportion of the initial price. K is calculated as:

$$K = \left(\frac{E(Y)}{\varepsilon}\right) - \left(\frac{E(C)}{1+E(Y)}\right) \times p \times A \times (1 - d) \qquad (2)$$

where E(Y) is the expected proportionate yield increase per hectare after the adoption of the new technology; E(C) is the expected proportionate change in variable input cost per hectare; p is the probability of success associated with the research; A is the adoption rate for the technology; and d is the depreciation rate of the new technology [89].

"Economic benefits" is the change in total economic surplus for each year, and the costs are the expenditures on the research projects plus estimated after-project costs related to developing and disseminating the new varieties. The annual costs and benefits are netted and totaled using the discount rate to calculate a net present value (NPV) using the standard NPV formula:

$$NPV = \sum_{t=1}^{T} \frac{R_t - C_t}{(1+i)^t} \quad (3)$$

where: R_t is the benefits in year t; C_t is research, development, and dissemination cost in year t; and i is the discount rate. In other words, the NPV is calculated as the sum of future benefits, minus the costs associated with the project discounted over time.

Data and Key Assumptions

Several factors must be considered to estimate the economic benefits of an agricultural innovation using the economic surplus model: (1) the proportion of farmers who adopt the innovation over time; (2) the price of the commodity; (3) the change in yield of the commodity with the new technology; (4) the nature of the market, as products that are traded may not experience price declines if production increases [37,89]; (5) the time it takes to develop the innovation, and the number of years for maximum adoption to be reached; and (6) the discount rate for future benefits compared to current benefits, since for example, a 5% discount rate implies that after 20 years the calculated benefits of the new technology are not sensitive to the usual depreciation of a new technology that occurs with time.

Dey [37] found that the adoption of the GIFT strain in Bangladesh, China, the Philippines, Thailand, and Vietnam led to yield increases of between 24% and 61%. Since this information is not available for Ghana, we used the mean yield change across these five countries (41.4%), bounded by 24% and 61% for our calculations. The Food and Agricultural Organization estimated Ghana's Nile tilapia aquaculture production at 18,200 tons in 2011 [93]. Increases in Nile tilapia aquaculture production in the four preceding years were 1600 t, 1576 t, 2748 t, and 8776 t, respectively. Therefore, we extrapolated the aquaculture production of tilapia in 2012 to be 28,200 t, an increase of 10,000 t over 2011's production value. We calculated the price per ton of production as $2,646 by dividing FAO's 2011 estimate of the value of production ($48,159,000) by the annual production that year (18,200 t).

Experience from tilapia culture indicates that the only production cost variable expected to be directly impacted by the adoption of the GIFT strain is fingerling price. Fingerling cost constitutes approximately 14% of total costs in a typical enterprise budget for a tilapia farm [94]. Also, according to the Asian Development Bank [95], one of the consequences of the introduction of the GIFT strain in the Philippines was the increase in the price of GIFT fingerlings by 150% compared to non-GIFT fingerlings. Assuming this same increase in fingerling price with the adoption of the GIFT strain in Ghana, total production costs will be increased by 20%. This increase in total costs is expected to incorporate increases and decreases in all input costs resulting from the adoption of the new strain. We also assumed that the technology will be available to Ghanaian farmers from 2013, since the country took delivery of the strain officially in 2012 [96], although primarily for research purposes.

Dey [37] defined "early" GIFT adoption rates to be 30%–40% during the initial dissemination of the technology in Asia, peaking at 70% at the latter stages. We are not aware of any data on supply and demand elasticities of tilapia in Ghana. In the absence of such studies, a supply elasticity of 0.5 usually is applied to perennial crops and other livestock, and demand elasticity of 1 can be used for most livestock [89]. Since the GIFT technology exists already, the probability of success of GIFT research can be assumed to be 100%. Research and development costs also can be assumed to be zero.

Following Ponzoni *et al.* [23], we anticipate that a government department, such as the Aquaculture Research and Development Center in Ghana, will have to invest in the establishment and running of a nucleus to continuously maintain and improve the strain. Hatchery operators in Ghana will replace their brood stock annually with the latest version of the strain from the nucleus. This arrangement will ensure that production will benefit from the greatest amount of genetic gain. However, this arrangement will have cost implications. Ponzoni *et al.* ([23], Table 4) provide values for annual or recurrent costs of such a program, with the most likely value being $60,000. We assume that the initial investment cost component is not necessary in our case, since the strain already has been developed. Recurrent costs include dissemination costs, and the costs of continuous improvement and maintenance. We acknowledge that continuous improvement could result in the continuous increase in growth rate and other traits, such as survival and disease resistance, and this expectation should be captured in the sensitivity analysis (see next section).

Analysis

We included the data above in a modified version of a spreadsheet that incorporates all the model formulas obtained from George W. Norton (Department of Agricultural and Applied Economics, Virginia Polytechnic Institute and State University, Blacksburg, VA, USA). To the typical or most likely value of each key variable, we added a possible minimum and a possible maximum to create a triangular distribution, based on the assumptions or extrapolations from the preceding sub-section, in order to incorporate uncertainty (Table 1), and then conducted economic risk analysis by running Monte Carlo simulations using the @Risk 6 software [97]. We ran the model to calculate the net present values of the annual economic benefits of the GIFT technology for 20 years, at a discount rate of 5%, in Microsoft Excel. The total net present value (NPV) then was calculated as the sum of these annual benefits. We also conducted an analysis to determine how sensitive the average NPV was to the key variables in Table 1.

Table 1: Values of key variables used in the Monte Carlo simulation to estimate net present value of adopting the GIFT strain in Ghana

Variable	Minimum	Most likely	Maximum
Recurrent costs ($)	30,000	60,000	90,000
Increase in production costs (%)	10	20	30
2012 Nile tilapia aquaculture production (metric tons)	26,200	28,200	30,200
Peak adoption rate (%)	60	70	80
Yield change (%)	24	41.4	61

Results

The estimated cumulative net benefits of adoption of the Genetically Improved Farmed Tilapia strain in Ghana (discounted at 5%) over twenty years ranged from over $130 million to about $650 million, with a mean close to $400 million (Figure 2). There is a probability of about 40% that the NPV is greater than $400 million (Figure 3). Sensitivity analysis indicated that the variable with the biggest impact on the mean

NPV (across the variable's range of values from Table 1) was the level of change in the aquaculture yield of Nile tilapia resulting from the adoption of the GIFT strain (Figure 4). Not as important were the peak rate of adoption of the strain, level of change in production costs with the adoption of GIFT, the production of Nile tilapia in the year preceding the introduction of the GIFT strain, and the recurrent costs associated with maintaining, improving, and disseminating the new strain, in that order.

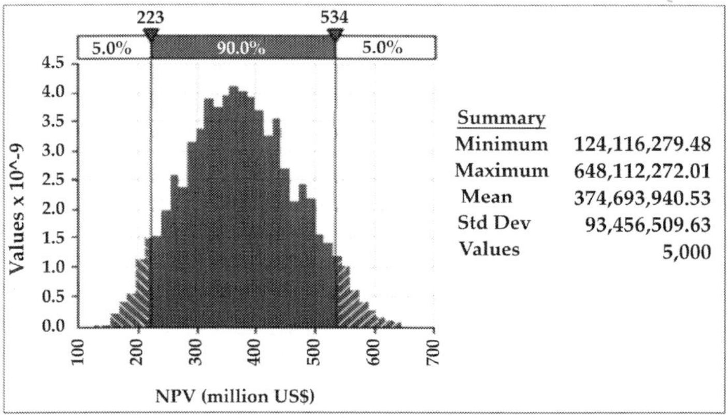

Figure 2: Summary of results of Monte Carlo simulation to determine net present value (NPV) in US$ for the adoption of GIFT-strain tilapia in Ghana, showing a 90% confidence interval.

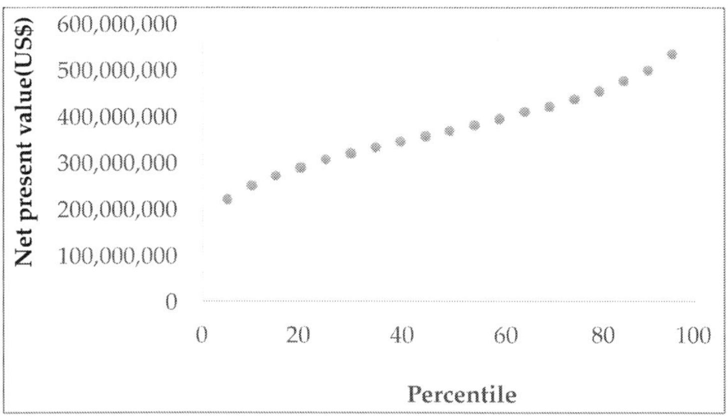

Figure 3: Net present values (NPV) at different percentiles for the adoption of GIFT-strain tilapia in Ghana.

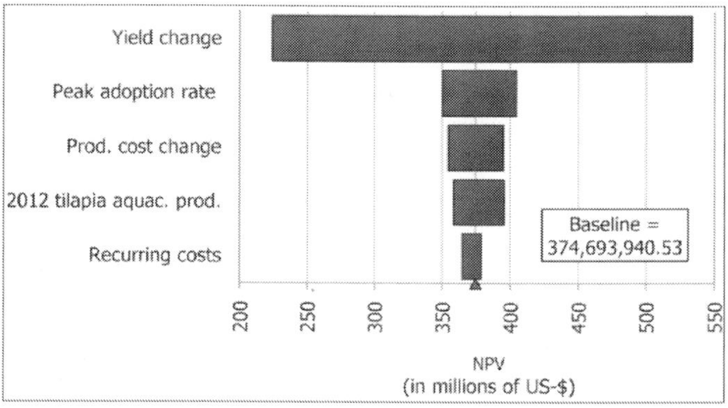

Figure 4: Relative impacts of key variables on the mean net present value of adoption of GIFT-strain tilapia in Ghana across the range of key variables.

Discussion

According to the World Bank [98], the gross domestic product (GDP) of Ghana was $40.71 billion in 2012. This implies that Ghana stands to benefit by approximately 1% of the country's GDP from the introduction of the GIFT strain. The calculated NPV is also approximately 2.5% of the component of the GDP contributed by the country's agriculture sector [99]. This percentage is significant, given that the total farmed fish share of the Ghana fish market is currently 3% [100]. The Ghana National Aquaculture Development Plan [100] quotes the value of commercially-farmed fish as $28.44 million in 2010. This value should be close to an optimistic $40 million at the time of developing this manuscript (May 2014) if any progress in growth has been made. Our calculated NPV ($400 million over 20 years) indicates that Ghana could achieve or add on about 50% of the annual farmed fish value from the introduction of the GIFT strain alone. Ponzoni *et al.* [23] conducted an investment appraisal of a genetic improvement program in Nile tilapia and found the most conservative estimates of economic benefits ranging between 4 and 32 million US$. Their simulated timeframe was 10 years and included the cost of developing the strain from scratch. However, our analysis was done for a period of 20 years with the assumption that the strain was already developed, which is the case with both the GIFT and Akosombo strains.

The version of the economic surplus model that we used assumed that Ghana exports little to no tilapia, as is the case currently. Therefore, the computed NPV values could be higher if this assumption no longer holds true, as might be the case when there is an abundant supply of tilapia. Clearly Ghana, and hence Africa, stands to gain substantial socio-economic benefits from the adoption of the GIFT strain. Some of the possible key benefits include the increased availability of relatively low-cost, high-quality animal protein from the increased yield; increased employment within the expanded aquaculture sector; and possible foreign exchange earnings in the long term.

The Monte Carlo simulation technique allowed us to estimate ranges for variables with uncertain values and to analyze economic risk stemming from changes in these values on net present value. This approach incorporated the element of uncertainty into our computations. Of the five assumed or extrapolated key variables, the cost of maintenance, continuous improvement, and dissemination of the new strain in Ghana (recurring costs) had the least impact on changes in mean NPV. It is intuitive that not much investment will be needed in the dissemination of the GIFT strain in Ghana to cage fish farmers on Volta Lake and to the most progressive pond farmers in other parts of the country. We have observed that this group of tilapia farmers has widely adopted the locally improved Akosombo strain of *O. niloticus*. However, more work needs to be done to disseminate these relatively new strains of tilapia to the majority of pond farmers, who are still farming inferior, mixed, and unknown stocks of tilapia more than a decade after the development and dissemination of the Akosombo strain by the ARDEC started [28,85]. Our analysis shows how small the investment cost of maintaining, continuous improvement, and dissemination of improved strains of tilapia could be relative to the economic benefits of these strains reaching all farmers. It is a sound policy for the government of Ghana to invest in the dissemination of the improved Akosombo and/or the GIFT strain of tilapia.

On the other hand, the variable with the largest impact on mean NPV was the difference in aquaculture yield between the GIFT and local strains of Nile tilapia. A number of production bottlenecks exist currently in the Ghanaian and African aquaculture industry. The AquaFish Innovation Lab, formerly the Aquaculture and Fisheries Collaborative Research Support Program (AquaFish CRSP) has observed, through on-farm demonstrations in Ghana, that for pond farmers growing the Akosombo strain of *O. niloticus*, improving pond construction and maintenance, supplementary use of commercial feeds, water quality management, and control of excessive reproduction in ponds (by hormonal sex-reversal and polyculture with a predator) could result in an increase of 2–4 times the current average yields from unimproved local strains. Other significant bottlenecks observed are institutional, including, inadequate access to urban markets due to bad roads, and unreliable electricity supply leading to absence of cold-storage in the rural tilapia value-chain and therefore inefficient and risky post-harvest handling [101]. Addressing these problems would increase the yield and income from improved local strains of tilapia to the extent that the differences in yield between local and GIFT strains could diminish the calculated NPV for the GIFT strain. Notably, some of these bottlenecks have the potential to limit the production and profitability of the GIFT strain, irrespective of its superior growth rate. For example, a farmer who invests in the necessary better management practices and the extra cost of fingerlings of the GIFT strain but who cannot market his or her fish effectively may become skeptical about the value of investing in increased production at an increased cost. Therefore, the problems currently impeding the production of local strains could also affect the realization of the attractive economic benefit computed for GIFT in this study.

Other socio-economic impacts are possible with the adoption of improved strains (see Figure 1). These include higher incomes, better nutrition (more protein), reduced poverty, improved health and welfare. Also, since women are traditionally involved more in processing and marketing of fish, they too will benefit from the adoption of improved strains. This analysis focused generally on the GIFT strain as a representation of an improved tilapia strain. This analysis will still hold true with the Akosombo or any other improved strain. At this time, we are not aware of any completed biological studies comparing the performance of these two strains. These extensions are recommended for future studies.

CONCLUSIONS

Clearly, the introduction of the GIFT strain in Africa has the potential for substantial economic benefits. However, the potential ecological impacts of the GIFT strain on African aquatic ecosystems cannot be overlooked. After a country makes the decision to introduce the GIFT tilapia strain, each facility that would like to culture this strain must be able to show the structures it has put in place to prevent escapes and both intentional and inadvertent release into the wild. Of course, facilities prone to such events as flooding should not be allowed to culture the GIFT strain. Possible methods to achieve this goal of non-release of the GIFT strain from farms include physical confinement, reproductive confinement, and operations management [63]. Effective physical confinement involves a combination of measures, such as mechanical barriers (e.g., standpipe screens, gravel traps, *etc.*). Also, effluents could be filtered to exclude any life-stage of the strain. Reproductive confinement is particularly important where physical confinement alone cannot be relied on. Production of GIFT tilapia could be limited to either monosex or sterile individuals. Operations management includes those measures put in place on a tilapia farm to ensure that the activities of all workers are in conformance with the goal of effective confinement, to prevent any unauthorized human access to the facility, and to routinely inspect and maintain all physical barriers to ensure they are constantly operational [63]. Operations management also could ensure that only dead fish on ice are marketed.

African governments, such as Ghana, which have obtained, or are in the process of obtaining, the GIFT germplasm should commission rigorous baseline and ecological risk analyses. Since each country has unique conditions, exhaustive and baseline ecological, genetic, and economic benefit and risk analyses should be conducted separately for each country. Efforts also must be made to improve the culture of locally available strains of tilapia. Improvements in management practices and infrastructure could go a long way to increase the yield of local strains even if genetically-improved strains are not introduced. These improvements also will ensure that the full potential of introduced, improved strains are realized.

Governments also must develop the political will to refuse, or to discontinue the adoption of, the GIFT strain if results of environmental, genetic and economic risk analyses point them in that direction. Each country should develop the capacity to prevent the needless exposure of native aquatic ecosystems to preventable risk, while reaping the substantial economic benefits of adopting the GIFT or similar improved strains of tilapia.

Acknowledgments

This research was partly funded by the Aquaculture and Fisheries (AquaFish) Innovation Lab, supported by the U.S. Agency for International Development (USAID) award number CA/LWA No. EPP-A-00-06-0012-00, and by contributions from participating institutions. The AquaFish Innovation Lab accession number is 1421. We would like to thank Raul Ponzoni (formerly of the WorldFish Center) for making important factual corrections to the review section 8 on dissemination of GIFT/GST Strains in Africa. The opinions expressed herein are those of the authors and do not necessarily reflect the views of the U.S. Agency for International Development.

Author Contributions

The study was conceived by Emmanuel A. Frimpong, and designed with Eric M. Hallerman. Yaw B. Ansah did the desk study, data collection and economic analysis, and a first draft of results. Emmanuel A. Frimpong extended interpretation and discussion of economic analysis results. Each author contributed to writing and editing of specific portions of the manuscript as follows: Yaw B. Ansah (Economics and Ecology); Emmanuel A. Frimpong (Ecology and Economics); and Eric M. Hallerman (Genetics and Ecology).

REFERENCES AND NOTES

1. Subasinghe, R.; Soto, D.; Jia, J. Global aquaculture and its role in sustainable development. *Rev. Aquac.* 2009, *1*, 2–9.
2. Béné, C.; Heck, S. Fish and food security in Africa. *NAGA WorldFish Center Q.* 2005, *28*, 8–13.
3. Brummett, R.E.; Angoni, D.E.; Pouomogne, V. On-farm and on-station comparison of wild and domesticated Cameroonian populations of *Oreochromis niloticus*. *Aquaculture* 2004, *242*, 157–164.
4. Lind, C.E.; Brummett, R.E.; Ponzoni, R.W. Exploitation and conservation of fish genetic resources in Africa: Issues and priorities

for aquaculture development and research. *Rev. Aquac.* 2012, *4*, 125–141.

5. Ponzoni, R.W.; Nguyen, N.H.; Khaw, H.L.; Hamzah, A.; Bakar, K.R.A.; Yee, H.Y. Genetic improvement of Nile tilapia (*Oreochromis niloticus*) with special reference to the work conducted by the Worldfish Center with the GIFT strain. *Rev. Aquac.* 2011, *3*, 27–41.
6. Greer, D.S.; Harvey, B.J. *Blue Genes: Sharing and Conserving the World's Aquatic Biodiversity*; Earthscan: London, UK, 2004.
7. Teichert-Coddington, D.; Popma, T.; Lovshin, L. Attributes of tropical pond-cultured fish. In *Dynamics of Pond Aquaculture*; Egna, H.S., Boyd, C.E., Eds.; CRC Press: Boca Raton, FL, USA, 1997; pp. 183–198.
8. El-Sayed, A.F.M. *Tilapia Culture*; CABI: Cambridge, MA, USA, 2006.
9. Fitzsimmons, K. Potential to increase global tilapia production. In *Global Outlook for Aquaculture Leadership*; GOAL Conference: Kuala Lumpur, Malaysia, 2010.
10. Food and Agriculture Organization. *The State of World Fisheries and Aquaculture*; Food and Agriculture Organization (United Nations): Rome, Italy, 2010; p. 218.
11. Shelton, W.L.; Popma, T.J. Biology. In *Tilapia: Biology, Culture, and Nutrition*; Lim, C.E., Webster, C.D., Eds.; Haworth Press: Binghamton, NY, USA, 2006; pp. 1–49.
12. Food and Agriculture Organization. *State of World Fisheries and Aquaculture*; Food and Agriculture Organisation: Rome, Italy, 2012; p. 209.
13. Fitzsimmons, K. Prospect and potential for global production. In *Tilapia: Biology, Culture, and Nutrition*; Lim, C., Webster, C., Eds.; Harworth Press: Binghamton, NY, USA, 2006; pp. 51–72.
14. Pullin, R. Tilapia: Everyman's fish. *Biologist* 1985, *32*, 84–88.
15. Gjoen, H. GIFT Program Continues: Distribution of Fast-Growing Tilapia to Expand. Available online: http://pdf.gaalliance.org/pdf/GAA-Gj%F8en-Dec01.pdf (accessed on 30 May 2014).
16. McAndrew, B.J. Evolution, phylogenetic relationships and biogeography. In *Tilapias: Biology and Exploitation*; Beveridge,

M.C.M., McAndrew, B.J., Eds.; Kluwer Academic: Dordrecht, The Netherlands, 2000; pp. 1–32.

17. Pillay, T.V.R.; Kutty, M. *Aquaculture: Principles and Practices*; Blackwell Publishing: Oxford, UK, 2005.
18. Canonico, G.C.; Arthington, A.; McCrary, J.K.; Thieme, M.L. The effects of introduced tilapias on native biodiversity. *Aquat. Conserv.: Mar. Freshw. Ecosyst.* **2005**, *15*, 463–483.
19. Fish Base *Tilapia spp*. Available online: http://www.fishbase.org/Nomenclature/ScientificNameSearchList.php?(accessed on 14 December 2013).
20. Courtenay, W. Tilapias as non-indigenous species in the Americas: Environmental, regulatory and legal issues. *Tilapia Aquac. Am.* **1997**, *1*, 18–33.
21. Pullin, R.S.V.; Palomares, M.; Casal, C.; Dey, M.; Pauly, D. Environmental impacts of tilapia. In *Fourth International Symposium on Tilapia in Aquaculture*; Fitzsimmons, K., Ed.; Northeast Regional Aquacultural Engineering Services: Ithaca, NY, USA, 1997; pp. 554–570.
22. De Silva, S.S.; Subasinghe, R.P.; Bartley, D.M.; Lowther, A. *Tilapias as Alien Aquatics in Asia and the Pacific: A Review*; Food and Agriculture Organization: Rome, Italy, 2004; Volume 453.
23. Ponzoni, R.W.; Nguyen, N.H.; Khaw, H.L. Investment appraisal of genetic improvement programs in Nile tilapia (*Oreochromis niloticus*). *Aquaculture* **2007**, *269*, 187–199.
24. Lutz, C.G. Recent directions in genetics. In *Tilapia: Biology, Culture, and Nutrition*; Lim, C.E., Webster, C.D., Eds.; Haworth Press: Binghamton, NY, USA, 2006; pp. 139–180.
25. Poompuang, S.; Hallerman, E.M. Toward detection of quantitative trait loci and marker-assisted selection in fish. *Rev. Fish. Sci.* **1997**, *5*, 253–277.
26. Attipoe, F.Y.K.; Agyakwah, S.K. Status of Catfish Farming and Research in Ghana. In Proceedings of the a Workshop on the Development of a Genetic Improvement Program for African Catfish *Clarias gariepinus*, Accra, Ghana, 5–9 November 2007; Ponzoni, R.W., Nguyen, N.H., Eds.; p. 23.

27. Attipoe, F.Y.; Blay Jnr, J.; Agyakwah, S.; Ponzoni, R.W.; Khaw, H.L.; Abban, E.K. Genetic parameters and response to selection in the development of Akosombo strain of the Nile tilapia (*Oreochromis niloticus*) in the Volta Basin, Ghana. In Proceedings of the International Symposium on Tilapia in Aquaculture, Jerusalem, Palestine, 6–10 October 2013.
28. Ofori, J.; Abban, E.; Karikari, A.; Brummett, R. Production parameters and economics of small-scale tilapia cage aquaculture in the Volta Lake, ghana. *J. Appl. Aquac.* **2010**, *22*, 337–351.
29. Bentsen, H.B.; Gjerde, B.; Nguyen, N.H.; Rye, M.; Ponzoni, R.W.; Palada de Vera, M.S.; Bolivar, H.L.; Velasco, R.R.; Danting, J.C.; Dionisio, E.E. Genetic improvement of farmed tilapias: Genetic parameters for body weight at harvest in Nile tilapia (*Oreochromis niloticus*) during five generations of testing in multiple environments. *Aquaculture* **2012**, *338*, 56–65.
30. Dey, M.M.; Gupta, M.V. Socioeconomics of disseminating genetically improved Nile tilapia in Asia: An introduction. *Aquac. Econ. Manag.* **2000**, *4*, 5–11.
31. Eknath, A.; Acosta, B. *Genetic Improvement of Farmed Tilapias (GIFT) Project: Final Report, March 1988 to December 1997*; World Fish Center: Penang, Malaysia, 1998.
32. Asian Development Bank. *An Impact Evaluation of the Development of Genetically Improved Farmed Tilapia and Their Dissemination in Selected Countries*; Asian Development Bank: Manila, Phillippines, 2005.
33. Asian Development Bank. *Overview of Freshwater Aquaculture of Tilapia in the Philippines*; Asian Development Bank: Manila, Phillippines, 2004.
34. Eknath, A.E.; Tayamen, M.M.; Palada-de Vera, M.S.; Danting, J.C.; Reyes, R.A.; Dionisio, E.E.; Capili, J.B.; Bolivar, H.L.; Abella, T.A.; Circa, A.V. Genetic improvement of farmed tilapias: The growth performance of eight strains of *Oreochromis niloticus* tested in different farm environments. *Aquaculture* **1993**, *111*, 171–188.
35. Gupta, M.V.; Acosta, B.O.; Eknath, A.E.; Dey, M.M. Breakthrough in genetic improvement of tropical finfish through partnership between ICLARM, ASI and developing country NARS. Available

online: http://www.fao.org/docs/eims/upload/206603/3_8_cases.PDF (accessed on 30 May 2014).

36. Acosta, B.O.; Gupta, M.V. The genetic improvement of farmed tilapias project: Impact and lessons learned. In *Success Stories in Asian Aquaculture*; De Silva, S.S., Davy, F.B., Eds.; Springer: London, UK, 2010; pp. 149–171.
37. Dey, M.M. The impact of genetically improved farmed Nile tilapia in Asia. *Aquac. Econ. Manag.* **2000**, *4*, 107–124.
38. Gupta, M.; Acosta, B. From drawing board to dining table: The success story of the GIFT project. *NAGA WorldFish Center Q.* **2004**, *27*, 4–14.
39. Dey, M.M.; Eknath, A.E. *Current Trends in the Asian Tilapia Industry and the Significance of Genetically Improved Tilapia Breeds*; INFOFISH: Kuala Lumpur, Malaysia, 1997.
40. Gupta, M.V.; Bartley, D.M.; Acosta, B.O. *Use of Genetically Improved and Alien Species for Aquaculture and Conservation of Aquatic Biodiversity in Africa*; The WorldFish Center: Penang, Malaysia, 2004; Volume 68.
41. Gupta, M.V. Genetic enhancement and conservation of aquatic biodiversity in Africa. *NAGA WorldFish Center Q.***2002**, *25*, 48–53.
42. WorldFish Center. Policy on the transfer of genetically improved farm tilapia (GIFT) from Asia to Africa by the Worldfish Center. In *The WorldFish Center Working Papers*; WorldFish Center: Penang, Malaysia, 2007.
43. Convention on Biological Diversity. The Nagoya Protocol on Access and Benefit-Sharing. Available online: http://www.cbd.int/abs/ (accessed on 21 March 2014).
44. Ansah, Y.B.; Frimpong, E.A.; Amisah, S. Biological assessment of aquaculture effects on effluent-receiving streams in Ghana using structural and functional composition of fish and macroinvertebrate assemblages. *Environ. Manag.* **2012**, *50*, 166–180.
45. Volpe, J.P.; Taylor, E.B.; Rimmer, D.W.; Glickman, B.W. Evidence of natural reproduction of aquaculture-escaped Atlantic salmon in a coastal British Columbia river. *Conserv. Biol.* **2000**, *14*, 899–903.
46. Naylor, R.L.; Williams, S.L.; Strong, D.R. Aquaculture—A gateway

for exotic species. *Science* **2001**, *294*, 1655–1656.

47. Breder, C.M.; Rosen, D.E. *Modes of Reproduction in Fishes*; The American Museum of Natural History: New York, NY, USA, 1966.
48. Balon, E.K. Reproductive guilds of fishes: A proposal and definition. *J. Fish. Res. Board Can.* **1975**, *32*, 821–864.
49. Arthington, A.; Blüdhorn, D.; Kennard, M. Food resource partitioning by the introduced cichlid, *Oreochromis mossambicus*, and two native fishes in a subtropical Australian impoundment. In Proceedings of the Third Asian Fisheries Forum, Singapore, 26–30 October 1994; pp. 425–428.
50. Crutchfield, J.U., Jr. Establishment and expansion of redbelly tilapia and blue tilapia in a power plant cooling reservoir. *Am. Fish. Soc. Symp.* **1995**, *15*, 452–461.
51. McCrary, J.; van den Berghe, E.; McKaye, K.; Lopez Perez, L. Tilapia cultivation: A threat to native fish species in Nicaragua. *Encuentro* **2001**, *58*, 9–19.
52. Starling, F.; Lazzaro, X.; Cavalcanti, C.; Moreira, R. Contribution of omnivorous tilapia to eutrophication of a shallow tropical reservoir: Evidence from a fish kill. *Freshw. Biol.* **2002**, *47*, 2443–2452.
53. Moreau, J. A review of introductions of tilapia in open waters of Africa, their influence on ecology and fisheries. In Proceedings of the International Symposium on Tilapia in Aquaculture; Fishelson, L., Yaron, Z., Eds.; Tel Aviv University Press: Nazareth, Israel, 1983; pp. 77–85.
54. Vareschi, E.; Jacobs, J. The ecology of Lake Nakuru. *Oecologia* **1985**, *65*, 412–424.
55. Noga, E.J. *Fish Disease: Diagnosis and Treatment*, 2nd ed.; John Wiley & Sons: Ames, IA, USA, 2010; p. 519.
56. Zambrano, L.; Martínez-Meyer, E.; Menezes, N.; Peterson, A.T. Invasive potential of common carp (*Cyprinus carpio*) and Nile tilapia (*oreochromis niloticus*) in American freshwater systems. *Can. J. Fish. Aquat. Sci.* **2006**, *63*, 1903–1910.
57. OIE Aquatic animal health. Available online: http://www.oie.int/en/international-standard-setting/aquatic-code/access-online/ (accessed on 20 March 2014).

58. Carvalho, G.R.; Hauser, L. Genetic impacts of fish introductions: A perspective on African lakes. In *The Impact of Species Changes in African Lakes*; Chapman and Hall: London, UK, 1995; pp. 457–493.
59. Ryman, N.; Laikre, L. Effects of supportive breeding on the genetically effective population size. *Conserv. Biol.* **1991**, *5*, 325–329.
60. Hallerman, E.M. *Population Genetics: Principles and Applications for Fisheries Scientists*; Hallerman, E.M., Ed.; American Fisheries Society: Bethesda, MD, USA, 2003; Hallerman, E.M., Ed.; pp. 239–259.
61. Ferguson, A.; Fleming, I.A.; Hindar, K.; Skaala, O.; McGinnity, P.; Cross, T.; Prodohl, P. The Atlantic Salmon: Genetic Conservation and Management. Blackwell Publishing: Oxford, UK, 2007; Verspoor, E., Stradmeyer, L., Nielsen, J.L., Eds.; pp. 357–398.
62. McGinnity, P.; Prodöhl, P.; Ferguson, A.; Hynes, R.; ó Maoiléidigh, N.; Baker, N.; Cotter, D.; O'Hea, B.; Cooke, D.; Rogan, G. Fitness reduction and potential extinction of wild populations of Atlantic salmon, *Salmo salar*, as a result of interactions with escaped farm salmon. *Proc. R. Soc. Lond. Ser. B: Biol. Sci.* **2003**, *270*, 2443–2450.
63. Araki, H.; Cooper, B.; Blouin, M.S. Genetic effects of captive breeding cause a rapid, cumulative fitness decline in the wild. *Science* **2007**, *318*, 100–103.
64. Araki, H.; Cooper, B.; Blouin, M.S. Carry-over effect of captive breeding reduces reproductive fitness of wild-born descendants in the wild. *Biol. Lett.* **2009**, *5*, 621–624.
65. Hindar, K.; Ryman, N.; Utter, F. Genetic effects of cultured fish on natural fish populations. *Can. J. Fish. Aquat. Sci.***1991**, *48*, 945–957.
66. Tufto, J. Effects of releasing maladapted individuals: A demographic-evolutionary model. *Am. Nat.* **2001**, *158*, 331–340.
67. Tufto, J. Gene flow from domesticated species to wild relatives: Migration load in a model of multivariate selection. *Evolution* **2010**, *64*, 180–192.
68. Garant, D.; Forde, S.E.; Hendry, A.P. The multifarious effects of dispersal and gene flow on contemporary adaptation. *Funct. Ecol.* **2007**, *21*, 434–443.

69. Storfer, A. Gene flow and endangered species translocation. *Biol. Conserv.* **1999**, *87*, 173–180.

70. Tallmon, D.A.; Luikart, G.; Waples, R.S. The alluring simplicity and complex reality of genetic rescue. *Trends Ecol. Evol.* **2004**, *19*, 489–495.

71. Hallerman, E. Application of risk analysis to genetic issues in aquaculture. In *Understanding and Applying Risk Analysis in Aquaculture*; FAO Fisheries and Aquaculture Technical Paper; Bondad-Reantaso, M.G., Arthur, J.R., Subasinghe, R.P., Eds.; Food and Agriculture Organization: Rome, Italy, 2008; Volume 519, pp. 47–66.

72. Seward, N.W.; VerCauteren, K.C.; Witmer, G.W.; Engeman, R.M. Feral swine impacts on agriculture and the environment. *Sheep Goat Res. J.* **2004**, *19*, 34–40.

73. McLeod, R.; Norris, A. *Counting the Cost: Impact of Invasive Animals in Australia, 2004*; Cooperative Research Centre for Pest Animal Control Canberra: Canberra, Australia, 2004.

74. Muir, W.; Howard, R. Fitness components and ecological risk of transgenic release: A model using Japanese medaka (*Oryzias latipes*). *Am. Nat.* **2001**, *158*, 1–16.

75. Pimm, S.L. The complexity and stability of ecosystems. *Nature* **1984**, *307*, 321–326.

76. Moyle, P.B.; Light, T. Biological invasions of fresh water: Empirical rules and assembly theory. *Biol. Conserv.* **1996**, *78*, 149–161.

77. Brummett, R.E.; Ponzoni, R.W. Concepts, alternatives, and environmental considerations in the development and use of improved strains of tilapia in African aquaculture. *Rev. Fish. Sci.* **2009**, *17*, 70–77.

78. Hallerman, E.; Hilsdorf, A.W.S. Conservation genetics of tilapias: Seeking to define appropriate units for management. *Isr. J. Aquac.—Bamidgeh* **2014**. in press.

79. Trewavas, E. *Tilapiine Fishes of the Genera Sarotherodon, Oreochromis and Danakilia*; British Museum (Natural History): London, UK, 1983.

80. Philippart, J.-C.; Ruwet, J.-C. Ecology and distribution of

tilapias. *Biol. Cult. Tilapias* **1982**, *7*, 15–60.

81. Bezault, E.; Balaresque, P.; Toguyeni, A.; Fermon, Y.; Araki, H.; Baroiller, J.-F.; Rognon, X. Spatial and temporal variation in population genetic structure of wild Nile tilapia (*Oreochromis niloticus*) across Africa. *BMC Genet.* **2011**, *12*.
82. Nyingi, D.; de Vos, L.; Aman, R.; Agnèse, J.-F. Genetic characterization of an unknown and endangered native population of the Nile tilapia (*Oreochromis niloticus* (Linnaeus, 1758) (Cichlidae; Teleostei) in the Loboi Swamp (Kenya). *Aquaculture* **2009**, *297*, 57–63.
83. Navarro-Martín, L.; Vinas, J.; Ribas, L.; Diaz, N.; Gutierrez, A.; di Croce, L.; Piferrer, F. DNA methylation of the gonadal aromatase (*cyp19a*) promoter is involved in temperature-dependent sex ratio shifts in the European sea bass. *PLoS Genet.* **2011**, *7*, e1002447.
84. Hassanien, H.A.; Gilbey, J. Genetic diversity and differentiation of Nile tilapia (*Oreochromis niloticus*) revealed by DNA microsatellites. *Aquac. Res.* **2005**, *36*, 1450–1457.
85. Eknath, A.E.; Hulata, G. Use and exchange of genetic resources of Nile tilapia (*Oreochromis niloticus*). *Rev. Aquac.***2009**, *1*, 197–213.
86. Convention on Biological Diversity—The Cartegena Protocol on Biosafety. Available online: http://bch.cbd.int/protocol/default.shtml (accessed on 20 March 2014).
87. Kapuscinski, A.R.; Hayes, K.R.; Li, S.; Dana, G. *Environmental Risk Assessment of Genetically Modified Organisms Volume 3: Methodologies for Transgenic Fish*; CABI: Cambridge, MA, USA, 2007.
88. Masters, A.; Coulibaly, B.; Sanogo, D.; Sidibé, M.; Williams, A. *The Economic Impact of Agricultural Research: A Practical Guide*; Department of Agricultural Economics, Purdue University: West Lafayette, IN, USA, 1996.
89. Alston, J.M.; Norton, G.W.; Pardey, P.G. International Service for National Agricultural Research. In *Science under Scarcity: Principles and Practice for Agricultural Research Evaluation and Priority Setting*; Cornell University Press: Ithaca, NY, USA, 1998.
90. Alpuerto, V.L.E.B.; Norton, G.W.; Alwang, J.; Ismail, A.M. Economic impact analysis of marker-assisted breeding for tolerance

to salinity and phosphorous deficiency in rice. *Appl. Econ. Perspect. Policy* **2009**, *31*, 779–792.

91. Rudi, N.; Norton, G.W.; Alwang, J.; Asumugha, G. Economic impact analysis of marker-assisted breeding for resistance to pests and post-harvest deterioration in cassava. *Afr. J. Agric. Resour. Econ.* **2010**, *4*, 110–122.
92. Food and Agriculture Organization (FAO). *FAOSTAT*; FAO: Rome, Italy, 2012.
93. Food and Agriculture Organization (FAO). *Global Aquaculture Production 2013*; FAO: Rome, Italy, 2013.
94. Engle, C.R.; Neira, I. *Tilapia Farm Business Management and Economics*; Aquaculture CRSP, Oregon State University: Corvallis, OR, USA, 2005.
95. Asian Development Bank. *Special Evaluation Study on Small-Scale Freshwater Rural Aquaculture Development for Poverty Reduction*; Asian Development Bank: Manila, Phillippines, 2004; p. 67.
96. Padi, J.; ARDEC, Water Research Institute, Akosombo, Ghana. Personal communication, 2012.
97. Palisade Corporation. *@Risk*, Edition 6 ed; Palisade Corporation: Ithaca, NY, USA, 2013.
98. World Bank Data by Country; Ghana. The World Bank Group, 2013. Available online: http://data.worldbank.org/country/ghana (accessed on 6 July 2013).
99. Overseas Development Institute. *Ghana's Sustained Agricultural Growth: Putting Underused Resources to Work*; ODI Publications: London, UK, 2011; p. 27.
100. Ghana Fisheries Commission. *Ghana National Aquaculture Development Plan*; Ghana Ministry of Food and Agriculture: Accra, Ghana, 2012; p. 85.
101. Frimpong, E.A.; Anane-Taabeah, G. *Social and Economic Performance of Tilapia Farming in Ghana*; Junning, C., Quagrainie, K.K., Hishamunda, N., Eds.; FAO: Rome, Italy, 2014; in press.

SECTION II

MANAGEMENT, TECHNOLOGIES AND MAIN SELECTED TRAITS

CHAPTER

8

DEVELOPMENT AND APPLICATION OF BIOLOGICAL TECHNOLOGIES IN FISH GENETIC BREEDING

Kang Xu, Wei Duan, Jun Xiao, Min Tao, Chun Zhang, Yun Liu and Shao-Jun Liu

Key Laboratory of Protein Chemistry and Fish Developmental Biology of Ministry of Education of China, College of Life Sciences, Hunan Normal University, Changsha 410081, China

ABSTRACT

Fish genetic breeding is a process that remolds heritable traits to obtain neotype and improved varieties. For the purpose of genetic improvement, researchers can select for desirable genetic traits, integrate a suite of traits from different donors, or alter the innate genetic traits of a species. These improved varieties have, in many cases, facilitated the development of the aquaculture industry by lowering costs and increasing both quality and yield. In this review, we present the pertinent literatures and summarize the biological bases and application of selection breeding technologies (containing traditional selective breeding, molecular marker-assisted breeding, genome-wide selective breeding and breeding by controlling single-sex groups), integration breeding technologies (containing cross breeding, nuclear transplantation, germline stem cells and germ cells transplantation, artificial gynogenesis, artificial androgenesis and polyploid breeding) and modification breeding technologies (represented by transgenic breeding) in fish genetic breeding. Additionally, we discuss the progress our laboratory has made in the field of chromosomal ploidy breeding of fish, including distant hybridization, gynogenesis, and androgenesis. Finally, we systematically summarize the research status and known problems associated with each technology.

Keywords: Fish genetic breeding genetic improvement biological method traits new variety

INTRODUCTION

Fish is high in protein, unsaturated fatty acids, vitamins, and microelements, so it is an important component of a high-protein, low-calorie diet. Recent outbreaks of mad cow disease, foot-and-mouth disease, and bird flu suggest that fish is one of the safer animal-based foods for human consumption. Unfortunately, however, the increased demand for fish has resulted in rapid depletion of fish stocks throughout the world. Concurrently, water resources are becoming scarcer. Given these trends, there is an urgent need to improve the efficiency and sustainability of the aquaculture industry. To address this, researchers have used a variety of methods to produce new varieties that have desirable traits such as rapid growth, high meat quality, and stress resistance.

The various approaches to genetic enhancement can be divided into three categories: selective breeding, integration breeding, and modification breeding. Selective breeding involves selection and breeding of individuals in a population that have desirable traits. Integration breeding involves the combination of two or more groups to obtain a mix of desirable traits from the donors. Modification breeding involves the creation of new genetic traits. Selective breeding can be further divided into traditional selective breeding and molecular marker-assisted selective breeding. The classical approach to traditional selective breeding is to choose and breed only individuals that exhibit desirable characteristics for one or multiple traits such as growth rate, meat quality, and stress resistance. Methods that yield all-female and all-male groups by treatments with hormones, temperature, or other factors are also included in this category. The research and application of molecular markers associated with functional genes in genetic breeding can also be classified as molecular marker-assisted selection.

The essence of integration breeding is to produce individuals whose genetic materials are altered from those of the parents. For this purpose, hybridization, hydrostatic pressure, colchicine treatment and other biological, physical, or chemical methods are used to obtain hybrid or polyploid individuals. During the processes of distant hybridization, gynogenesis, and androgenesis, chromosome ploidy is reflected in the gametes and zygotes. Application of nucleus transplantation and stem cells transplantation (SCT) results in the recombination of genetic materials, and these methods are categorized into integration breeding. It is worth noting that distant hybridization can not only generate heterosis in new varieties, but also result in the formation of diploid or tetraploid hybrid strains, in which both the females and males are fertile. Under these conditions it is possible to form a new species. Modification breeding involves the creation of transgenic fish by the transfer of genetic material from a donor to a recipient via micromanipulation. The purpose of integration breeding and modification breeding is to alter the genotype and phenotype of offspring. Selection breeding can be used subsequent to integration or modification breeding to ensure the persistence and emphasis of desirable inherited characteristics.

FISH GENETIC BREEDING

Selection Breeding Techniques

Directional filtering of traits is an important and indispensable step in genetic breeding programs. Recently, research has focused on methods to rapidly and efficiently screen for important economic traits. Advances in the fields of genetics, molecular biology, and other biological techniques have resulted in a shift from single traditional selective breeding to more diverse selective breeding methods. The following discussion summarizes selective breeding based on the divisions between traditional selective breeding, markerassisted selection, genomic technology breeding, and monosex breeding.

Traditional selective breeding

In fish, traditional selective breeding is the classical method of genetic breeding. The primary objective is to screen and select for desirable genetic traits in individuals or groups. Common methods of traditional selective breeding include population selective breeding, pedigree selective breeding, parental selective breeding, and integrated selective breeding. Additionally, best linear unbiased prediction (BLUP) has been used for rainbow trout (Oncorhynchus mykiss) [1], coho salmon (Oncorhynchus kisutch) [2], black sea bass (Dicentrarchus labrax L.) [3] and other species.

Selective breeding has been successfully applied to enhance desirable traits in multiple species, including rainbow trout [1], silver carp (Hypophthalmichthys molitrix) [4], channel catfish (Ictalurus punctutatus) [5], and so on. In early 1919, Embody and Hyford [6] used selective breeding to increase survival from 2% to 69% in brook trout (Salvelinus fontinalis) infected with furunculosis. Similarly, a number of selected strains of rainbow trout have been bred for desirable traits, such as fast growth, increased fecundity, and early spawning. In China, traditional selective breeding is frequently applied in freshwater fish breeding. For example, purse red carp (Cyprinus carpio var. wuyuanensis) were selected using a combination of population selection and family selection methods [7]. Xingguo red carp (Cyprinus carpio var singuonensis.) were obtained using population selection breeding [7] and Molong carp (Cyprinus carpio var Molong.) were obtained using a combination of comprehensive and directional selection [7]. A targeted group breeding method was used to obtain Gansu golden trout (Oncorhynchus mykiss) [8]. Additionally, "Pujiang No.1" blunt snout bream (Pujiang No.1 Megalobrama amblycephala) [9], new Gift tilapia (GIFT, Oreochromis niloticus) [10] and Matsuura mirror carp (Cyprinus carpio Songpu carp) [7] were also obtained by selective methods. Among marine fish, bastard halibut (Paralichthys olivaceus) [11], large yellow croaker (Pseudosciaen acrocea) [12], etc. [13] were bred using selection methods and the BLUP breeding method was used for the selection of turbot (Scophthalmus maximus) [14].

Molecular marker-assisted breeding

Molecular markers are a direct reflection of the genetic diversity at the DNA level. As such, these markers have a number of applications, including the identification of population structure or commercially important traits [15,16], identification and cloning of genes [17], construction of genetic maps [18], analysis of genetic relationships [16], prediction of heterosis and molecular-assisted selection breeding [19]. A number of methods have been developed to obtain molecular markers, including Variable Number of Tandem Repeats (VNTR), Random Amplified Polymorphic DNA (RAPD), DNA Amplification Fingerprinting (DAF), Single Nucleotide Polymorphisms (SNP), Amplified Fragment Length Polymorphism (AFLP), Inter-simple Sequence Repeat (ISSR), Simple Sequence Repeat (SSR), Single-strand Conformation Polymorphism (SSCP), Restriction Fragment Length Polymorphism (RFLP) and Sequenced Characterized Amplified Region Marker (SCAR), and others [20].

The development of molecular marker has improved the accuracy and efficiency of selection for specific traits. As a result, molecular-assisted breeding technologies are increasingly used in aquaculture. For example, genetic linkage maps have been constructed for zebrafish [21], medaka [22], pufferfish [23], swordtail fish (Xiphophorus helleri) [24], three thorns fish (Gasterosteus aculeatus) [25], rainbow trout [26], Nile tilapia [27] and several other economically important fish species. These genetic linkage maps have been used to direct selective breeding for many traits [28,29]. In China, researchers have completed genome sequencing in common carp and have subsequently developed genetic linkage maps to identify quantitative trait loci (QTLs) [30], including those linked to muscle fiber related traits and cold resistance, using a BAC library [31]. BAC libraries have been constructed for grass carp (Ctenopharyngodon idellus) [32], blunt snout bream, red crucian carp (Carassius auratus var.), silver crucian carp [33], half smooth tongue sole (Cynoglossus semilaevis Gunther) [34] and others. These BAC libraries lay the foundation for the physical map building and genome sequencing [31].

Genome-wide selective breeding technology

Recently, researchers have decoded the genome of several fish species, including zebrafish (Danio rerio), medaka (Oryzias latipes), pufferfish

(Takifugu rubripes), green pufferfish (Tetraodon nigroviridis), Nile tilapia (Oreochromis niloticus), channel catfish (Ictalurus punctatus), rainbow trout, Atlantic salmon (Salmo salar), bass (Dicentrarchus labrax), and Atlantic cod (Gadus morhua) [31,35–37]. The availability of these genomic data has opened up a number of opportunities to advance basic research and develop commercial applications. For example, genetic linkage maps and molecular genetic marker techniques have been used to explore the location of genes associated with growth, sex decision, disease resistance, and other traits. Additionally, quantitative trait loci are used to design DNA markers for assisted breeding technology. In 2010, Chinese researchers obtained the whole genome sequences of half smooth tongue sole [38], Pacific oysters (Crassostrea gigas) [39], large yellow croaker, epinephelus bleekeri (Epinephelus bleekeri), common carp (Cyprinus carpio) [40], and bastard halibut [31]. In 2011, researchers at the Chinese Academy of Sciences Institute of Aquatic Organisms and other research institutions also started whole genome sequencing of bighead carp (Aristichthys nobilis), grass carp, and silver carp (Hypophthalmichthys molitrix), species that form part of the "four famous fish" in China [31]. Additionally, our laboratory has basically completed the comprehensive genome sequencing of red crucian carp. The data from this latter effort are critical to improving our understanding of allotetraploid heredity, and will also provide important insights into ways to advance molecular genetic breeding in cyprinidae. At the same time, our laboratory has started the whole genome sequencing of Erythroculter ilishaeformis, which is one of the original parents of a new type of hybrid bream that is obtained by our laboratory and has been approved by National Certification Committee for Aquatic Varieties. The obtaining of the genome sequencing of Erythroculter ilishaeformis is beneficial for explaining the molecular mechanism of heterosis and other traits of this new type of hybrid bream. In summary, genomic information allows for more rapid and complete understanding of economically important performance traits and provides a basis for the development of new varieties for aquaculture.

More recently, next-generation sequencing (NGS) has been used in our laboratory to obtain transcriptome data from different tissues in red crucian carp, common carp (Cyprinus carpio L.), blunt snout bream, Xenocypris davidi bleeker, grass carp, and some hybrid fish (e.g., allotetraploid and allotriploid). These transcriptome data are essential for comprehensive analyses of the genetic characteristics of cyprinid fish and also can be used to obtain SSR and SNP markers which are helpful for seeking specific molecular markers for selective breeding.

Breeding by controlling single-sex groups

Sex-specific differences in morphology and physiological function are relatively common in animals. For example, male and female individuals of several fish species exhibit differences in growth rate, maturation age, reproductive pattern, body color, and body size. Aquaculturists can exploit these differences by controlling the sex of fry, typically by sex reversal, to conduct monosex cultivation. There are two types of sex determination mechanisms in fish: chromosomal and environmental. In chromosomal sex determination, the sex is determined by genes and the sex of an embryo depends on its allosome composition or other sex-determined genes. Under environmental sex determination, sex is influenced by external factors such as temperature, humidity, and pH [41].

Since the 1950s, researchers and culturists have tested the utility of hormone treatments to alter fish sex in a number of species, including Oryzias latipes, Carassius auratus auratus, Carassius aruatus, Cyprinus carpio, Brachydanio rerio, Onchorynchus mykiss, Salmo salar, Oncorhynchus kisutch, O. tshawytscha, Oreochromis mossambicus, Epinephelus akaara, Oreochromis niloticus, Poecilia reticulates Peters, Mollienisia velifera, Xiphophorous hellerii, Mugil cephalus, Epinephelus coioides, E. tauvina, and Carassius auratus gibelio [41–43].

The mechanism of action and efficacy of these treatments varies among sex hormones. Among androgens, 17α-methyltestosterone is most widely used as it can be added to the feed relatively easily. Conversely, 17β- estradiol and oestrone are the most commonly used estrogens. In one of the earliest experiments, Yamamoto [44] used estrogen to transform male goldfish (Carassius auratus auratus) and medaka into females, which were then crossed with wild male fish to generate YY males. In 1975, Guerrero tested the efficacy of adding methyltestosterone to the feed to convert sex in Oreochromis aureus [45]. In our laboratory, Liu et al. [46] noted that feeding fry with estradiol for an extended period resulted in conversion of males to females with only remnants of testis-like tissue on the ovary surface in leather moustache catfish (Clarias lazera). Conversely, feeding with methyltestosterone (MT) induced sex reversed male and hermaphrodites fish with female genotype. The authors subsequently acquired all-female offspring by self-fertilization of bisexual fish, which helped determine the genetic pattern of sex choice (the type of ♀ XX-♂XY) in Leather moustache catfish and raised a new means of obtaining homozygotes [46,47].

Chen et al. [48] successfully transformed gynogenetic Cyprinus carpio into functional male fish by feeding with methyltestosterone, then crossed these functional male fish with pure red common carp and generated all-female common carp. In addition, our laboratory [49] fed gynogenetic Carassius auratus cuvieri fry with MT and obtained functional sex-reversed male individuals. Crossing these individuals with common female C. auratus cuvieri and female allotetraploid C. aruatus (♀)×Cyprinus carpio (♂), we obtained all-female diploid C. auratus cuvieri and all-female triploid fish. The population of all-female diploid C. auratus cuvieri was crossed with male allotetraploid fish to generate triploid carp. Using this approach, we were able to eliminate the need to artificially remove male individuals, resulting in reduced costs for culturists [50]. Liu et al. [33] obtained functional XY female Pelteobagrus fulvidraco using hormone-mediated reversal, then used gynogenesis to generate XX, XY and YY individuals. On this basis, Wang et al. isolated two pairs of Y and X chromosome-specific markers from yellow catfish (Pelteobagrus fulvidraco) by AFLP and SCAR screening, which then were applied on screening YY super-male individuals and thereby developed a Y- and X-specific allele marker-assisted sex control technical route for cultivating all-male populations [31,51,52]. Following this technical route, all-male Pelteobagrus fulvidraco with stable genetic traits has been approved as a novel variety "yellow catfish all-male No. 1" by National Certification Committee for Aquatic Varieties, because yellow catfish males grow faster than females, and generate difference in individual ultimate size about 2–3 fold.

Integration Breeding Technologies

Cross breeding

The cross breeding (hybridization) of two species is one of the most basic methods of integrating desirable traits (e.g., growth rate, disease resistance). Hybridization aims to get improved new varieties of better growth rate, reproduction rate, disease resistance, yield and quality through integrating excellent parental characters, which plays an important role in fish breeding improvement and production value.

The process of hybridization is divided into two categories, distant and close hybridization, based on the parental genetic relationship. Distant hybridization characterizes crosses between parents that differ by species or higher classifications. Close hybridization characterizes crosses between parents of the same species, but different strains, different varieties, different ecological types, or different populations of individuals.

Because of the close evolutionary relationship between parents, close hybridization generally results in fertile hybrid progeny, so is commonly used in hybrid breeding. The most representative example of close hybridization in fish is the carp [53]. The intraspecific hybridization of carp has occurred throughout the world, resulting in multiple geographic strains or domesticated strains [53,54]. In China, researchers have developed 15 superior carp strains of common carp (e.g., Feng carp, black dragon carp, Xingde carp and so on [7,53,55]. In our laboratory, we have developed Yue carp, and more recently, we have obtained two new hybrid fish from crossing combinations of Japanese crucian carp and red crucian carp. The new strains not only integrate many desirable characteristics of their parents, exhibiting significant heterosis, but also have significantly improved fertilization and hatching rates1). Using a combination of hybridization and selection breeding, we obtained Hefang crucian carp which was produced by crossing female Japanese crucian carp (Carassius auratus cuvieri) with male red crucian carp (Carassius auratus red var.), a strain that has rapid growth, a desirable shape, and improved meat quality. We are in the process of having this strain certified as a national farmed fish (Figure 1). At the same time, there is controversy about the biological categories of Carassius auratus cuvieri and Carassius auratus red var., and put them into two different species, so hybridization between these two fish can also be recognized as distant hybridization. In addition to carp hybrids, researchers at the Shanghai Fisheries University and other institutions have used mixed hybrid methods to obtain Gift tilapia, which has both growth and size advantages over the parent varieties [56].

Figure 1: Carassius auratus cuvieri (♀)×Carassius auratus red var. (♂).

The filial generation of close hybridization places a limitation on the available genetic variability. As a result, there is currently increased interest in distant hybridization. Distant hybridization involves a much larger gene pool so has a greater potential for breeding new groups and even a new species [57]. Between 1558 and 1980, 1,080 species from 56 families of fish were used in hybrid tests [58]. In the 1950s, Chinese researchers began to experiment with distant hybridization, using fish from three orders (Cypriniformes, Perciformes, Siluriformes), seven families (Cyprinidae, Serranidae, Cichidae, Sparidae, Siluridae, Clariidae, Bagridae), more than 40 species, and >100 hybrid combinations, with the majority involving Cyprinidae.

Distant hybridization has been successfully achieved between different orders, including between M. amblycephala (♀) in Cypriniformes and Siniperca chuatsi (♂) in Perciformes, and between Hypophthalmichthys molitrix (♀) in Cypriniformes and Pagrosomus major (♂) in Perciformes. Fingerlings were successfully hatched from these crosses, but these were not subsequently reared for production [59]. Inter-familial hybridization has been reported between perciform cichlids O. aureus (♀) and Siniperca chuatsi (♂), with fingerling survival rates of 0.3%–0.5%, but this level is not adequate for production [59]. In contrast, the hybrid offspring of H. molitrix and Aristichys nobilis have higher survival and are less aggressive than their parents [60]. Similarly, the filial generations of O. mossambicus (♀) and O. niloticus (♂) have a faster growth rate than the parents [61]. Last, the hybrid offspring of Morone sexatilis (♀) and Morone chrysops (♂) males exhibit obvious heterosis [62]. In recent years, the domestic scientists processed much interspecific hybridization of fish in Cyprinidae, Paralichthyidae [63], Siluridae, Sparidae, Cichlid,

Percichthyidae, etc. [59].

Hybrid combinations are often chosen between different species, different genus and even different subfamilies in Cyprinidae fish. In Cyprinidae fish, common carp, crucian carp, black carp (Mylopharyngodon piceus), grass carp, bighead carp, silver carp, blunt snout bream, Xenocypris davidi Bleeker, Topmouth Culter (Erythroculter ilishaeformis), etc. are often used in hybrids. Over the past several years, our laboratory has carried out numerous hybridization experiments between different subfamilies such as Carassius auratus red Var. (♀)×Elopichthys bambusa (♂) [61] and other combinations; between different genus such as Carassius auratus red Var. (♀)×Cyprinus carpio Xiangjiangnensis (♂) [64] and other combinations (Table 1).

The utilization of distant heterosis of hybridization is primarily limited to the F1 generation in fish at present. However, the ability to maintain heritable traits for multiple generations will be important to realize the full benefits of heterosis from distant hybridization. To address this need, we obtained an allotetraploid hybrid strain by crossing Carassius auratus red Var. (♀) with Cyprinus carpio Xiangjiangnensis (♂) [57,64], and obtained allotetraploid hybrid groups (4n=148) from the F1 of Carassius auratus red Var. (♀)×Megalobrama amblycephala (♂). After continued breeding we acquired an autotetraploid hybrid group from the F_2 and their self-crossed progenies (F_2-F_9, 4n=200) [69]. The characteristics of these two tetraploid hybrid strains are heritable and they can be used in the production of triploid fish [61]. Similarly, from the hybridization between Megalobrama amblycephala (♀)×Erythroculter ilishaeformis (♂) and Erythroculter ilishaeformis (♀)× Megalobrama amblycephala (♂), we obtained fertile male and female individuals in the F_1 and F_2 generations, which allowed development of an F_3 generation. Subsequently, by using female diploid offspring of the cross between Megalobrama amblycephala (♀)×Erythroculter ilishaeformis (♂) and backcrossing with male Megalobrama amblycephala, we produced a new type of hybrid bream that has rapid growth (20% greater than for M. amblycephala), increased disease resistance, improved meat quality, and desirable shape characteristics.

Parent selection is an important component in the success of hybridization experiments. A number of factors must be taken into account, including the chromosome number of the parents, phylogenetic relationships, reproductive behaviors, feeding, appearance, and growth rates. It is important to obtain the different ploidy fish offspring with different special traits and through breeding to develop a variety of different types of excellent fish strains. After the experiments of above hybridization (Table 1), the importance of the combination parental choice is showed ubiquitously. Meanwhile, each design for a good combination of distant hybridization research is a systematic project, and needs long-term unremitting efforts.

Nuclear transplantation

Nuclear transplantation involves the transplantation of a donor-cell nucleus into an enucleated unfertilized egg by micromanipulation. The new cell will continue to split and develop as usual. This technology can be used to break the reproductive isolation among species and change the mode of breeding. Tong [70] was the first to test the possibility of nuclear transplantation between conspecifics using the goldfish and Rhodeus sinensis Gunther, and firstly demonstrated that nucleus of cell in mid-blastocyst of teleostean had totipotency to guide the enucleated eggs developing into embryos and adults.

After transplantation, the success of embryo development is determined by the totipotency of the transplanted nucleus and the degree of fusion with the recipient cytoplasm. Chen et al. [71] transplanted the renal cell nuclei of triploid crucian carp into the enucleated oocytes of diploid crucian carp by way of serial nuclear transfer, and obtained fertile cloned fish. The authors verified the totipotency of the differentiated somatic cells. Lee et al. [72] demonstrated that somatic cells from zebrafish maintained totipotency even after long-term culture.

Table 1: Distant hybridization experiment in Cyprinidae

Genetic relationship		Serial number	Hybridized combination	Food habit combination	Generation number
Subfamily	The parents with equal chromosomal numbers	1	*Megalobrama amblycephala* (♀) × *Xenocypris davidi* Bleeker (♂) [65] ($2n$=48)×($2n$=48)	herbivority×omnivory	F_1–F_2
		2	*Xenocypris davidi* Bleeker (♀) × *Megalobrama amblycephala* (♂) [59] ($2n$=48)×($2n$=48)	omnivory×herbivority	F_1
		3	*Ctenopharyngodon idellus* (♀) × *Megalobrama amblycephala* (♂) [66] ($2n$=48)×($2n$=48)	herbivority×herbivority	F_1
		4	*Xenocypris davidi* Bleeker (♀) × *Erythroculter ilishaeformis* (♂) [59] ($2n$=48)×($2n$=48)	omnivory×predacity	F_1
		5	*Erythroculter ilishaeformis* (♀) × *Xenocypris davidi* Bleeker (♂) [59] ($2n$=48)×($2n$=48)	predacity×omnivory	F_1
		6	*Megalobrama amblycephala* (♀) × *Elopichthys bambusa* (♂) [59] ($2n$=48)×($2n$=48)	herbivority×predacity	F_1

	The parents with unequal chromosomal numbers	1	*Carassius auratus* red Var. (♀) × *Megalobrama amblycephala* (♂) [67] (2*n*=100)×(2*n*=48)	omnivory×herbivority	F_1–F_9
		2	*Carassius auratus* red Var. (♀) × *Xenocypris davidi* Bleeker (♂) [59] (2*n*=100)×(2*n*=48)	omnivory×omnivory	F_1
		3	*Carassius auratus* red Var. (♀) × *Erythroculter ilishaeformis* (♂) [68] (2*n*=100)×(2*n*=48)	omnivory×predacity	F_1
		4	*Carassius auratus* red Var. (♀) × *Elopichthys bambusa* (♂) [59] (2*n*=100)×(2*n*=48)	omnivory×predacity	F_1
Genus	The parents with equal chromosomal numbers	1	*Carassius auratus* red Var. (♀) × *Cyprinus carpio* (♂) [64] (2*n*=100)×(2*n*=100)	omnivory×omnivory	F_1–F_{23}
		2	*Cyprinus carpio* (♀) × *Carassius auratus* red Var. (♂) [59] (2*n*=100)×(2*n*=100)	omnivory×omnivory	F_1–F_2
		3	*Megalobrama amblycephala* (♀) × *Erythroculter ilishaeformis* (♂) [59] (2*n*=48)×(2*n*=48)	herbivority×predacity	F_1–F_3
		4	*Erythroculter ilishaeformis* (♀) × *Megalobrama amblycephala* (♂) [59] (2*n*=48)×(2*n*=48)	predacity×herbivority	F_1–F_3

Yan et al. [73] suggested that nucleoplasm hybridization by nuclear transplantation can be used to facilitate crosses between two species that are distantly related, though the success of the cross is correlated with the evolutionary distance. If true, this ability may be a function of the evolutionary history of fish within the animal line or a low level of "incompatibility" among fishes with respect to their ability to hybridize. Yu et al. [74] demonstrated that the evolutionary relationship between the donor and receptor influenced the ease of nuclear transplantation. The survival rate of nucleoplasm hybridized common carp and crucian carp was higher than the hybrids of grass carp and blunt snout bream or silver carp and blunt snout bream.

Although there are still some problems associated with nuclear transplantation in fish (e.g., low survival, physiological deficiency, immune deficiency), researchers have used this approach to successfully conduct nuclear transplants between different genera, subfamilies, families and orders. For example, Yan et al. [75] transplanted nuclei from blastula cells of crucian carp, tilapia, and grass carp into the enucleated eggs of carp, blunt snout bream, and carp, respectively. The authors obtained crucian carp nucleoplasm hybrids, which were of significant value for breeding as their excellent traits in faster growth rate, higher protein content and lower fat content. Additionally, Yu et al. [76] transplanted nuclei from the blastula cells of carp, tilapia, and grass carp into the non-enucleated eggs of crucian carp, bream, and goldfish, respectively. Lin et al. [77] transplanted nuclei from the head kidney cells of crucian carp, Cirrhinus molitorella, and tilapia into the enucleated unfertilized eggs of carp. Using a combination of androgenesis and nuclear transplantation,

Liu et al. [78] obtained five diploid loaches (Misgurnus anguillicaudatus) that developed from arrhenokaryon and could be bred for 2–7 months.

Researchers in China have made a number of advances in the use of nuclear transplantation in fish. However, a number of problems still persist, such as low survival rate of nucleoplasm hybrid fish, physiological deficiencies, and immune defects in some fish. Thus, there is a need for further study of this approach and a better understanding of the mechanisms controlling embryo development. Additionally, the means of detection of nuclear transplantation require improvement.

Transplantation of germline stem cells and germ cells

Fish germ cell transplantation is a form of cell engineering technology that can be used to regulate fish reproduction. Primordial germ cell (PGC) transplantation technology was first applied successfully in chickens [79], and has since been used in chickens, goats, and pigs [80–82]. In fish, Takeuchi et al. [83] developed the first PGC transplantation system in 2003. Hong et al. [84] successfully used adult fish testis to develop a normal medaka spermatogonial stem cell line (SG3) that had the ability to produce spermatozoids. This not only confirmed that adult spermatogonial cells had the ability to form cell lines, but also provided a new way to obtain germline stem cells and simplified the fish germ cell transplantation program. In 2006, Okutsu et al. [85] developed a spermatogonial transplantation system of rainbow trout. In 2009, Yi et al. [86,87] established haploid ES cell lines from medaka gynogenetic haploid embryos, and then created the semi-cloned female medaka which shows normal fertility by using the semi-cloning technology. Currently, most germ cells used for transplantation are derived from juvenile and adult males in a process that typically involves harvest of PGCs in spermatogonial stem cells or spermatogonia. Fish reproductive cell transplant methods can be divided into three categories: intraspecific transplantation of fish germ cells, interspecific transfer of fish reproductive cells, and transplantation of germ cells into a sterile triploid receptor.

The intraspecific transplantation of fish germ cells is primarily used in studies of fish germ cells and reproductive development mechanisms. Nobrega et al. [88] observed spermatogenesis and oogenesis after transplant of zebrafish spermatogonial stem cells to male and female zebrafish, respectively, thereby demonstrating that germ cell differentiation is affected by the environment. In 2004, Takeuchi et al. [89] transplanted rainbow trout PGCs to O. masou and obtained O. masou gametes that developed normally. Similarly, Majhi et al. [90] transplanted germ cells from Odontesthes bonariensis to mature O. hatcheri and obtained viable sperm within 6 months of the transfer. In these instances, the parent PGCs reached sexual maturity more rapidly and differentiated to sperm faster than the donor's own in the heterologous receptor.

Using an alternate approach, Okutsu et al. [91] obtained O. masou that could produce 100% rainbow trout offspring following transplant of homozygous orange rainbow trout mutant germ cells into sterile triploid O. masou. Fish reproductive stem cell and germ cell transplantation methods have opened up new possibilities for the protection and restoration of endangered germplasm resources and genetic breeding. However, there remains a need for more theoretical and practical understanding of the mechanisms to decrease the development time between laboratory and commercial application.

Artificial gynogenesis and androgenesis

(1) Artificial gynogenesis technology. Gynogenesis refers to a form of reproduction in which the eggs are activated by sperm and the development of zygotes is controlled by matrilineal genes. Gynogenesis can be divided into natural and artificial gynogenesis. To date, natural gynogenesis has been documented in at least 10 species, including Poecilia formosa, Menidia clarkhubbsi, and Misgurnus anguillicaudatus [92]. It should be noted that Carassius auratus gibelio were cultivated by Chinese aquaculture researchers according to the reproductive mode of natural gynogenesis in Heilongjiang Fangzheng silver crucian carp [93]. Our laboratory has obtained bisexual fertile natural gynogenetic red crucian carp from the offspring of red crucian carp (♀) and M. amblycephala (♂). These natural gynogenetic red crucian carps have been stably bred for many generations. The formation of bisexual fertile natural gynogenetic red crucian carp will allow development of a new strain of Carassius auratus [94].

Artificially induced gynogenesis refers to a breeding technique that uses inactivated sperm to activate eggs. The development of zygotes is controlled by maternally derived genes. For fish species with female homogamety, artificially induced gynogenetic progeny are all females.

In the process of artificial gynogenesis, heterogenous spermatozoas are usually used to induce the development of eggs. Artificial gynogenesis can purify genetic material of gametes and result in new genetic material and new genetic traits as a result of allogynogenesis. In summary, gynogenesis can be used to control the sex of fish and obtain pure inbred lines to purify and rejuvenate the fish species.

Artificially induced gynogenesis has been carried out in common carp, carassius auratus gibellio, grass carp, rare minnow (gobiocypris rarus), blunt snout bream, large yellow croaker, sterlet (Acipenser ruthenus), tilapia, rainbow trout, Atlantic halibut (Hippoglossoides platessoides), European seabass (Dicentrarchus labrax), south flounder (Paralichthys lethostigma), turbot (Scophthalmus maximus), bass (Perca fluviatilis), Esox masquinongy and >100 other fish species [92,95]. Gynogenetic strains of red sea bream (Pagrosomus major), rainbow trout, weever (Lateolabrax japonicus), calico salmon (Oncorhynchus keta), olive flounder (Paralichthys olivaceus) have subsequently been used in production settings [96–102].

Our group has developed gynogenetic strains from natural origin red crucian carp, Carassius auratus cuvieri, goldfish, common carp, orange ornamental carp (Cyprinus carpio), grass carp, Erythroculter ilishaeformis, Xenocypris davidi Bleeker, and artificially cultivated strains including tetraploid hybrids of common carp with red crucian carp and tetraploid hybrids of red crucian carp with blunt snout bream [103–107]. We have developed gynogenetic strains of red crucian carp, Japanese crucian carp, and goldfish following activation using heterogenous sperms (M. amblycephala). These gynogenetic strains have a higher survival rate and are easily identified. The combination of artificial gynogenesis and sex reversal technology sets the stages for the production of all-female diploids and triploids [103,104]. It is worth noting that Zhang et al. [105] have built disease-resistant gynogenetic strains of grass carp by the activation of inactivated sperm of tetraploids. These gynogenetic grass carp demonstrate that gynogenesis can be used to obtain all-female individuals that have desirable genetic traits. Lived artificial gynogenetic fish have strong resistance and tolerance, which is due to the process of selection, such as cold shock and the inactivated genetic material of sperm. The survived fish that underwent the adversity were easy to obtain traits that resist harsh conditions. Liu et al. [106] have obtained diploid gynogenetic offspring by using the diploid eggs from allotetraploid hybrids by gynogenesis without chromosome-doubling treatment. Through long- term research, we established a strain of diploid gynogenetic hybrids (G_1–G_8) that were able to produce fertile diploid eggs. This strain is not only a model for research into the mechanism of diploid gamete formation, but also can provide high-quality diploid eggs for developing new types of modified tetraploids. Our laboratory has obtained improved allotetraploid hybrids by fertilizing diploid eggs from the diploid gynogenetic hybrids with diploid sperm from allotetraploid hybrids, an important step in the mass production of improved triploid fish.

The artificial induction of gynogenesis in fish has a wide range of applications from theoretical research to commercial production. However, the methods used to date have low survival rates and complex identification of gynogenetic fish. Recent improvements in methods of inactivating sperm and doubling egg chromosomes have improved the survival rate of gynogenetic fish. Furthermore, research suggests that using sperm from distant fish can improve the survival rate of gynogenetic fish and simplify their identification. We have obtained gynogenetic crucian carp using sperm from blunt snout bream. This variety has higher survival rate than gynogenetic crucian carp created using common carp sperm, and is more easily identified [103].

(2) Artificial androgenesis technology. In contrast to gynogenesis, androgenesis involves the use of genetically inactivated eggs and normal sperm to form a “zygote”. The “zygote” develops into an individual that is primarily under the control of nuclear genetic material from the sperm. The steps involved in the artificial induction of androgenesis and gynogenesis are similar, and include the inactivation of gynogenesis genetic material and the induction of diploid androgenesis. This method has been used for the rapid establishment of pure lines, to determine mechanisms of sex determination, to create unisexual populations, and for the protection of endangered fish [92].

Natural androgenesis individuals are rare in nature but occur in the hybrid offspring common carp (female) and grass carp (male) [108]. The artificial induction of androgenesis fish has primarily been attempted in O. mykiss, Platichthys flesus, Salvelinus japonicus, Oreochromis niloticus, Danio rerio, Oncorhynchus masou, Ctenopharyngodon idellu, Esox reticulatu, Misgurnus anguillicaudrus, Pelteobagrus fulvidraco, and so on [92,109–111]. It is worth noting that because male tilapia grows more rapidly than female tilapia, androgenesis could be used to rapidly obtain super-male individuals with a male nucleus (YY). Then, crossing with normal female (XX) hybrids would yield all-male tilapia (XY) offspring.

To date, haploid sperm has been rarely used to induce artificial androgenesis in fish for industry. The primary limiting factor is the difficulty in inactivating the DNA in the egg and the negative impact on the sperm nucleus genome resulting from artificial doubling of androgenesis individual embryos and later life activities that lead to a low rate of individual survival in androgenesis. The use of tetraploid fish, with two sets of chromosomes in the diploid sperm, for artificial androgenesis would yield diploid androgenetic offspring without the need for chromosome-doubling treatment, thereby improving the success rate of androgenesis. Indeed, researchers have carried out androgenesis using diploid sperm that was produced by artificial tetraploid rainbow trout and natural tetraploid loach without chromosome-doubling treatment, and obtained diploid androgenetic fish that have higher survival rate than those when using haploid sperm [112–114].

We have successfully obtained bisexual fertile diploid androgenetic fish (A0) using diploid sperm produced by allotetraploid hybrids, without male nuclear chromosome-doubling treatment. Diploid androgenetic fish were self-crossed to obtain bisexual fertile tetraploids. When compared with ordinary allotetraploid hybrids, the new-type tetraploid fish exhibited significant growth advantages and increased disease resistance [115,116]. Additionally, using diploid sperms produced by autotetraploid males, we obtained super-male diploid fish from offspring that can produce unreduced diploid sperm. Subsequent crossing with haploid or diploid eggs yields all-male triploid and tetraploid fish, respectively. The all-male tetraploid fish can be used for the subsequent production of triploid fish, eliminating the artificial selection steps needed to produce all-male fish, reducing the cost and improving breeding efficiency.

Polyploid breeding

Polyploids contain three or more complete chromosome sets, which are widespread in plants and animals and are important for speciation [117]. Polyploid breeding refers to doubling cell genomes by artificial or natural mutagenesis to produce polyploid individuals, and then selected and fostered by persons to meet requirements. Fish genomes have greater plasticity, and could be doubled, so research about fish polyploidy is flourishing.

Triploids typically exhibit rapid growth, strong resistance, and high yield, so are of high economic value [118]. Additionally, the majority of allotriploid fish are infertile, which reduces issues associated with diversion of energy to gametic growth, mixing of genetic lines, and interbreeding with wild stocks. Triploids are primarily produced using one of two methods: by direct induction or by the hybridization of tetraploid parent groups and hybrid triploids to produce diploid groups. Using the first method, it is difficult to guarantee a high proportion of triploids in offspring. In contrast, the second method yields more predictable results.

Currently, the methods used to induce polyploidy fall into one of three categories: biological, physical, and chemical. Biological methods include hybridization, nuclear transfer, and cell fusion. Physical methods include temperature shock, hydrostatic pressure, and high-salt and highalkaline electric shock treatment. Chemical methods involve the use of different chemicals (e.g., cytochalasin, caffeine, polyethylene glycol, colchicine, 6-dimethyl aminopurine) to induce embryonic polyploidy.

Physical and chemical methods artificially induce polyploidy work by inhibiting the exclusion of the first or second polar body in the oocytes or inhibiting the first cleavage of zygotes [117]. Gui et al. [119] succeeded in inducing triploid Carassius auratus transparent colored var. by hydrostatic pressure. Similarly, Chen et al. [120] used hydrostatic pressure to prevent loss of the second polar body of zygotes and obtain triploid fry in Cynoglossus semilaevis Gunther. Hybridization is one of the most commonly used biological methods to induce polyploidy in fish. Wu et al. [121] crossed the female descendants of Cyprinus carpio (♀)×Carassius auratus (♂) with male mirror carp (Cyprinus carpio) and obtained triploid carp. The proportion of triploid offspring is often low and variable using these methods; therefore they are not widely used.

Interestingly, Carassius auratus gibelio is a naturally occurring triploid that differs from the other single-sex line gynogenetic species as a certain proportion is males, making them a triploid amphoteric group. When Carassius auratus gibelio eggs are activated by heterogeneous sperms, the offspring are all females. Using molecular marker technology, Gui and Zhou [122] developed different clonal strains of Carassius auratus gibelio. Using these clone lines, the researchers obtained a novel clone A^+ with obvious growth advantage by sexual mating between clone D female and clone A male in polyploid gibel carp. Genetic composition analysis showed that clone A^+ was a nucleocytoplasmic hybrid between clones A and D, because its entire nuclear genome was from the paternal clone A and its mtDNA genome (cytoplasm) was from the maternal clone D [123]. Therefore, the formation of clone A^+ was suggested to be the androgenesis of clone A sperm in clone D ooplasm. Significantly, selected female in clone A^+ still possessed gynogenetic ability, and then produced a novel clone variety "CAS III" which had been approved by National Certification Committee for Aquatic Varieties. This new variety has been cultured throughout China [123].

Researchers have used heat shock to obtain tetraploid rainbow trout [124], channel catfish [125], tench (Tinea tinea L.) [126], and loach [127]. Similarly, hydrostatic pressure was used to obtain tetraploid tongue sole (Cynoglossus semilaevis. Günther) [128] and a combination of hydrostatic pressure and cold shock was used to obtain tetraploid Oreochromis spp [129] and Carassius auratus transparent colored var. (C. auratus) [130]. Zou et al. [131] successfully obtained autotetraploid Megalobrama amblycephala by heat shock inhibition of first cleavage of the zygote derived from self-crosses of Pujiang No. 1 M. amblycephala. While using hybridization to obtain tetraploid fish, Gui et al. [132] noted the occurrence of a small number of tetraploid individuals in an artificially cultured population of C. auratus gibelio. The chromosomes in these individuals were derived from silver crucian carp and fused with a set of haploid chromosome from C. carpio var. However, these authors were unsuccessful at obtaining a bisexual fertile tetraploid fish population, so these approaches have not been applied to the production of triploids.

Beginning in the 1970s, our laboratory experimented with hybridization between red crucian carp (♀, 2n=100) and common carp (♂, 2n=100) and noted the presence of some fertile diploid individuals in the F_1 hybrids. Using these, we obtained the F_2 generation, and in the F_3 (4n=200 or 2n=4X=200) we obtained tetraploid hybrids, called allotetraploids [57,64]. This allotetraploid population has a stable genetic characteristic and we have since continued breeding this to the F_{24} generation. Thus, we are able to mass-produce infertile triploid fish.

Currently, we produce infertile triploid fish by crossing allotetraploid male carp and diploid female crucian carp. Infertile triploid carps are in demand by culturists because they have significant social, economic, and ecological benefits. The allotetraploid carp populations have heterosis of genetic stability and characteristics that are needed [57] to form a new species. Tetraploid crucian carp also provide an important research model to evaluate the origin and evolution of polyploid fish.

The red carp have 100 chromosomes whereas M. amblycephala have 48 chromosomes. Using red carp as a female parent and Megalobrama amblycephala as a male parent, we carried out a hybridization experiment [67,133]. We obtained fertile allotetraploid and infertile triploid F1 hybrids. The fertile allotetraploid was able to produce two types of gametes concurrently, one type having 148 chromosomes and the other having 100 chromosomes. To date, we have cultured nine generations of autotetraploid progeny by self-crossing and have formed a stable autotetraploid line. Autotetraploids (4n=200) are used to produce triploid hybrids by crossing with red carp for use in aquaculture [69]. The first generation of tetraploid crucian carp bream (♀, 4n=148) was backcrossed with its female parent M. amblycephala (♂, 2n=48) to obtain a pentaploid hybrid (5n=172). A backcross of the pentaploid with its male parent red carp yielded a new pentaploid hybrid (5n=198).

Research in our laboratory has also yielded triploid (3n=124) and tetraploid (4n=148) hybrids from distant hybridization between Carassius auratus red var. (♀, 2n=100) and Xenocypris davidi (♂, 2n=48) and diploid (2n=74), triploid (3n=124), and tetraploid (4n=148) hybrids from distant hybridization between Carassius auratus red var. (♀, 2n=100) and M. amblycephala (♂, 2n=48) [68]. In the progeny of distant hybridization between grass carp (♀, 2n=48) and blunt snout bream (♂, 2n=48), we detected diploid (2n=48) and triploid (3n=72) hybrids [66].

Liu [134] summarized the experiments and achievements in fish distant hybridization in about 20 years, and wrote Fish Distant Hybridization. This book systematically described the processes and mechanism of the formation of autotetraploid and allotetraploid fish, and also described the processes and mechanism of new diploid varieties by using distant hybridization and the merits and value of triploid fish in production and applications.

Modification Breeding Technologies

The essence of fish genetic breeding is a cyclic process, involving the selection or development of variable strains from stable strains, and then the culture of stable strains from variable strains. The ideal mutant strains are rapidly obtained by reasonable trait transformation methods, and decide the efficiency and effectiveness of fish genetic breeding. Transgenic breeding is an effective means of trait transformation that can be used to rapidly acquire new variant strains.

Transgenic fish are a category of fish that were developed using gene transfer technology whereby a foreign gene is incorporated into the genome of the recipient fish and it can be passed on to future generations. Fish are a relatively primitive vertebrate taxa, with large genetic plasticity and fecundity. Furthermore, the eggs can easily be developed in vitro so can be micro-manipulated. The earliest reported study of transgenic fish was by a Chinese researcher who developed the world's first transgenic fish in 1985 [135]. Since then, the team led by Zhu has developed a number of transgenic fish models, and has laid a theoretical foundation for fish gene transfer research and development. Subsequently, a number of countries have experimented with transgenic fish, developing transgenic medaka, coho salmon, rainbow trout and others [136–139].

A feature of transgenic technology is the introduction of exogenous genes into the recipient fish. Using this approach, it is possible to obtain new varieties that have desirable and predictable traits. Traditional genetic breeding requires repeated breeding over many generations whereas transgenic technology can cut the time course of evolution and create new breeds or strains in a short time (single generation). At present, the application of fish transgenic technology in fish genetic breeding centers around three needs. The first is the cultivation of high-yield varieties that have increased growth rate and food conversion rates. The second is the cultivation of resistant strains (disease, cold). The third is the basic study of the mechanisms of animal growth, development, reproduction and other life activities.

With respect to the cultivation of high-yielding varieties, Zhu [140–142] obtained transgenic carp that had individual growth rates 3–4 times faster than the original control. Likewise, a cooperative effort between our laboratory and the Sciences Institute of Chinese

Academy of Aquatic Organisms developed transgenic allotetraploid hybrids that were transfected with the fish growth hormone gene using a microinjection technique. These transgenic tetraploids exhibit a rapid growth rate and strong disease resistance [143]. Additionally, Feng et al. [144] successfully transferred a black carp recombinant growth hormone gene into transgenic allotetraploids by microinjection. The transgenic fish had a growth rate that was 2.65 times faster than the control group. In summary, transfer of fish growth hormone genes appears to yield significant growth-promoting benefits. In addition to growth hormone genes, researchers have experimented with the transfer of genes promoting resistance to factors such as cold, disease, and hypoxia. Fletcher et al. [145] transferred an antifreeze protein gene from winter flounder (Pseudopleuronectes americanus) into Atlantic salmon, so that the Atlantic salmon had increased resistance to low temperature before entering salt water. A number of genes are known to increase disease resistance in fish, including interferon and lysozyme genes. Transfer of these genes has been associated with increased survival by improving disease resistance, resulting in improved yield [146]. Dunham et al. [147] demonstrated that transfer of the cecropin B gene to channel catfish enhanced immunity against pathogenic bacteria. Mao et al. [148] found that some genetically modified grass carp anti-aeromonas' hydrophilia capacity significantly increased by transferring recombinant plasmid containing the human lactoferrin into the grass carp. The transfer of genes coding for increased resistance appears to improve the ability of transgenic fish to cope with their environment. These new varieties will significantly improve the efficiency of fish farming.

Transgenic breeding has been frequently used to quickly establish new variant strains. However, concerns over bio-safety/security have prevented these strains from entering the mass production application stage [139]. To address these concerns, research will likely focus on the development of transgenic sterile triploid fish. Additionally, the application of "whole fish" genes as exogenous genes for transgenosis would reduce the food safety risks of transgenic fish. To date, efforts by our laboratory and the Sciences Institute of Chinese Academy of Aquatic Organisms have resulted in two generations of transgenic triploid carp (positive fish) and non-transgenic triploid carp (negative fish), by mating between Yellow River carp (C. carpio) transfected with diploid grass carp growth hormone gene and improved allotetraploid crucian carp. We have since collected data on the growth, reproductive and developmental characteristics of these two triploid carp, providing important baseline information for the industrialization of trans-growth-hormone gene carp [149].

DEFICIENCIES AND FUTURE OF FISH GENETIC BREEDING

After years of efforts, there has been significant progress, evidenced by the establishment of hybrid, ploidy, cell nuclear transfer, chromosome engineering, and genetic engineering technologies. These efforts have significantly improved the productivity of the aquaculture industry and increased the genetic diversity of fish.

On the whole, however, good varieties of the main aquaculture fish are comparatively few. One of the primary reasons is that the contact between basic research of fish genetic breeding and the breeding practices is not close enough. How to quickly apply these new technologies to genetic breeding practices is another question. In addition, conventional cellular engineering technologies, such as distant hybridization and artificial gynogenesis, have been proved to be effective methods, but new varieties obtained by using these methods are limited, because of the demands for good conditions (larger aquaculture sites and perfect aquaculture facilities) and systematic detection methods of genetic breeding. Nevertheless, it is promising to resolve the above issues and stick to targeted research with long-term unremitting efforts in these aspects. Thus, there is a need to link basic research with industry demand and integrate the use of genetic breeding techniques into the industry. Additionally, there is a need to foster a climate of innovation to accelerate development of new technologies.

REFERENCES

1. Kause A, Ritola O, Paananen T, Wahlroos H, Mantysaan EA. Genetic trends in growth, sexual maturity and skeletal deformations, and rate of inbreeding in a breeding programme for rainbow trout (*Oncorhynchus mykiss*). Aquaculture, 2005, 247: 177–187
2. Neira R, Díaz NF, Gall GAE, Gallardo JA, Lhorente JP, Alert A.

Genetic improvement in coho salmon (*Oncorhynchus kisutch*). II: Selection response for early spawning date. Aquaculture, 2006, 257: 1–8

3. Saillant E, Dupont-Nivet M, Haffray P, Beatrice C. Estimates of heritability and genotype environment interactions for body weight in sea bass (*Dicentrarchus labrax* L.) raised under communal rearing conditions. Aquaculture, 2006, 254: 139–147
4. Gheyas AA, Woolliams JA, Taggart JB, Sattar MA, Das TK, McAndrew BJ, Penman DJ. Heritability estimation of silver carp (*Hypophthalmichthys molitrix*) harvest traits using microsatellite based parentage assignment. Aquaculture, 2009, 294: 187–193
5. Rezk MA, Smitherman RO, Williams JC, Nichols A, Kucuktas H, Dunham RA. Response to three generations of selection for increased body weight in channel catfish, *Ictalurus punctatus*, grown in earthen ponds. Aquaculture, 2003, 228: 69–79
6. Embody GC, Hayford CO. The advantage of rearing brook trout fingerlings from selected breeders. Trans Am Fish Soc, 1925, 55: 135–142
7. Main KL, Reynolds E. Selective breeding of fishes in Asia and the United States. In: Proceedings of a workshop in Honolulu, Hawaii. 1993, 3-7: 206–213
8. Wang HF, Liu X, Su JH, Wang GF, Wei YM. Effects of starvation and subsequent refeeding on growth and biochemical compositions of Gansu Golden Trout. J Anim Vet Adv, 2013, 12: 289–294
9. Tang SJ, Li SF, Cai WQ, Zhao Y. Microsatellite analysis of variation among wild, domesticated, and genetically improved populations of blunt snout bream (*Megalobrama amblycephala*). Zool Res, 2014, 35: 108–117
10. Li SF, Tang SJ, Cai WQ. RAPD-SCAR markers for genetically improved NEW GIFT Nile Tilapia (*Oreochromis niloticus niloticus* L.) and their application in strain identification. Zool Res, 2010, 31: 147–53
11. Liu YG, Chen SL, Li BF, Wang ZJ, Liu ZJ. Analysis of genetic variation in selected stocks of hatchery flounder, *Paralichthys olivaceus*, using AFLP markers. Biochem Syst Ecol, 2005, 33: 993–1005

12. Ning Y, Liu XD, Wang ZY, Guo W, Li YY, Xie FJ. A genetic map of large yellow croaker *Pseudosciaena crocea*. Aquaculture, 2007, 264: 16–26
13. Zhang TS, Kong J, Liu BS, Wang QY, Cao BX, Luan S, Wang WJ. Genetic parameter estimation for juvenile growth and upper thermal tolerance in turbot (*Scophthalmus maximus* Linnaeus). Acta Oceanol Sin, 2014, 33: 106–110
14. Hong WS, Zhang QY. Review of captive bred species and fry production of marine fish in China. Aquaculture, 2003, 227: 305–318
15. Yang WJ, Kang XL, Yang QF, Lin Y, Fang MY. Review on the development of genotyping methods for assessing farm animal diversity. Journal of Animal Science and Biotechnology, 2013, 4: 2
16. Arif IA, Khan HA. Molecular markers for biodiversity analysis of wildlife animals: a brief review. Anim Bioiv Conserv, 2009, 32: 9–17
17. Jander G. Gene identification and cloning by molecular marker mapping. Methods Mol Biol, 2006, 323: 115–126
18. Luo ZW, Hackett CA, Bradshaw JE, McNicol JW, Milbourne D. Construction of a genetic linkage map in tetraploid species using molecular markers. Genetics, 2001, 157: 1369–1385
19. Montaldo HH, Meza-Herrera CA. Use of molecular markers and major genes in the genetic improvement of livestock. Electron J Biotechn, 1998, 1: 1–7
20. Agarwal M, Shrivastava N, Padh H. Advances in molecular marker techniques and their applications in plant sciences. Plant Cell Rep, 2008, 27: 617–631
21. Postlethwait JH, Johnson SL, Midson CN, Talbot WS, Gates M, Ballinger EW, Africa D, Andrews R, Carl T, Eisen JS. A genetic linkage map for the zebrafish. Science, 1994, 264: 699–703
22. Naruse K, Tanaka M, Mita K, Shima A, Postlethwait J, Mitani H. A medaka gene map: the trace of ancestral vertebrate protochromosome revealed by comparative gene mapping. Genome Res, 2004, 14: 820–828

23. Kai W, Kikuchi K, Fujita M, Suetake H, Fujiwara A, Yoshiura Y, Ototake M, Venkatesh B, Miyaki K, Suzuki Y. A genetic linkage map for the tiger pufferfish, *Takifugu rubripes*. Genetics, 2005, 171: 227–238
24. Walter RB, Rains JD, Russell JE, Guerra TM, Daniels C, Johnston DA, Kumar J, Wheeler A, Kelnar K, Khanolkar VA, Williams EL, Hornecker JL, Hollek L, Mamerow MM, Pedroza A, Kazianis S. A microsatellite genetic linkage map for *Xiphophorus*. Genetics, 2004, 168: 363–372
25. Peichel CL, Nereng KS, Ohgi KA, Cole BLE, Colosimo PF, Buerkle CA, Schluter D, Kingsley DM. The genetic architecture of divergence between three spine stickleback species. Nature, 2001, 414: 901–905
26. Young WP, Wheeler PA, Coryell VH. A detailed linkage map of rainbow trout produced using doubled haploids. Genetics, 1998, 148: 839–850
27. Lee BY, Lee WJ, Streelman JT, Carleton KL, Howe AE, Hulata G, Slettan A, Stern JE, Terai Y, Kocher TD. A second generation genetic linkage map of tilapia (*Oreochromis spp*.). Genetics, 2005, 170: 237–244
28. Liu ZJ, Cordes JF. DNA marker technologies and their applications in aquaculture genetics. Aquaculture, 2004, 238: 1–37
29. Chauhan T, Rajiv K. Molecular markers and their applications in fisheries and aquaculture. Adv Biosci Biotech, 2010, 1: 281–291
30. Laghari MY, Lashari P, Zhang X, Xu P, Narejo NT, Xin B, Zhang Y, Sun X. QTL mapping for economically important traits of common carp (*Cyprinus carpio* L.). J Appl Genet, 2014, doi: 10.1007/s13353-014-0232-y
31. Gui JF, Zhu ZY. Molecular basis and genetic improvement of economically important traits in aquaculture animals. Chin Sci Bull, 2012, 57: 1751–1760
32. Jang SH, Liu H, Su JG, Dong F, Xiong F, Liao LJ, Wang YP, Zhu ZY. Construction and characterization of two bacterial artificial chromosome libraries of grass carp. Mar Biotechnol, 2010, 12: 261–266

33. Geng FS, Zhou L, Gui JF. Construction and characterization of a BAC library for *Carassius auratus gibelio*, a gynogenetic polyploid fish. Anim Genet, 2005, 36: 535–536

34. Zhang JJ, Shao CW, Zhang LY, Liu K, Gao FT, Dong ZD, Xu P, Chen SL. A first generation BAC-based physical map of the half-smooth tongue sole (*Cynoglossus semilaevis*) genome. BMC Genomics, 2014, 15: 215

35. Zhao SY. Genomics and life sciences industry. Chin Sci Bull, 1999, 11: 1–5

36. Sarropoulou E, Nousdili D, Magoulas A, Kotoulas G. Linking the genomes of nonmodel teleosts through comparative genomics. Mar Biotechnol, 2008, 10: 227–233

37. Star B, Nederbragt AJ, Jentoft S, Grimholt U, Malmstrom M, Gregers TF, Rounge TB, Paulsen J, Solbakken MH, Sharma A, Wetten OF, Lanzen A, Winer R, Knight J, Vogel JH, Aken B, Andersen Q, Lagesen K, Tooming-Klunderud A, Edvardsen BO, Moum T, Skage M, Berg PR, Gjqen T, Kuhl H, Thorsen J, Malde K, Reinhardt R, Du L, Johansen SD, Searle S, Lien S, Nilsen F, Jonassen I, Omholt SW, Stenseth NC, Jakobsen KS. The genome sequence of Atlantic cod reveals a unique immune system. Nature, 2011, 477: 207–210

38. Chen SL, Zhang GJ, Shao CW, Huang QF, Liu G, Zhang P, Song WT, An N, Chalopin D, Volff JN, Hong YH, Li QY, Sha ZX, Zhou HL, Xie MS, Yu QL, Liu Y, Xiang H, Wang N, Wu K, Yang CG, Zhou Q, Liao XL, Yang LF, Hu QM, Zhang JL, Meng L, Jin LJ, Tian YS, Lian JM, Yang JF, Miao GD, Liu SS, Liang Z, Yan F, Li YZ, Sun B, Zhang H, Zhang J, Zhu Y, Du M, Zhao YW, Schartl M, Tang QS, Wang J. Whole-genome sequence of a flatfish provides insights into ZW sex chromosome evolution and adaptation to a benthic lifestyle. Nat Genet, 2014, 46: 253–260

39. Zhang GF, Fang XD, Guo XM, Li L, Luo RB, Xu F, Yang PC, Zhang LL, Wang XT, Qi HG, Xiong ZQ, Que HY, Xie YL, Holland PWH, Paps J, Zhu YB, Wu FC, Chen YX, Wang JF, Peng CF, Meng J, Yang L, Liu J, Wen B, Zhang N, Huang ZY, Zhu QH, Feng Y, Mount A, Hedgecock D, Xu Z, Liu YJ, Domazet-Loso T, Du YS, Sun XQ, Zhang SD, Liu BH, Cheng PZ, Jiang XT, Li J, Fan DD, Wang W, Fu WJ, Wang T, Wang B, Zhang JB, Peng ZY, Li YX, Li N, Wang

JP, Chen MS, He Y, Tan FJ, Song XR, Zheng QM, Huang RL, Yang HL, Du XD, Chen L, Yang M, Gaffney PM, Wang S, Luo LH, She ZC, Ming Y, Huang W, Zhang S, Huang BY, Zhang Y, Qu T, Ni PX, Miao GY, Wang JY, Wang Q, Steinberg CEW, Wang HY, Li N, Qian LM, Zhang GJ, Li YR, Yang HM, Liu X, Wang J, Yin Y, Wang J. The oyster genome reveals stress adaptation and complexity of shell formation. Nature, 2012, 490: 49–54

40. Xu P, Zhang XF, Wang XM, Li JT, Liu GM, Kuang YY, Xu J, Zheng XH, Ren LF, Wang GL, Zhang Y, Huo LH, Zhao ZX, Cao DC, Lu CY, Li C, Zhou Y, Liu ZJ, Fan ZH, Shan GL, Li XG, Wu SX, Song LP, Hou GY, Jiang YL, Jeney Z, Yu D, Wang L, Shao CJ, Song L, Sun J, Ji PF, Wang J, Li Q, Xu LM, Sun FY, Feng JX, Wang CH, Wang SL, Wang BS, Li Y, Zhu YP, Xue W, Zhao L, Wang JT, Gu Y, Lv WH, Wu KJ, Xiao JF, Wu JY, Zhang Z, Yu J, Sun XW. Genome sequence and genetic diversity of the common carp, *Cyprinus carpio*. Nat Genet, 2014, doi: 10.1038/ng.3098
41. Devlin R, Nagahama Y. Sex determination and sex differentiation in fish: an overview of genetic, physiological, and environmental influences. Aquaculture, 2002, 208: 191–364
42. Pandian TJ, Sheela SG. Hormonal induction of sex reversal in fish. Aquaculture, 1995, 138: 1–22
43. Piferrer F. Endocrine sex control strategies for the feminization of teleost fish. Aquaculture, 2001, 197: 229–281
44. Yamamoto T. Progeny of artificially induced sex-reversals of male genotype (XY) in the medaka (*Oryzias latipes*) with special reference to YY male. Geneties, 1955, 40: 406–429
45. Guerrero RD. Use of Androgens for the production of all male Tilapia aurea (steindacher). Trans Amer Fish Soc, 1975, 104: 342–348
46. Liu SJ, Yao ZZ, Wang YQ. Sex hormone induction of sex reversal in the teleost clarias lazera and evidence for female homogmety and male heterogamety. J Exp Zool, 1996, 276: 432–438
47. Liu SJ, Yao ZZ. Self-fertilization of hermaphrodites of the teleost *Clarias lazera* after oral administration of 17-α-methyltestosterone and their offspring. J Exp Zool, 1995, 273: 527–532

48. Chen RD, Huang WY, Wu QJ, Ye YZ. Cultivation and breeding effects of all-female carp (in Chinese). Reserv Fish, 1990, 3: 21–23
49. Luo KK, Xiao J, Liu SJ, Wang J, He WG, Hu J, Qin QB, Zhang C, Tao M, Liu Y. Massive production of all-female diploids and triploids in the crucian carp. Int J Biol Sci, 2011, 7: 487–495
50. Liu HQ, Cui SQ, Hou CC, Xu J, Chen HX. YY supermale generated gynogenetically from XY female in *Pelteobagrus fulvidraco* (Richardson) (in Chinese). Acta Hydrobiol Sin, 2007, 31: 718–725
51. Wang D, Mao HL, Chen HX, Liu HQ, Gui JF. Isolation of Y- and X-linked SCAR markers in yellow catfish and application in the production of all-male populations. Anim Genet, 2009, 40: 978–981
52. Dan C, Mei J, Wang D, Gui JF. Genetic differentiation and efficient sex-specific marker development of a pair of Y- and X-linked markers in Yellow Catfish. Int J Biol Sci, 2013, 9: 1043–1049
53. Hulata G. A review of genetic improvement of the common carp (*Cyprinus carpio* L.) and other hybrids by crossbreeding, hybridization and selection. Aquaculture, 1995, 129: 143–155
54. Kirpichnikov VS, Ilyasov JI, Shart LA, Vikhman AA, Ganchenko MV, Ostashevsky AL, Simonov VM, Tikhonov GF, Tjurin VV. Selection of krasnodar common carp (*Cyprinus carpio* L.) for resistance to dropsy: principal results and prospects. Aquaculture, 1993, 111: 7–20
55. Zhou J, Wu Q, Wang Z, Ye Y. Genetic variation analysis within and among six varieties of common carp (*Cyprinus carpio* L.) in China using microsatellite markers. Rus J Genet, 2004, 40: 1144–1148
56. Li SF, He XJ, Han FJ. Third-fifth generation selective evaluation of GIFT strain Nile tilapia. Beijing: World Aquaculture, 2002. 410
57. Liu SJ. Distant hybridization leads to different ploidy fishes. Sci China Life Sci, 2010, 53: 416–425
58. Schwartz FJ. World literature to fish hybrids with an analysis by family, species and hybrid: supplement. NOAA Technical Report NMFS SSRF, 1981, 2: 750
59. Zhang ZH, Chen J, Li L, Tao M, Zhang C, Qin QB, Xiao J, Liu Y, Liu SJ. Research advances in animal distant hybridization. Science China Life Sci, 2014, 44: 161–174

60. Beck ML, Biggers CJ, Dupree HK. Karyological analysis of *Hypophthalmichthys molitrix*, *Aristichthys nobilis* and their F1 hybrid. Trans Am Fish Soc, 1980, 109: 433–438

61. Sulaiman AH. Toxicity of malathion to red tilapia (Hybrid *Tilapia mossambica*×*Tilapia nilotica*): behavioural, histopathological anti-cholinesterase studies. Mal Appl Biol (Malaysia), 1989, 18: 163–170

62. Harms CA, Kennedy-Stoskopf S, Horne WA, Fuller FJ, Tompkins WA. Cloning and sequencing hybrid striped bass (*Morone saxatilis* ×*M. chrysops*) transforming growth factor-beta (TGF-beta), and development of a reverse transcription quantitative competitive polymerase chain reaction (RT-qcPCR) assay to measure TGF-beta mRNA of teleost fish. Fish Shellfish Immunol, 2000, 10: 61–85

63. Sui J, Liu QH, Xiao ZZ, Ma DY, Xu SH, He T, Liu YF, Xiao YS, Li J. The viability, melanophore and embryo genesis of first- and second-generation hybrids between *Paralichthys olivaceus* and *P. dentatus*. Mar Biol Res, 2013, 9: 220–226

64. Liu SJ, Liu Y, Zhou GJ, Zhang XJ, Luo C, Feng H, He XX, Zhu GH, Yang H. The formation of tetraploid stocks of red crucian carp×common carp hybrids as an effect of interspecific hybridization. Aquaculture, 2001, 192: 171–186

65. Hu J, Liu SJ, Xiao J, Zhou Y, You CP, He WG, Zhao RR, Song C, Liu Y. Characteristics of diploid and triploid hybrids derived from female *Megalobrama amblycephala* Yih×male *Xenocypris davidi* Bleeker. Aquaculture, 2012, (364–365): 157–164

66. He WG, Xie LH, Li TL, Liu SJ, Xiao J, Hu J, Wang J, Qin QB, Liu Y. The formation of diploid and triploid hybrids of female grass carp × male blunt snout bream and their 5S rDNA analysis. BMC Genetics, 2013, 14: 110

67. Liu SJ, Qin QB, Xiao J, Lu W, Shen J, Li W, Liu J, Duan W, Zhang C, Tao M, Zhao R, Yan J, Liu Y. The formation of the polyploid hybrids from different subfamily fish crossing and its evolutionary significance. Genetics, 2007, 176: 1023–1034

68. He WG, Qin QB, Liu SJ, Li TL, Wang J, Xiao J, Xie LH, Zhang C, Liu Y. Organization and variation analysis of 5S rDNA in different ploidy-level hybrids of red crucian carp×topmouth culter. PLoS

One, 2012, 6: e38976

69. Qin QB, Wang YD, Wang J, Dai J, Xiao J, Hu FZ, Luo KK, Tao M, Zhang C, Liu Y, Liu SJ. The autotetraploid fish derived from hybridization of *Carassius auratus* red var. (female)×*Megalobrama amblycephala* (male). Biol Reprod, 2014, 91: 93, 1–11
70. Tung TC, Wu SQ, Ye YF, Yan SY, Du M, Lu DY. Nuclear transplantation of fish. Sci Sin B, 1963, 7: 60–61
71. Chen HX, Yi YL, Chen MR, Yang XQ. Studies on the developmental potentiality of cultured cell nuclei of fish (in Chinese). Acta Hydrobiol Sin, 1986, 10: 1–7
72. Lee KY, Huang HG, Ju BS, Yang Z, Lin S. Cloned zebrafish by nuclear transfer from long-term cultured cells. Nature, 2002, 20: 795–799
73. Yan SY. A historical review and some comments on the nuclear transplantation in fish (in Chinese). Chin J Biotechnol, 2000, 16: 541–547
74. Yu LN, Zuo WG, Fang YL, Zheng WD. Cell-engineering grass carp produced by the combination of electric fusion and nuclear transplantation (in Chinese). J Fish China, 1996, 20: 312–318
75. Yan SY. The nucleo-cytoplasmic interaction as revealed by nuclear transplantation in fish. In: Gytoplasm Organization System. New York: McGraw-Hill Pub. Co., 1989. 61–81
76. Yu LN, Yang YQ, Liu L, Zheng WD, Fang YL. Study on the fish cell nucleus transplant with egg not take off the nucleus as a receptor (in Chinese). Freshw Fish, 1989, 3: 3–7
77. Lin LT, Xia SL, Zhu XP. Studies on nuclear transplantation of somatic cells in teleost. Zool Res, 1996, 17: 337–340
78. Liu HQ, Yi YL, Chen HX. The birth of the androgenetic homozygous diploid loach (in Chinese). Acta Hydrobiol Sin, 1987, 11: 241–246
79. Tajima A, Naito M, Yasuda Y, Kuwana T. Production of germ line chimera by transfer of primordial germ cells in the domestic chicken (*Gallus domesticus*). Theriogenology, 1993, 40: 509–519
80. Lee YM, Jung JG, Kim JN, Park TS, Kim TM, Shin SS, Kang DK, Lim JM, Han JY. A testis-mediated germline chimera production based on transfer of chicken testicular cells directly into heterologous

testes. Biol Reprod, 2006, 75: 380–386
81. Hill JR, Dobrinski I. Male germ cell transplantation in livestock. Reprod Fertil Dev, 2006, 18: 13–18
82. Dobrinski I. Germ cell transplantation and testis tissue xenografting in domestic animals. Anim Reprod Sci, 2005, 89: 137–145
83. Takeuchi Y, Yoshizaki G, Takeuchi T. Generation of live fry from intraperitoneally transplanted primordial germ cells in rainbow trout. Biol Reprod, 2003, 69: 1142–1149
84. Hong YH, Liu T, Zhao H, Xu H, Wang W, Liu R, Chen T, Deng J, Gui J. Establishment of a normal medakafish spermatogonial cell line capable of sperm production in vitro. Proc Natl Acad Sci USA, 2004, 101: 8011–8016
85. Okutsu T, Suzuki K, Takeuchi Y, Takeuchi T, Yoshizaki G. Testicular germ cells can colonize sexually undifferentiated embryonic gonad and produce functional eggs in fish. Proc Natl Acad Sci USA, 2006, 103: 2725–2729
86. Yi MS, Hong N, Hong YH. Generation of medaka fish haploid embryonic stem cells. Science, 2009, 326: 430–433
87. Yi MS, Hong N, Li ZD, Yan Y, Wang D, Zhao H, Hong YH. Medaka fish stem cells and their applications. Sci China Life Sci, 2010, 53: 426–434
88. Nobrega RH, Greebe CD, Kant HX, Bogerd J, Franca LR, Schulz RW. Spermatogonial stem cell niche and spermatogonial stem cell transplantation in zebrafish. PLoS One, 2010, 5: e12808
89. Takeuchi Y, Yoshizaki G, Takeuchi T. Surrogate broodstock produces salmonids. Nature, 2004, 430: 629–630
90. Majhi SK, Hattori RS, Yokota M, Watanabe S, Strussmann CA. Germ cell transplantation using sexually competent fish: an approach for rapid propagation of endangered and valuable germlines. PLoS One, 2009, 4: e6132
91. Okutsu T, Shikina S, Kanno M, Takeuchi Y, Yoshizaki G. Production of trout offspring from triploid salmon parents. Science, 2007, 317: 1517
92. Komen H, Thorgaard GH. Androgenesis, gynogenesis and the production of clones in fishes: a review. Aquaculture, 2007, 269:

150–173

93. Liu J, Yang G. Changes in copper content of allogynogenetic silver crucian carp after application of copper sulfate to fishponds. Isr J Aquacult-Bamid, 2009, 61: 351–355
94. Liu SJ, Qin QB, Wang YQ, Zhang H, Zhao R, Zhang C, Wang J, Li W, Chen L, Xiao J, Luo K, Tao M, Duan W, Liu Y. Evidence for the formation of the male gynogenetic fish. Mar Biotechnol, 2010, 12: 160–172
95. Felip A, Zanuy S, Carrillo M, Piferrer F. Induction of triploidy and gynogenesis in teleost fish with emphasis on marine species. Genetica, 2001, 111: 175–195
96. Komen J, Wiegertjes GF, Van Ginneken VJT, Eding EH, Richter CJJ. Gynogenesis in common carp (*Cyprinus carpio* L.) III. The effects of inbreeding on gonadal development of heterozygous and homozygous gynogenetic offspring. Aquaculture, 1992, 104: 51–66
97. Fopp-Bayat D, Kolman R, Woznicki P. Induction of meiotic gyno genesis in sterlet (*Acipenser ruthenus*) using UV-irradiated bester sperm. Aquaculture, 2007, 264: 54–58
98. Morgan AJ, Murashige R, Woolridge CA, Luckenbach JA, Watanabe WO, Borski RJ, Godwin J, Daniels HV. Effective UV dose and pressure shock for induction of meiotic gynogenesis in southern flounder *Paralichthys lethostigma* using black seabass (*Centropristis striata*) sperm. Aquaculture, 2006, 259: 290–299
99. Piferrer F, Cal RM, Gsmez C, Alvarez-Blazquez B, Castro J, Martinez P. Induction of gynogenesis in the turbot (*Scophthalmus maximus*): effects of UV irradiation on sperm motility, the Hertwig effect and viability during the first 6 months of age. Aquaculture, 2004, 238: 403–419
100. Rougeot C, Ngingo JV, Gillet L, Vanderplasschen A, Melard C. Gynogenesis induction and sex determination in the Eurasian perch, *Perca fluviatilis*. Aquaculture, 2005, 43: 411–415
101. Lin F, Dabrowski K. Induction of gynogenesis in muskellunge (*Esox masquinongy*). Aquaculture, 1995, 137: 153–154
102. Li Q, Kijima A. Microsatellite analysis of gynogenetic families in the Pacific oyster, *Crassostrea gigas*. J Exp Mar Biol Ecol, 2006, 331: 1–8

103. Sun YD, Tao M, Liu SJ, Zhang C, Duan W, Shen JM, Wang J, Zeng C, Long Y, Liu Y. Induction of gynogenesis in red crucian carp using spermatozoa of blunt snout bream. Prog Nat Sci, 2007, 17: 163–167

104. Sun YD, Zhang C, Liu SJ, Tao M, Zeng C, Liu Y. Induction of gynogenesis in Japanese crucian carp (*Carassius cuvieri*). Acta Genetica Sinica, 2006, 33: 405–412

105. Zhang H, Liu SJ, Zhang C, Tao M, Peng L, You CP, Xiao J, Zhou Y, Zhou G, Luo KK, Liu Y. Induced gynogenesis in Grass Carp (*Ctenopharyngodon idellus*) using irradiated sperm of allotetraploid hybrids. Mar Biotechnol, 2010, 13: 1017–1026

106. Liu SJ, Duan W, Tao M, Zhang C, Sun YD, Shen J, Wang J, Luo KK, Liu Y. Establishment of the diploid gynogenetic hybrid clonal line of red crucian carp×common carp. Sci China Ser C-Life Sci, 2007, 50: 186–193

107. Xiao J, Zou TM, Chen YB, Chen L, Liu SJ, Tao M, Zhang C, Zhao RR, Zhou Y, Long Y, You CP, Yan JP, Liu Y. Coexistence of diploid, triploid and tetraploid crucian carp (*Carassius auratus*) in natural waters. BMC Genetics, 2011, 12: 20

108. Stanley JG. Production of hybrid, androgenetic and gynogenetic grass carp and carp. T Am Fish Soc, 1976, 105: 10–16

109. Arai K, Onozato H, Yamazaki F. Artificial androgenesis induced with gamma irradiation in masou salmon (*oncorhynchus masou*). Bull Fac Fish Hokkaido Univ, 1979, 30: 181–186

110. Araki K, Shinma H, Nagoya H, Nakayama I, Onozato H. Androgenetic diploids of rainbow trout (*Oncorhynchus mykiss*) produced by fused sperm. Can J Fish Aquat Sci, 1995, 52: 892–896

111. Babiak I, Dobosz S, Goryczko K, Kuzminski H, Brzuzan P, Ciesielski S. Androgenesis in rainbow trout using cryopreserved spermatozoa: the effect of processing and biological factors. Theriogenology, 2002, 57: 29249

112. Thorgaard GH, Scheerer PD, Hershberger WK, Myers JM. Androgenetic rainbow trout produced using sperm from tetraploid males show improved survival. Aquaculture, 1990, 85: 215–221

113. Arai K, Ikeno M, Suzuki R. Production of androgenetic diploid loach *Misgurnus anguillicaudatus* using spermatozoa of natural tetraploids. Aquaculture, 1995, 137: 131–138

114. Fujimoto T, Yasui GS, Hayakawa M, Sakao S, Yamaha E, Arai K. Reproductive capacity of neo-tetraploid loaches produced using diploid spermatozoa from a natural tetraploid male. Aquaculture, 2010, 308: s133–139

115. Sun YD, Zhang C, Liu SJ, Duan W, Liu Y. Induced interspecific androgenesis using diploid sperm from allotetraploid hybrids of common carp×red crucian carp. Aquaculture, 2007, 264: 47–53

116. Duan W, Qin QB, Chen S, Liu SJ, Wang J, Zhang C, Sun YD, Liu Y. The formation of improved tetraploid population of red crucian carp×common carp hybrids by androgenesis. Sci China Ser C-Life Sci, 2007, 50: 753–761

117. Song C, Liu SJ, Xiao J, He WG, Zhou Y, Qin QB, Zhang C, Liu Y. Polyploid organisms. Sci China Life Sci, 2012, 55: 301–311

118. Benfey TJ. The physiology and behavior of triploid fishes. Rev Fish Sci, 1999, 1: 39–67

119. Gui JF, Xiao WH, Liang SC, Jiang YG. Preliminary study on the cytological mechanism of triploidy and tetraploidy induced by hydrostatic pressure shock in transparent colored crucian carp (in Chinese). Acta Hydrobiol Sin, 1995, 19: 49–55

120. Li WL, Xu Y, Deng H, Chen SL, Xie MS, Ji XS. Induction and identification of artificial triploid fry in *Cynoglossus semilaevis* (in Chinese). J Fish China, 2011, 35: 925–931

121. Wu C, Ye Y, Chen R, Liu X. An artificial multiple triploid carp and its biological characteristics. Aquaculture, 1993, 111: 255–262

122. Gui JF, Zhou L. Genetic basis and breeding application of clonal diversity and dual reproduction modes in polyploid *Carassius auratus gibelio*. Sci China Life Sci, 2010, 53: 409–415

123. Wang ZW, Zhu HP, Wang D, Jiang FF, Guo W, Zhou L, Gui JF. A novel nucleo-cytoplasmic hybrid clone formed via androgenesis in polyploid gibel carp. BMC Res Notes, 2011, 4: 82

124. Chourrout D. Tetraploidy induced by heat shocks in the rainbow trout (*Salmo gairdneri*R.). Reprod Nutr Dev, 1982, 22: 569–574

125. Bidwell CA, Chrisman CL, Libey G. Polyploidy induced by heat shock in channel catfish. Aquaculture, 1985, 51: 25–32

126. Flajshans M, Linhart O, Kvasnicka P. Genetic studies of tench (*Tinea*

tinea L.): induced triploidy and tetraploidy and first performance data. Aquaculture, 1993,113: 301–312

127. Nam YK, Choi GC, Park DJ, Kim DS. Survival and growth of induced tetraploid mud loach. Aquacult Int, 2001, 9: 61–71
128. Li WL, Chen SL, Ji XS, Xie MS, Xu Y, Deng H. Induction and identification of tetraploid fry in *Cynoglossus semilaevis* (in Chinese). J Fish Sci China, 2012, 19: 196–201
129. Myers JM. Tetraploid induction in *Oreochromis* spp. Aquaculture, 1986, 57: 281–287
130. Gui JF, Sun JM, Liang SC, Huang WY, Jiang YG. Studies on genome manipulation in fish II. Tetraploidy induced by hydrostatic pressure treatment and a combination of hydrostatic pressure and cold treatments in transparent colored crucian carp (in Chinese). Acta Hydrobiol Sin, 1991, 15: 333–341
131. Zhu HP, Gui JF. Identification of genome organization in the unusual allotetraploid form of *Carassius auratus gibelio*. Aquaculture, 2007, 265: 109–117
132. Zou SM, Li SF, Cai WQ, Zhao JL, Yang HY. Establishment of fertile tetraploid population of blunt snout bream (*Megalobrama amblycephala*). Aquaculture, 2004, 238: 155–164
133. Qin QB, He WG, Liu SJ, Wang J, Xiao J, Liu Y. Analysis of 5S rDNA organization and variation in polyploid hybrids from crosses of different fish subfamilies. J Exp Zool (Mol Dev Evol), 2010, 314: 403–411
134. Liu SJ. Fish Distant Hybridization (in Chinese). Beijing: Science Press, 2015
135. Zhu ZY, Li G, He L, Chen S. Novel gene transfer into fertilized eggs of gold fish (*Carassius auratus* L.1758). Z Angew Ichthyol, 1985, 1: 31–34
136. Houdebine LM, Chourrout D. Transgenesis in fish. Experientia, 1991, 47: 891–897
137. Rembold M, Lahiri K, Foulkes NS, Wittbrodt J. Transgenesis in fish: efficient selection of transgenic fish by co-injection with a fluorescent reporter construct. Nat Prot, 2006, 1: 1133–1139
138. Devlin RH. Production and evaluation of transgenic fish for

aquaculture. Australas Biotechnol, 1998, 8: 222–226

139. Maclean N, Laight RJ. Transgenic fish: an evaluation of benefits and risks. Fish Fish, 2000, 1: 146–172
140. Fu C, Hu W, Wang Y, Zhu Z. Developments in transgenic fish in the People's Republic of China. Rev Sci Tech Off Int Epiz, 2005, 24: 299–307
141. Fu C, Cui Y, Hung SSO, Zhu Z. Growth and feed utilization by F4 human growth hormone transgenic carp fed diets with different protein levels. J Fish Biol, 1998, 53: 115–129
142. Wu G, Sun Y, Zhu ZY. Growth hormone gene transfer in common carp. Aquat Living Resour, 2003, 16: 416–420
143. Feng H, Zeng ZQ, Liu SJ, Zhang XJ, Zhou GJ, Li JZ, Liu Y, Wang YP, Chen SP, Hu W, Zhu ZY. Studies of F1 of transgenic allotraploid hybrids of *Carassius auratus* red var. (♀)×*Cyprinus carpio* (♂) (in Chinese). Acta Genet Sin, 2002, 29: 434–437
144. Feng H, Fu YM, Luo J, Wu H, Liu Y, Liu SJ. Black carp GH gene transgenic allotetraploid hybrids of *Carassius auratus* red var. (♀)×*Cyprinus carpio* (♂). Sci China Life Sci, 2011, 41: 202–209
145. Hew CL, Davies PL, Fletcher G. Antifreeze protein gene transfer in Atlantic salmon. Mol Mar Biol Biotechnol, 1992, 1: 309–317
146. Dunham RA. Transgenic fish resistant to infectious diseases, their risk and prevention of escape into the environment and future candidate genes for disease transgene manipulation. Comp Immunol Microbiol Infect Dis, 2009, 32: 139–161
147. Dunham RA, Warr GW, Nichols A, Duncan PL, Argue B, Middleton D, Kucuktas H. Enhanced bacterial disease resistance of transgenic channel catfish, *Ictalarus punctatus*, possessing cecropin genes. Mar Biotechnol, 2002, 4: 338–344
148. Mao WF, Wang YP, Wang WB, Wu B, Feng JX, Zhu ZY. Enhanced resistance to *Aeromonas hydrophila* infection and enhanced phagocytic activities in human lactoferrin-transgenic grass carp (*Ctenopharyngodon idellus*). Aquaculture, 2004, 242: 93–103
149. Yu F, Xiao J, Liang XY, Liu SJ, Zhou GJ, Luo KK, Liu Y, Hu W, Wang YP, Zhu ZY. Rapid growth and sterility of growth hormone gene transgenic triploid carp. Chin Sci Bull, 2011, 56: 1679–1684

CHAPTER

9

GENETIC AND GENOMIC ANALYSES FOR ECONOMICALLY IMPORTANT TRAITS AND THEIR APPLICATIONS IN MOLECULAR BREEDING OF CULTURED FISH

JinGou Tong[1] and XiaoWen Sun[2]

[1] State Key Laboratory of Freshwater Ecology and Biotechnology, Institute of Hydrobiology, Chinese Academy of Sciences, Wuhan 430072, China;

[2] Heilongjiang Fisheries Research Institute, Chinese Academy of Fishery Sciences, Harbin 150070, China

ABSTRACT

The traits of cultured fish must continually be genetically improved to supply high-quality animal protein for human consumption. Economically important fish traits are controlled by multiple gene quantitative trait loci (QTL), most of which have minor effects, but a few genes may have major effects useful for molecular breeding. In this review, we chose relevant studies on some of the most intensively cultured fish and concisely summarize progress on identifying and verifying QTLs for such traits as growth, disease and stress resistance and sex in recent decades. The potential applications of these major-effect genes and their associated markers in marker-assisted selection and molecular breeding, as well as future research directions are also discussed. These genetic and genomic analyses will be valuable for elucidating the mechanisms modulating economically important traits and to establish more effective molecular breeding techniques in fish.

Keywords: Aquaculture fish economic traits QTL major genes molecular markers molecular breeding

INTRODUCTION

Fish are produced to supply high-quality protein to the market through breeding of new fish strains with better traits, such as fast growth, high resistance to disease and stress, and a high feed conversion rate (FCR). Fish-breeding methodologies have developed rapidly in past decades, from traditional selection and hybridization to modern biotechnologies, such as marker-assisted selection and molecular breeding. Although selection is the basis for all breeding techniques, traditional family or population-based selection is usually less efficient and the breeding programs may take a long period of time to develop. Most economically important traits in fish, such as growth rate, disease resistance, and sex, are controlled by multiple genes, known as quantitative trait loci (QTL). Most of these genes have minor effects, but several may have major effects on traits. Theoretically, if genes and genetic markers associated with traits of interest are identified, the genetic variants could be used as tools in marker-assisted selection (MAS) analyses. The idea of using genetic markers during selective breeding has been proposed since the 1960s, but the genetic basis for economically important traits in animals including fish has not been analyzed effectively due to a shortage in high-resolution and powerful techniques. However, genetic mechanisms underlying the economically important traits in fish have been studied more effectively with the development of quantitative genetics, molecular genetics, structural and functional genomics, and molecular marker technologies. In particular, an understanding of QTL from zero to the present provides a solid basis for developing fish molecular breeding strategies.

The tools to genetically analyze economic traits in fish include (i) developing a large number of molecular markers; (ii) constructing medium and high-density genetic linkage maps and conducting QTL/

expression quantitative trait locus (eQTL) analyses throughout the whole genome; (iii) using a candidate gene approach or an association study; (iv) conducting a bulked segregant analysis (BSA); (v) linkage disequilibrium (LD), or LD associate mapping; and (vi) genome-wide association study (GWAS).

Molecular markers are one of the most powerful tools for genetic analyses of fish's economically important traits. In the early stages of this field, genetic variations were represented by a limited number of less polymorphic allozyme loci. Then dominant markers, such as restriction fragment length polymorphism (RFLP), randomly amplified polymorphic DNA (RAPD), and amplified fragment length polymorphism (AFLP), were developed one after another. In recent years, two types of DNA markers—microsatellite or simple sequence repeat (SSR) and single nucleotide polymorphisms (SNPs)—have been widely used in genetic analyses due to their wide genome distribution and high rate of polymorphisms [1]. Molecular breeding includes molecular marker-assisted selection, transgenic breeding, and computer-based molecular design breeding [2]. In the late 1990s, the United States, Japan, Norway, Canada, and Australia initiated genome projects in several aquatic animals. Chinese scientists also started genome sequencing projects for oyster, scallop, shrimp, half-smooth tongue sole, common carp, yellow croaker, grass carp, and grouper. Silver carp, bighead carp, bluntsnout bream, and gibeio carp will also be decoded one after another in the near future. The great advances in genomics, including structural, functional, and comparative genomics, have provided new insights into molecular breeding studies [3–5]. In this review, we will concisely summarize current genetic and genomic analyses for economically important traits and discuss future directions and prospects for fish molecular breeding.

STATUS OF GENETIC AND GENOMIC STUDIES ON ECONOMICALLY IMPORTANT TRAITS IN FISH

Genetic Linkage Maps

Since the first genetic linkage map was published for tilapia in 1998 [6], genetic maps of various densities have been published on some important aquaculture species, and many of these maps used co-dominant genetic markers, such as microsatellites and SNPs. More high-density maps of aquatic species have been reported with the advances in molecular marker technology and increased financial input. Up to 2010, about 30 aquatic animals had publicly accessible genetic maps [7]. However, more aquaculture fish have been added in the past three years, including three of the four major Chinese carp, among them grass carp, silver carp, and bighead carp, and other important aquaculture fish (Table 1).

QTL Analysis for Growth Traits

Growth is one of the most important economic traits of all aquaculture species. Up to 2012, QTL analyses have been conducted in more than 20 aquatic species [7], and growth is the most popular trait studied. Wang et al. [21] used 380 F1 Asian seabass to identify five major QTLs and 27 potential QTLs. Of them, three major QTLs for body weight, total length, and body length were located at a similar linkage group 2 (LG2) position with the nearby Lca287 microsatellite and accounted for 28.8%, 58.9%, and 59.7% of the phenotypic variations. The other two major QTLs for body weight were located at another LG2 position. These five major QTLs have been confirmed in two other Asian seabass populations [22]. Further QTL fine mapping of the Asian seabass growth trait identified three candidate "growth genes" (cathepsin D, KCTD15, and csmd2) affecting body weight, body length, and total length [10]. The function of the cathepsin D gene in humans involves cell proliferation and cell growth; therefore, cathepsin D may also be a major "growth gene" in Asian seabass.

Table 1: Medium and high-density genetic linkage maps for some economically important fish species

Species	Type of markers	No. of markers	Average distance (cM)	Year	No. of linkage groups	Chromosome No. of haploid genome	References
Pseudosciaena crocea	AFLP and SSR	375	>9	2007	24	24	[8]
Ctenopharyngodon idella	SSR and SNP	279	4.2	2010	24	24	[9]
Lates calcarifer	SSR and SNP	790	3.4	2011	20	20	[10]
Oncorhynchus mykiss	SSR and SNP	1,459	2.29	2012	29	30	[11]
Paralichthys olivaceus	SSR	1,487	1.22	2012	24	24	[12]
Cynoglossus semilaevis	SSR	1,009	1.67	2012	21	21	[13]
Scophthalmus maximus	SSR and SNP	487	3.3	2013	24	22	[14]
Oreochromis spp	SSR	401	3.3	2013	22	22	[15]
Hypophthalmichehys molitrix	SSR	703	2.2	2013	24	24	[16]
Cyprinus carpio	SSR and SNP	1,209	3.5	2013	50	50	[17]
Aristichthys nobilis	SSR	659	2.9	2014	24	24	[18]
Ictalurus punctatus	SNP	62,870	0.4	2014	29	29	[19]
Salmo salar	SNP	6,458	–	2014	29	29	[20]

O'Malley et al. [23] identified QTLs for body weight in rainbow trout on 10 different LGs. Wringe et al. [24] used additional backcrossed families and SSR markers to confirm the O'Malley et al.'s results and found several major candidate growth genes (e.g., GH2 and Pax7). Reid et al. [25] identified a QTL for body weight in two LGs (AS8 and 11) of Atlantic salmon, and reported that it was homologous to the growth QTL in rainbow trout. Houston et al. [26] identified QTLs for body weight in LG1 and LG5 of Atlantic salmon. Gutierrez et al. [27] further used a 6.5 K SNP chip to identify QTLs in six LGs at the genomic level. Cnaani et al. [28,29] identified a QTL for tilapia growth on LG23, which is the linkage group with the genetic sex-determining region. Song et al. [12] used 1487 SSRs to produce a high-density genetic linkage map and successfully identified a QTL affecting body weight in LG14 of Japanese flounder.

Some reports have used a candidate gene approach to identify growth-related genes and molecular markers in fish. Tao and Boulding [30] found polymorphisms in the growth hormone gene (GH) that were significantly associated with growth rate of Arctic charr (Salvelinus alpinus). Li et al. [31] reported an SNP in the insulin-like growth factor- (IGF)1 gene 5′ untranslated region (UTR) of largemouth bass (Micropterus salmoides). Sun et al. [32] reported that two SNPs in exon 3 of the myostatin (MSTN) gene were significantly related to body weight and Fulton's factor in common carp. Liu et al. [33] also found that a SNP in the MSTN 3′ UTR was very significantly associated with total length, body length, and body weight of bighead carp.

QTL for Disease Resistance

Diseases strongly influence the aquaculture industry. Infectious hematopoietic necrosis (IHN) disease, for instance, is an acute infectious virus disease that affects salmonids in the northern hemisphere. IHN can result in fingerling and larval fish mortality rates of 70%–100%. Identifying immune or disease-resistant molecular markers would provide theoretical and technical bases for breeding anti-disease characteristics. Rodriguez et al.[34] used rainbow trout and steelhead trout hybrids in a backcross with rainbow trout to produce 70 BC1 families. The offspring were exposed to an IHN infection, and the phenotypes were categorized as: 1 (death) and 0 (survival). As a result, six SSR markers were related to IHN resistance based on BSA screening.

Ozaki et al. [35] used 51 SSRs across all rainbow trout LGs in a BC1 family QTL analysis and found two QTLs for anti-infectious pancreatic necrosis virus (IPNV) disease on LG21 in Atlantic salmon. Baerwald et al. [36] used 143 SSRs and AFLP markers in a large F2 family (n=480) and found a QTL on Omy9 associated with whirling disease in young salmonids. This QTL was confirmed in three other families. Several authors have reported QTLs for IPNV, and most show that the anti-IPNV loci are located on LG21 [37–40], indicating that rainbow trout and Atlantic salmon anti-IPNV QTLs may be evolutionarily conserved. The genetic effects of Atlantic salmon anti-IHN QTLs have been validated in other studies [41]. Fuji et al. [42] used a segregated population and located a single major QTL controlling resistance to lymphocystis disease on LG15 in Japanese flounder with a phenotypic variation value of 50%. Further study confirmed that an SSR locus within that QTL was closely related with a phenotype for lymphocystis disease resistance [43].

Molecular markers associated with disease resistance are also found on immune-related genes. For example, Palti et al. [44] reported that MHCII alleles are significantly associated with anti-IHN in rainbow trout. Zhang et al. [45] and Xu et al. [46] successfully screened for anti-disease molecular markers by comparing the polymorphisms of Japanese flounder groups susceptible and resistant to Vibrio anguillarum.

QTL for Feed Conversion Rate

FCR is one of the most important economic traits in fish, as fish with a better FCR increase profits. QTLs for FCR have been reported in cattle [47], pigs [48] and chickens [49]. Some SNPs in the growth hormone receptor (GHR), neuropeptide Y (NPY), uncoupling protein 3 (UCP3), ghrelin (GHRL), and IGF-2 genes of cattle affect FCR. [50]. Liu [51] used AFLP markers to construct a catfish genetic map and found a QTL associated with FCR. Zimmerman et al. [52] revealed three QTLs for the number of pyloric caeca in three LGs of rainbow trout, and this is an important index associated with FCR. Although FCR studies have been reported in common carp from the Heilongjiang Fisheries Research Institute of the Chinese Academy of Fishery Sciences [53], QTL analyses for FCR in aquaculture fish are rarely reported.

QTL for Sex Determination

Sex phenotype and sex determination in fish have specific evolutionary status and diversity. Males and females of some species have significant differences in growth rate or commercial value; therefore, monosex fish culture is a promising strategy. The sex-determining (SD) loci and QTLs have been studied in a limited number of fish, such as tilapia [54], rainbow trout [55], and salmonids [56]. Previous studies have demonstrated that sex QTLs are located on LG1, 2, 3, 6, and 23 of tilapia [29,54,57]. Eshel et al.[58,59] reported a major candidate sex QTL that is considered the sexdetermining region in tilapia. Fifty-one genes in this region have been annotated, and 10 have been confirmed. The anti-Müllerian hormone gene is the most differentially expressed gene in male and female tilapia. Sun et al. [60] recently published several sex-specific markers, and one is tightly linked with the sex-determining region discovered by Eshel et al. The sex-determining locus in rainbow trout is located on the LG of RT10 [61], and this locus also significantly affects thermo-resistance and body length [62]. The sex-determining regions in Artic charr [63], brown trout [64], and Atlantic salmon [65] are located on the LGs of AC4, BT28, and AS1, respectively. Woram et al. [66] compared LGs of sex-determining loci in four salmonids and found that although the nucleotide sequences flanking the sex-determining loci were well-conserved, the SD LGs were diverse, suggesting that the regions underwent different recombination events. Loukovitis et al. [67] located growth and sex-determining QTLs in gilthead sea bream and showed that these two traits have similar genetic control in LG21. Martínez et al. [68] located a sex QTL on LG5 of turbot and proposed a ZZ/ZW sex-determining mechanism. Viñas et al. [69] also found a major sex QTL on turbot LG5. These findings suggest that the sex-determining genes may occur on turbot LG5. Song et al. [13] used high-density genetic maps to locate seven sex QTLs on the half-smooth tongue sole LG1f, LG14f, and LG1m. Additional study by Chen et al. [70] provided insight into ZW sex chromosome evolution and identified sex-determining genes, such as dmrt1 and neurl3.

QTL Analysis for Anti-Stress Traits

Recent evidence shows that some fish may have sufficient genetic variation to adapt to environmental changes, such as changes in temperature and salinity. For example, cold tolerance in the stickleback may be strongly selected [71]. Growth of tilapia slows markedly when temperature drops below the optimum, and can cause death. Cnaani et al. [28] located a marker associated with cold resistance and body size in a F2 hybrid between Mozambique and Nile tilapia. Moen et al. [72] confirmed that cold-tolerance QTL exists on LG23 in tilapia. Rainbow trout have weak tolerance to high temperature; however, Jackson et al. [73] located thermo-tolerant QTLs on two LGs and identified two tightly linked markers. Danzmann et al. [74] further reported that one microsatellite was associated with thermo-tolerance in rainbow trout, which was confirmed by Perry et al. [75]. Perry et al. [62] further found that the Ssa20.19NUIG sex marker is associated with body length and thermo-resistance in rainbow trout. Sun et al. [76] located a cold resistance locus on common carp LG5. Rengmark et al. [77] reported a high salinity-tolerance gene, while Norman et al. [78] compared QTLs for salinity tolerance among the salmonid genomes and found that the QTLs were distributed on nine different LGs.

APPLICATIONS AND DISCUSSION OF MOLECULAR BREEDING

Progress of Applied Molecular Breeding Studies

Many aquatic geneticists have used genetic or molecular markers as tools to perform parentage and molecular marker-assisted selection [4]. Common carp is a fish cultured worldwide and is one of the most important fish cultured in China. Chinese scientists have achieved good breeding results using genetic distances of female and male parents to design a mirror carp mating scheme. In common carp, bighead carp and other species, Chinese scientists confirmed a positive correlation between growth traits and the number of advantageous variants from candidate major genes. QTL analyses have reported good results from selective breeding programs for spindle-type mirror carp and slim-type common carp [4]. A multi-locus congruent breeding technique has been proposed, based mainly on common carp studies, and provides a basis for establishing a molecular breeding technique in fish (Figure 1).

Although many research programs are progressing, studies in a few species have shown favorable effects on breeding for anti-disease traits. Fuji et al. [42] found an SSR allele associated with IPN, and used it during large-scale anti-IPN selective breeding and fingerling production [43]. Selective breeding of gibeio carp has made good progress using anti-myxosporean-related MHC alleles as molecular markers [79]. Chen's group selected for anti-disease Japanese flounder using BSA and QTL analyses to identify anti-Vibrio-anguillarum-related markers [80,81].

Sex markers have been used in fish species with large sex differences. Chinese scientists have developed strategies for large-scale YY supermale and XY all-male yellow catfish based on the combination of artificial gynogenesis, sex reversal, and X- and Y-chromosome-specific molecular markers, and these methods have been used in commercial production [82-84]. Female-specific AFLP and sex-specific SSR markers have been isolated in half-smooth tongue sole [85,86], and facilitate sex identification during early growth. Sex determination and sex marker studies have been conducted in Nile tilapia [59,60], and sex-marker-assisted selection is expected to be commercialized in the near future.

Discussion of and Directions for Molecular Breeding

People continue to consume more fish protein with the increased demand for foods with beneficial health effects. Aquatic products have become increasingly popular in recent decades, but the increase in aquatic production relies heavily on aquaculture of a limited number of species, such as grass carp, silver carp, bighead carp, common carp, crucian carp, rainbow trout, and tilapia, which account for more than two-thirds of total annual world fish production.

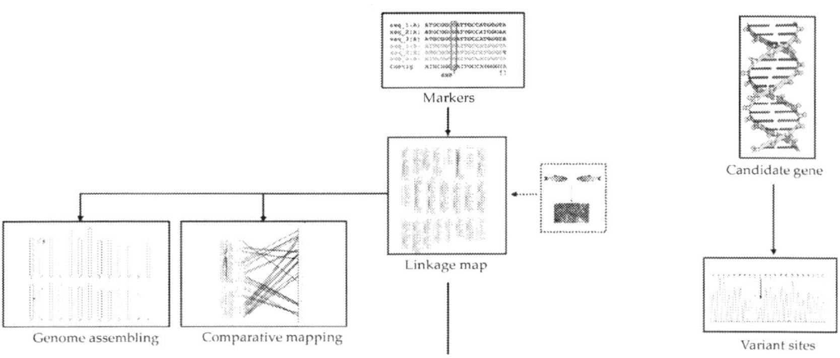

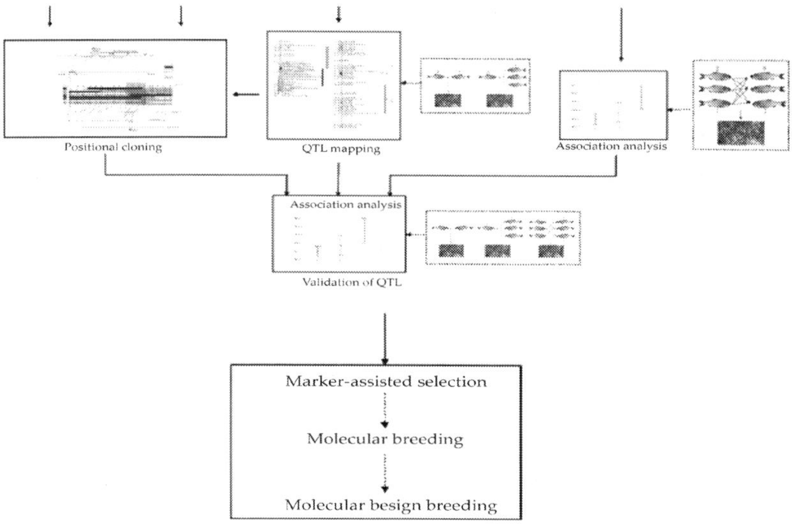

Figure 1: Fish molecular breeding strategies.

Some fish species have met poor traits after many years of aquaculture, such as early sex maturation, weak stress resistance, or frequent disease outbreaks. These features demand selection and genetic improvement for better strains and varieties. Up to now, more than 60 fish and shellfish have undergone selective breeding, and traits of interest have focused on growth and other phenotypes, such as anti-disease and anti-stress (e.g., cold- and thermo-resistant). However, other economically important traits, such as FCR and intramuscular bone content, have received less attention, probably because the phenotypes for these traits are difficult to measure and study [4]. It is expected that future scientists will consider these traits and elucidate the genetic mechanisms underlying them so that molecular breeding for composite traits would be possible.

Genomes of the half-smooth tongue sole and common carp have been sequenced and assembled at the Chinese Academy of Fishery Sciences [70,87]. After several years of efforts by the scientists there, hundreds of QTLs and molecular markers for growth traits have been found in various common carp strains, as well as markers for immuno-related and cold-resistance genes and markers. QTLs associated with sex and growth-related traits have been identified, and the sex chromosome sequences have provided a solid foundation for molecular and mono-sex breeding studies in half-smooth tongue sole. Scientists have also identified many QTLs for various traits in salmonids, tilapia, and flounder. Nevertheless, QTL fine mapping has been rarely reported in cultured fish. Currently, QTL analyses in fish have provided LG intervals, and those markers tightly linked with an interval may be suitable for MAS. However, a more powerful method is to identify genes from those intervals, screen the SNPs at the population level, and apply the SNPs in a molecular breeding program through quantitative trait nucleotides [88]. Genes in the QTL interval that affect traits of interest should generally meet the following criteria: (i) the function of the gene is known and it plays one of the most important roles affecting the trait(s); (ii) the gene has been subjected to functional studies in this or other species, such as gene transfer or knockdown; (iii) the gene is expressed in tissues or organs that involve traits of interest; and (iv) the gene modulates the trait throughout development. Further confirmation is needed for those genes and markers in different populations and families. Only confirmed genes and linked markers can be used for MAS or molecular breeding.

Although QTL analyses screen for trait-related loci throughout a fish genome, it is usually used for family material, which makes QTL results difficult to confirm in other populations or families. LD analysis or LD mapping (usually of population material) is used to localize QTLs for more polymorphic sites and at higher resolution. Therefore, LD mapping is a powerful method to identify the functions of major-effect genes or high-resolution QTL [89]. High-throughput DNA sequencing and SNP-detecting technologies have been used with LD mapping to identify possible advantageous alleles of genes in fish. As mentioned above, some reports have demonstrated that different QTLs for sex, anti-disease (or stress resistance), and growth traits may harbor the same LG on a similar or nearby position. Fine mapping of those QTL intervals with multiple phenotypes could be conducted with LD mapping, comparative genomics, and other cost-effective methods to identify and confirm novel genes and their associated markers for molecular breeding.

A candidate gene approach [90] together with the BSA method is an effective way to identify trait-associated genes and molecular markers in species with or without genetic maps. The BSA method has been applied to discover sex and disease-resistance genes in aquatic animals [34,91], but it has rarely been used to identify major candidate growth genes or those for other economically important traits. In the future, BSA should be more frequently used in eQTL to identify major genes and associated markers, particularly when used in combination with high-throughput next generation sequencing.

GWAS is a powerful technology to reveal trait-related genes and alleles. However, GWAS relies heavily on the development and genotyping of thousands of molecular markers throughout the target genome. As re-sequencing costs are high, GWAS has only been used in humans and a few domestic animals. Fortunately, whole-genome sequencing is complete or is nearly complete in several cultured fish species, and sequences of more fish genomes will be available in the near future. Thus, GWAS will be used for more frequent genetic analyses of economically important traits in cultured fish.

Scientists have proposed a molecular design breeding concept with the rapid accumulation of known genes and genetic variants that affect economically important traits of domestic animals and plants. This is to integrate major-effect genes and favorable variants to create new strains and varieties to improve agricultural performance [2]. Scientists from the Chinese Academy of Sciences also proposed a molecular module design breeding concept to establish new breeding biotechnology in the new era. Molecular design breeding or molecular module design breeding is based on molecular docking of three-dimensional biomacromolecular data, computational analysis, integration, and simulation to select the best breeding strategy for a computer virtual design before initiating an actual breeding program [92]. Further understanding of fish genomes and identifying large numbers of QTLs, candidate major genes, and associated markers will reveal the potentials of molecular breeding. Enriching omics information and molecular elements (genes and markers) for major traits in cultured fish will allow molecular (module) design breeding technology to achieve great progress in the near future.

REFERENCES

1. Liu ZJ, Cordes JF. DNA marker technologies and their applications in aquaculture genetics. Aquaculture, 2004, 238: 1–37
2. Peleman JD, van der Voort JR. Breeding by design. Trends Plant Sci, 2003, 8: 330–334PubMed

3. Varshney RK, Graner A, Sorrells ME. Genomics-assisted breeding for crop improvement. Trends Plant Sci, 2005, 10: 621–630
4. Sun XW. Molecular Breeding of Fish (in Chinese). Beijing: China Ocean Press, 2010
5. Gui JF, Zhu ZY. Molecular basis and genetic improvement of economically important traits in aquaculture animals. Chin Sci Bull, 2012, 57: 1751–1760
6. Kocher TD, Lee WJ, Sobolewska H, Penman D, McAndrew B. A genetic linkage map of a cichlid fish, the tilapia (*Oreochromis niloticus*). Genetics, 1998, 148: 1225–1232
7. Yue GH. Recent advances of genome mapping and marker-assisted selection in aquaculture. Fish Fisheries, 2014, 15: 376–396
8. Ning Y, Liu X, Wang ZY, Guo W, Li YY, Xie FJ. A genetic map of large yellow croaker *Pseudosciaena crocea*. Aquaculture, 2007, 264: 16–26
9. Xia JH, Liu F, Zhu ZY, Fu JJ, Feng JB, Li JL, Yue GH. A consensus linkage map of the grass carp (*Ctenopharyngodon idella*) based on microsatellites and SNPs. BMC Genomics, 2010, 11: 135
10. Wang CM, Bai Z Y, He XP, Lin G, Xia JH, Sun F, Lo LC, Feng F, Zhu ZY, Yue GH. A high-resolution linkage map for comparative genome analysis and QTL fine mapping in Asian seabass, *Lates calcarifer*. BMC Genomics, 2011, 12: 174
11. Palti Y, Genet C, Gao G, Hu Y, You FM, Boussaha M, Rexroad CE 3rd, Luo MC. A second generation integrated map of the rainbow trout (*Oncorhynchus mykiss*) genome: analysis of conserved synteny with model fish genomes. Mar Biotechnol, 2012, 14: 343–357
12. Song W, Pang R, Niu Y, Gao F, Zhao Y, Zhang J, Sun J, Shao C, Liao X, Wang L, Tian Y, Chen S. Construction of high-density genetic linkage maps and mapping of growth-related quantitative trait loci in the Japanese flounder (*Paralichthys olivaceus*). PLoS One, 2012, 7: e50404
13. Song W, Li Y, Zhao Y, Liu Y, Niu Y, Pang R, Miao G, Liao X, Shao C, Gao F, Chen S. Construction of a high-density microsatellite genetic linkage map and mapping of sexual and growth-related traits in half-smooth tongue sole (*Cynoglossus semilaevis*). PLoS One, 2012, 7: e52097

14. Hermida M, Bouza C, Fernández C, Sciara AA, Rodríguez-Ramilo ST, Fernández J, Martínez P. Compilation of mapping resources in turbot (*Scophthalmus maximus*): a new integrated consensus genetic map. Aquaculture, 2013, 414: 19–25
15. Liu F, Sun F, Li J, Xia JH, Lin G, Tu RJ, Yue GH. A microsatellite-based linkage map of salt tolerant tilapia (*Oreochromis mossambicus* × *Oreochromis spp.*)) and mapping of sex-determining loci. BMC Genomics, 2013, 14: 58
16. Guo W, Tong J, Yu X, Guo W, Wang X, Liu H, Feng X, Sun Y, Liu L, Fu B. A second generation genetic linkage map for silver carp (*Hypophthalmichehys molitrix*) using microsatellite markers. Aquaculture, 2013, 412: 97–106
17. Zhao L, Zhang Y, Ji P, Zhang X, Zhao Z, Hou G, Huo L, Liu G, Li C, Xu P, Sun X. A dense genetic linkage map for common carp and its integration with a BAC-based physical map. PLoS One, 2013, 8:e63928
18. Zhu C, Tong J, Yu X, Guo W, Wang X, Liu H, Feng X, Sun Y, Liu L, Fu B. A second-generation genetic linkage map for bighead carp (*Aristichthys nobilis*) based on microsatellite markers. Anim Genet, 2014, 45: 699–708
19. Li Y, Liu S, Qin Z, et al. A High-density genetic linkage map of channel catfish for the assembly of reference whole genome sequences. In: Plant and Animal Genome XXII Conference. Plant and Animal Genome, 2014
20. Gonen S, Lowe NR, Cezard T, Gharbi K, Bishop SC, Houston RD. Linkage maps of the Atlantic salmon (*Salmo salar*) genome derived from RAD sequencing. BMC Genomics, 2014, 15: 166
21. Wang CM, Lo LC, Zhu ZY, Yue GH. A genome scan for quantitative trait loci affecting growth-related traits in an F1 family of Asian seabass (*Lates calcarifer*). BMC Genomics, 2006, 7: 274
22. Wang CM, Lo LC, Feng F, Zhu ZY, Yue GH. Identification and verification of QTL associated with growth traits in two genetic backgrounds of Barramundi (*Lates calcarifer*). Anim Genet, 2008, 39: 34–39
23. O'malley KG, Sakamoto T, Danzmann RG, Ferguson MM. Quantitative trait loci for spawning date and body weight in rainbow

trout: testing for conserved effects across ancestrally duplicated chromosomes. J Hered, 2003, 94: 273–284

24. Wringe BF, Devlin RH, Ferguson MM, Moghadam HK, Sakhrani D, Danzmann RG. Growth-related quantitative trait loci in domestic and wild rainbow trout (*Oncorhynchus mykiss*). BMC Genet, 2010, 11: 63
25. Reid DP, Szanto A, Glebe B, Danzmann RG, Ferguson MM. QTL for body weight and condition factor in Atlantic salmon (*Salmo salar*): comparative analysis with rainbow trout (*Oncorhynchus mykiss*) and Arctic charr (*Salvelinus alpinus*). Heredity, 2004, 94: 166–172
26. Houston RD, Bishop SC, Hamilton A, Guy DR, Tinch AE, Taggart JB, Derayat A, McAndrew BJ, Haley CS. Detection of QTL affecting harvest traits in a commercial Atlantic salmon population. Anim Genet, 2009, 40: 753–755
27. Gutierrez AP, Lubieniecki KP, Davidson EA, Lien S, Kent MP, Fukui S, Withler RE, Swift B, Davidson WS. (2012) Genetic mapping of quantitative trait loci (QTL) for body-weight in Atlantic salmon (*Salmo salar*) using a 6.5 5 K SNP array. Aquaculture, 2012, 358: 61–70
28. Cnaani A, Hallerman EM, Ron M, Weller JI, Indelman M, Kashi Y, Gall GAE, Hulata G. Detection of a chromosomal region with two quantitative trait loci, affecting cold tolerance and fish size, in an F2 tilapia hybrid. Aquaculture, 2003, 223: 117–128
29. Cnaani A, Zilberman N, Tinman S, Hulata G, Ron M. Genome-scan analysis for quantitative trait loci in an F2 tilapia hybrid. Mol Genet Genomics, 2004, 272: 162–172
30. Tao WJ, Boulding EG. Associations between single nucleotide polymorphisms in candidate genes and growth rate in Arctic charr (*Salvelinus alpinus* L.). Heredity, 2003, 91:60–69
31. Li XH, Bai JJ, Ye X, Hu YC, Li SJ, Yu LY. Polymorphisms in the 5′ flanking region of the insulin-like growth factor I gene are associated with growth traits in largemouth bass *Micropterus salmoides*. Fish Sci, 2009, 75: 351–358
32. Sun YH, Yu X, Tong J. Polymorphisms in myostatin gene and associations with growth traits in common carp. Int J Mol Sci, 2012,

13: 14956–14961

33. Liu L, Yu X, Tong J. Molecular characterization of myostatin (*MSTN*) gene and association analysis with growth traits in the bighead carp (*Aristichthys nobilis*). Mol Biol Rep, 2012, 39: 9211–9221
34. Rodriguez MF, LaPatra S, Williams S, Famula T, May B. Genetic markers associated with resistance to infectious hematopoietic necrosis in rainbow and steelhead trout (*Oncorhynchus mykiss*) backcrosses. Aquaculture, 2004, 241:93–115
35. Ozaki A, Sakamoto T, Khoo S, Nakamura K, Coimbra MR, Akutsu T, Okamoto N. Quantitative trait loci (QTLs) associated with resistance/susceptibility to infectious pancreatic necrosis virus (IPNV) in rainbow trout (*Oncorhynchus mykiss*). Mol Genet Genomics, 2001, 265: 23–31
36. Baerwald MR, Petersen JL, Hedrick RP, Schisler GJ, May B. A major effect quantitative trait locus for whirling disease resistance identified in rainbow trout (*Oncorhynchus mykiss*). Heredity, 2011, 106: 920–926
37. Moen T, Baranski M, Sonesson AK, Kjøglum S. Confirmation and fine-mapping of a major QTL for resistance to infectious pancreatic necrosis in Atlantic salmon (*Salmo salar*): population-level associations between markers and trait. BMC Genomics, 2009, 10: 368
38. Houston RD, Guy DR, Hamilton A, Ralph J, Spreckley N, Taggart JB, McAndrew BJ, Haley CS, Bishop SC. Mapping QTL affecting resistance to infectious pancreatic necrosis (IPN) in Atlantic salmon (*Salmo salar*). Aquaculture, 2007, 272: S269
39. Houston RD, Haley CS, Hamilton A, Guy DR, Tinch AE, Taggart JB, McAndrew BJ, Bishop SC. Major quantitative trait loci affect resistance to infectious pancreatic necrosis in Atlantic salmon (*Salmo salar*). Genetics, 2008, 178, 1109–1115
40. Houston RD, Haley CS, Hamilton A, Guy DR, Mota-Velasco JC, Gheyas AA, Tinch AE, Taggart JB, Bron JE, Starkey WG, McAndrew BJ, Verner-Jeffreys DW, Paley RK, Rimmer GSE, Tew IJ, Bishop SC. The susceptibility of Atlantic salmon fry to freshwater infectious pancreatic necrosis is largely explained by a major QTL. Heredity, 2009, 105: 318–327

41. Gheyas AA, Haley CS, Guy DR, Hamilton A, Tinch AE, Mota-Velasco JC, Woolliams JA. Effect of a major QTL affecting IPN resistance on production traits in Atlantic salmon. Anim Genet, 2010, 41: 666–668

42. Fuji K, Kobayashi K, Hasegawa O, Coimbra MRM, Sakamoto T, Okamoto N. Identification of a single major genetic locus controlling the resistance to lymphocystis disease in Japanese flounder (*Paralichthys olivaceus*). Aquaculture, 2006, 254: 203–210

43. Fuji K, Hasegawa O, Honda K, Kumasaka K, Sakamoto T, Okamoto N. Marker-assisted breeding of a lymphocystis disease-resistant Japanese flounder (*Paralichthys olivaceus*). Aquaculture, 2007, 272: 291–295

44. Palti Y, Nichols KM, Waller KI, Parsons JE, Thorgaard GH. Association between DNA polymorphisms tightly linked to MHC class II genes and IHN virus resistance in backcrosses of rainbow and cutthroat trout. Aquaculture, 2001, 194: 283–389

45. Zhang YX, Chen SL, Liu YG, Sha ZX, Liu ZJ. Major histocompatibility complex class IIB allele polymorphism and its association with resistance/susceptibility to *Vibrio anguillarum* in Japanese flounder (*Paralichthys olivaceus*). Mar Biotechnol, 2006, 8: 600–610

46. Xu TJ, Chen SL, Zhang YX. MHC class IIα gene polymorphism and its association with resistance/susceptibility to *Vibrio anguillarum* in Japanese flounder (*Paralichthys olivaceus*). Dev Comp Immunol, 2010, 34: 1042–1050

47. Nkrumah JD, Sherman EL, Li C, Marques E, Crews Jr DH, Bartusiak R, Murdoch B, Wang Z, Basarab JA, Moore SS. Primary genome scan to identify putative quantitative trait loci for feedlot growth rate, feed intake, and feed efficiency of beef cattle. J Anim Sci, 2007, 85: 3170–3181

48. Houston RD, Haley CS, Archibald AL, Rance KA. A QTL affecting daily feed intake maps to Chromosome 2 in pigs. Mamm Genome, 2005, 16: 464–470

49. Van Kaam JB, Groenen MA, Bovenhuis H, Veenendaal A, Vereijken AL, Van Arendonk JA. Whole genome scan in chickens for quantitative trait loci affecting growth and feed efficiency. Poultry

Sci, 1999, 78: 15–23

50. Sherman EL, Nkrumah JD, Murdoch BM, Li C, Wang Z, Fu A, Moore SS. Polymorphisms and haplotypes in the bovine neuropeptide Y, growth hormone receptor, ghrelin, insulin-like growth factor 2, and uncoupling proteins 2 and 3 genes and their associations with measures of growth, performance, feed efficiency, and carcass merit in beef cattle. J Anim Sci, 2008, 86: 1–16
51. Liu ZJ. Gene mapping, marker-assisted selection, gene cloning, genetic engineering and integrated genetic improvement programs at Auburn University. In: Gupta MV, Acosta BO, eds. Fish Genetics Research in Member Countries and Institutions of the International Network on Genetics in Aquaculture, ICLARM Conf Proc, 2001, 64: 179
52. Zimmerman AM, Wheeler PA, Ristow SS, Thorgaard GH. Composite interval mapping reveals three QTL associated with pyloric caeca number in rainbow trout, *Oncorhynchus mykiss*. Aquaculture, 2005, 247: 85–95
53. Wang XP, Zhang XF, Li WS, Zhang TQ, Li C, Sun XW. Mapping and genetic effect analysis on quantitative trait loci related to feed conversion ratio of common carp (*Cyprinus Carpio* L.) (in Chinese). Acta Hydrobiol Sin, 2012, 36: 177–196
54. Lee BY, Hulata G, Kocher TD. Two unlinked loci controlling the sex of blue tilapia (*Oreochromis aureus*). Heredity, 2004, 92: 543–549
55. Alfaqih MA, Brunelli JP, Drew RE, Thorgaard GH. Mapping of five candidate sex-determining loci in rainbow trout (*Oncorhynchus mykiss*). BMC Genet, 2009, 10: 2
56. Davidson WS, Huang TK, Fujiki K, von Schalburg KR, Koop BF. The sex determining loci and sex chromosomes in the family Salmonidae. Sex Dev, 2009, 3: 78–87
57. Cnaani A, Lee BY, Zilberman N, Ozouf-Costaz C, Hulata G, Ron M, D'Hont A, Baroiller JF, D'Cotta H, Penman DJ, Tomasino E, Coutanceau JP, Pepey E, Shirak A, Kocher TD. Genetics of sex determination in tilapiine species. Sex Dev, 2008, 2: 43–54
58. Eshel O, Shirak A, Weller JI, Slossman T, Hulata G, Cnaani A, Ron M. Fine-mapping of a locus on linkage group 23 for sex determination in Nile tilapia (*Oreochromis niloticus*). Anim Genet,

2011, 42: 222–224

59. Eshel O, Shirak A, Weller JI, Hulata G, Ron M. Linkage and physical mapping of sex region on LG23 of Nile tilapia (*Oreochromis niloticus*). G 3: Genes-Genom-Genet, 2012, 2: 35–42

60. Sun YL, Jiang DN, Zeng S, Hu CJ, Ye K, Yang C, Yang SJ, Li MH, Wang DS. Screening and characterization of sex-linked DNA markers and marker-assisted selection in the Nile tilapia (*Oreochromis niloticus*). Aquaculture, 2014, 433: 19–27

61. Sakamoto T, Danzmann RG, Gharbi K, Howard P, Ozaki A, Khoo SK, Woram RA, Okamoto N, Ferguson MM, Holm L, Guyomard R, Hoyheim B. A microsatellite linkage map of rainbow trout (*Oncorhynchus mykiss*) characterized by large sex-specific differences in recombination rates. Genetics, 2000, 155: 1331–1345

62. Perry GML, Ferguson MM, Sakamoto T, Danzmann RG. Sex-linked quantitative trait loci for thermotolerance and length in the rainbow trout. J Hered, 2005, 96: 97–107

63. Moghadam HK, Poissant J, Fotherby H, Haidle L, Ferguson MM, Danzmann RG. Quantitative trait loci for body weight, condition factor and age at sexual maturation in Arctic charr (*Salvelinus alpinus*): comparative analysis with rainbow trout (*Oncorhynchus mykiss*) and Atlantic salmon (*Salmo salar*). Mol Genet Genomics, 2007, 277: 647–661

64. Gharbi K, Gautier A, Danzmann RG, Gharbi S, Sakamoto T, Høyheim B, Taggart JB, Cairney M, Powell R, Krieg F, Okamoto N, Ferguson MM, Holm L, Guyomard R. A linkage map for brown trout (*Salmo trutta*): chromosome homeologies and comparative genome organization with other salmonid fish. Genetics, 2006, 172: 2405–2419

65. Gilbey J, Verspoor E, McLay A, Houlihan D. A microsatellite linkage map for Atlantic salmon (*Salmo salar*). Anim Genet, 2004, 35: 98–105

66. Woram RA, Gharbi K, Sakamoto T, Hoyheim B, Holm L, Naish K, McGowan C, Ferguson MM, Phillips RB, Stein J, Guyomard R, Cairney M, Taggart JB, Powell R, Davidson W, Danzmann RG. Comparative genome analysis of the primary sex-determining locus in salmonid fishes. Genome Res, 2003, 13: 272–280

67. Loukovitis D, Sarropoulou E, Tsigenopoulos CS, Batargias C, Magoulas A, Apostolidis AP, Chatziplis D, Kotoulas G. Quantitative trait loci involved in sex determination and body growth in the gilthead sea bream (*Sparus aurata* L.) through targeted genome scan. PLoS One, 2011, 6: e16599
68. Martínez P, Bouza C, Hermida M, Fernández J, Toro MA, Vera M, Pardo B, Millán A, Fernández C, Vilas R, Viñas A, Sánchez L, Felip A, Piferrer F, Ferreiro I, Cabaleiro S. Identification of the major sex-determining region of turbot (*Scophthalmus maximus*). Genetics, 2009, 183: 1443–1452
69. Viñas A, Taboada X, Vale L, Robledo D, Hermida M, Vera M, Martínez P. Mapping of DNA sex-specific markers and genes related to sex differentiation in turbot (*Scophthalmus maximus*). Mar Biotechnol, 2012, 14: 655–663
70. Chen S, Zhang G, Shao C, Huang Q, Liu G, Zhang P, Song W, An N, Chalopin D, Volff J, Hong Y, Li Q, Sha Z, Zhou H, Xie M, Yu Q, Liu Y, Xiang H, Wang N, Wu K, Yang C, Zhou Q, Liao X, Yang L, Hu Q, Zhang J, Meng L, Jin L, Tian Y, Lian J, Yang J, Miao G, Liu S, Liang Z, Yan F, Li Y, Sun B, Zhang H, Zhang J, Zhu Y, Du M, Zhao Y, Schartl M, Tang Q, Wang J. Whole-genome sequence of a flatfish provides insights into ZW sex chromosome evolution and adaptation to a benthic lifestyle. Nat Genet, 2014, 46: 253–260
71. Barrett RD, Paccard A, Healy TM, Bergek S, Schulte PM, Schluter D, Rogers SM. Rapid evolution of cold tolerance in stickleback. Proc Biol Sci, 2011, 278: 233–238
72. Moen T, Agresti JJ, Cnaani A, Moses H, Famula TR, Hulata G, Gall GAE, May B. A genome scan of a four-way tilapia cross supports the existence of a quantitative trait locus for cold tolerance on linkage group 23. Aquac Res, 2004, 35: 893–904
73. Jackson TR, Ferguson MM, Danzmann RG, Fishback AG, Ihssen PE, O'Connell M, Crease TJ. Identification of two QTL influencing upper temperature tolerance in three rainbow trout (*Oncorhynchus mykiss*) half-sib families. Heredity, 1998, 80: 143–151
74. Danzmann RG, Jackson TR, Ferguson MM. Epistasis in allelic expression at upper temperature tolerance QTL in rainbow trout. Aquaculture, 1999, 173: 45–58

75. Perry GML, Danzmann RG, Ferguson MM, Gibson JP. Quantitative trait loci for upper thermal tolerance in outbred strains of rainbow trout (*Oncorhynchus mykiss*). Heredity, 2001, 86: 333–341
76. Sun XW, Liang LQ. A genetic linkage map of common carp (*Cyprinus carpio* L.) and mapping of a locus associated with cold tolerance. Aquaculture, 2004, 238: 165–172
77. Rengmark AH, Slettan A, Lee WJ, Lie ø, Lingaas F. Identification and mapping of genes associated with salt tolerance in tilapia. J Fish Biol, 2007, 71 (sc): 409–422
78. Norman JD, Robinson M, Glebe B, Ferguson MM, Danzmann RG. Genomic arrangement of salinity tolerance QTLs in salmonids: a comparative analysis of Atlantic salmon (*Salmo salar*) with Arctic charr (*Salvelinus alpinus*) and rainbow trout (*Oncorhynchus mykiss*). BMC Genomics, 2012, 13: 420
79. Zhai YH, Zhou L, Wang Y, Wang ZW, Li Z, Zhang XJ, Gui JF. Proliferation and resistance difference of a liver-parasitized myxosporean in two different gynogenetic clones of gibel carp. Parasitol Res, 2014, 113: 1331–1341
80. Xu TJ, Chen SL, Ji XS, Tian YS. MHC polymorphism and disease resistance to *Vibrio anguillarum* in 12 selective Japanese flounder (*Paralichthys olivaceus*) families. Fish Shellfish Immun, 2008, 25: 213–221
81. Wang L, Fan C, Liu Y, Zhang Y, Liu S, Sun D, Deng H, Xu Y, Tian Y, Liao X, Xie M, Li W, Chen S. A genome scan for quantitative trait loci associated with *Vibrio anguillarum*infection resistance in Japanese flounder (*Paralichthys olivaceus*) by bulked segregant analysis. Mar Biotechnol, 2014, 16: 513–521
82. Wang D, Mao HL, Chen HX, Liu HQ, Gui JF. Isolation of Y- and X-linked SCAR markers in yellow catfish and application in the production of all-male populations. Anim Genet, 2009, 40: 978–981
83. Liu HQ, Guan B, Xu J, Hou C, Tian H, Chen H. Genetic manipulation of sex ratio for the large-scale breeding of YY super-male and XY all-male yellow catfish (*Pelteobagrus fulvidraco* (Richardson)). Mar Biotechnol, 2013, 15: 321–328
84. Dan C, Mei J, Wang D, Gui JF. Genetic differentiation and efficient sex-specific marker development of a pair of Y- and X-linked

markers in yellow catfish. Int J Biol Sci, 2013, 9:1043–1049

85. Che SL, Li J, Deng SP, Tian YS, Wang QY, Zhuang ZM, Sha ZX, Xu JY. Isolation of female-specific AFLP markers and molecular identification of genetic sex in half-smooth tongue sole (*Cynoglossus semilaevis*). Mar Biotechnol, 2007, 9: 273–280
86. Chen SL, Ji XS, Shao CW, Li WL, Yang JF, Liang Z, Liao XL, Xu GB, Xu Y, Song WT. Induction of mitogynogenetic diploids and identification of WW super-female using sex-specific SSR markers in half-smooth tongue sole (*Cynoglossus semilaevis*). Mar Biotechnol, 2012, 14: 120–128
87. Xu P, Zhang X, Wang X, Li J, Liu G, Kuang Y, Xu J, Zheng X, Ren L, Wang G, Zhang Y, Huo L, Zhao Z, Cao D, Lu C, Li C, Zhou Y, Liu Z, Fan Z, Shan G, Li X, Wu S, Song L, Hou G, Jiang Y, Jeney Z, Yu D, Wang L, Shao C, Song L, Sun J, Ji P, Wang J, Li Q, Xu L, Sun F, Feng J, Wang C, Wang S, Wang B, Li Y, Zhu Y, Xue W, Zhao L, Wang J, Gu Y, Lv W, Wu K, Xiao J, Wu J, Zhang Z, Yu J, Sun X. Genome sequence and genetic diversity of the common carp, *Cyprinus carpio*. Nat Genet, 2014, 46: 1212–1219
88. Ron M, Weller JI. From QTL to QTN identification in livestock—winning by points rather than knock-out: a review. Anim Genet, 2007, 38: 429–439
89. Remington DL, Thornsberry JM, Matsuoka Y, Wilson LM, Whitt SR, Doebley J, Kresovich S, Goodman MM, Buckler ES. Structure of linkage disequilibrium and phenotypic associations in the maize genome. Proc Natl Acad Sci USA, 2001, 98: 11479–11484
90. De-Santis C, Jerry DR. Candidate growth genes in finfish—Where should we be looking? Aquaculture, 2007, 272: 22–38
91. Lee BY, Penman DJ, Kocher TD. Identification of a sex — determining region in Nile tilapia (*Oreochromis niloticus*) using bulked segregant analysis. Anim Genet, 2003, 34: 379–383
92. Moose SP, Mumm RH. Molecular plant breeding as the foundation for 21st century crop improvement. Plant Physiol, 2008, 147: 969–977

CHAPTER

10

MOLECULAR BASIS AND GENETIC IMPROVEMENT OF ECONOMICALLY IMPORTANT TRAITS IN AQUACULTURE ANIMALS

JianFang Gui and ZuoYan Zhu

State Key Laboratory of Freshwater Ecology and Biotechnology, Institute of Hydrobiology, Chinese Academy of Sciences, Wuhan 430072, China

ABSTACT

Aquaculture has been believed to be a major Chinese contribution to the world. In recent 20 years, genome and other genetic technologies have promoted significant advances in basic studies on molecular basis and genetic improvement of aquaculture animals, and complete genomes of some main aquaculture animals have been sequenced or announced to be sequenced since the beginning of this century. Here, we review some significant breakthrough progress of aquaculture genetic improvement technologies including genome technologies, somatic cell nuclear transfer and stem cell technologies, outline the molecular basis of several economically important traits including reproduction, sex, growth, disease resistance, cold tolerance and hypoxia tolerance, and present a series of candidate trait-related genes. Finally, some application cases of genetic improvement are introduced in aquaculture animals, especially in China, and several development trends are highlighted in the near future.

Keywords: Aquaculture genome technology somatic cell nuclear transfer stem cell technology reproduction-related gene sex determination gene growth-related gene disease resistance-related genes cold tolerance-related gene hypoxia tolerance-related gene genetic improvement

INTRODUCTION

Aquaculture has a history of about 4000 years, and China is the earliest country for raising fish in fishponds. However, it is only in the recent 50 years that aquaculture has rapidly developed and become a true production industry because artificial propagation technique has been successfully established since the early 1960s for the "four important domestic fish" including silver carp, bighead carp, grass carp and black carp [1] and aquaculture biotechnology for genetic breeding and disease control has been quickly developed since the 1980s [2]. In China, the annual aquaculture production capacity has increased to 38.3 million tons in 2010 from less than 1 million tons in 1960s, and possesses the fastest growing rate among agriculture and food industry. The aquaculture proportion has risen to more than 71.3% of the food fish, and reaches about 70% of the world's aquaculture production. Aquaculture industry has become a major power to promote sustainable, rapid and stable development of China's fishery, and played an important role in guaranteeing supply, stabilizing the market, ensuring national food security and promoting trade development. As an optimal method for exchanging animal proteins by minimum plant proteins, aquaculture has been believed as the most efficient technology by some authoritative economists, which is a major Chinese contribution to the world.

Just entering the 21st century, sustainable fisheries have become a common theme for the world, and the effect and contribution of aquaculture on world fish supplies have been widely recognized by many scientists in the developed countries[3,4]. Moreover, fish and other aquatic products as safer food sources have been well identified with most of the people [5]. At the same time, China's aquaculture is in a crucial change stage from traditional farming to modern industry, in which the severely lagged basic research has become the outstanding issue for restricting aquaculture development. Especially, the improved varieties were very few, and major aquaculture fish and shellfish were still de-pendent on the wild species. The frequent incidences of infectious diseases led to serious economic losses of about 15 billion RMB annually, and the unreasonable responses and non-standard medication to the diseases further resulted in drug residues, which affected quality and food safety of aquatic products, and in turn restricted the sustainable development of aquaculture [6]. In order to solve the above problems, the National Key Basic Research Program of China (973 Program) has successively supported 7 projects since 1999, such as "Fundamental research on the disease occurrence and disease resistance of the commercially important organisms in mariculture, "Studies on genetic and developmental biological basis for improvement of important aquaculture fishes", "Studies on the disease outbreak and immunological control of commercially important organisms in mariculture", "Investigation of crucial scientific issues in intensive aquaculture of freshwater ponds in China", "Basic studies on functional genomics and molecular design breeding in important aquaculture fishes", "The molecular basis for the important economic traits of the mollusks and the study on the molecular design breeding", and "Basic studies on outbreak mechanism and immunological control of main virus diseases in marine aquaculture animals", which have promoted significant advances in basic studies on genetic improvement and disease control of aquaculture animals. Especially in basic studies on genetic improvement in the aquaculture performance, the molecular basis of several economically important traits including reproduction, sex, growth, disease resistance, cold tolerance, and hypoxia tolerance have begun to elucidate, although these traits are highly complex and problematic. Here, we will attempt to retrace and review some significant breakthrough progress in the field for the recent 10 years, especially the major achievements obtained by Chinese scientists, and attempt to forecast developmental prospects for the future.

GENETIC IMPROVEMENT TECHNOLOGIES OF AQUACULTURE ANIMALS

Aquaculture Animals in Which Complete Genomes Have Been Sequenced

As complete genomes have been sequenced and released from human, model animals and plants, as well as several model fish including zebrafish (Danio rerio), fugu (Fugu rubripes), Tetraodon (Tetraodon nigroviridis), medaka (Oryzias latipes), and three-spined stickleback (Gasterosteus aculeatus) since this century, the genomes of some important aquaculture fish, such as channel catfish (Ictalurus punctatus), Nile tilapia (Oreochromis niloticus), Atlantic salmon (Salmo salar), rainbow trout (Oncorhynchus mykiss), gilthead seabream (Sparus aurata), European seabass (Dicentrarchus labrax), and Atlantic cod (Gadus morhua), have been also sequenced [7,8]. From 2010, the complete genomes of some main aquaculture animals including tongue sole (Cynoglossus semilaevis), Pacific oyster (Crassostrea gigas), large yellow croaker (Pseudosciaena crocea), orange-spotted grouper (Epinephelus coioides), common carp (Cyprinus carpio) and Japanese flounder (Paralichthys olivaceu) have been announced to be sequenced by Chinese fisheries-related institutes (Table 1). In 2011, Institute of Hydrobiology, Chinese Academy of Sciences and the collaborators have also started to sequence the complete genomes of grass carp (Ctenopharyngodon idellus), silver carp (Hypophthalmichthys molitrix) and bighead carp (Aristichthys nobilis) of the "four important domestic fish", and small RNA transcriptomes of bighead and silver carp have been characterized [9]. Therefore, we are convinced that the complete genome bioinformatics and the detailed molecule anatomy will provide significantly referent and even directive knowledge for their trait improvement and disease control of these major aquaculture animals.

Genome Technologies of Aquaculture Animals

In 2007, Prof. Zhanjiang Liu at Auburn University of USA, together with more than 40 aquaculture genomics experts in the world, edited a highly technical and comprehensive book Aquaculture Genome Technologies[10]. According to the book, the major aquaculture genome technologies include DNA marker technologies, such as amplified fragment length polymorphism (AFLP), microsatellites and single nucleotide polymorphism, novel DNA sequencing technologies, such as the SOLiD sequencing platform, the Solexa sequencing platform and the 454 sequencing platform, gene discovery technologies of expressed sequence tags and de novo sequencing of whole transcriptomes, genome mapping technologies, such as genetic linkage mapping, QTL mapping, radiation hybrid mapping and BACbased physical mapping, and genome expression analysis technologies, such as microarray technology and sequence tag-based technology.

Recently, numerous polymorphic DNA markers for aquaculture animals have been developed and significant advances have been made in China. In polyploid gibel carp (Carassius auratus gibelio), several significant molecular markers including transferrin alleles [11], SCAR [12], microsatellites [13], AFLP [14] and mitochondria DNA sequences [15,16] were identified, and dual modes of unisexual gynogenesis and sexual reproduction were revealed by these markers[17,18]. Significantly, the unique dual reproduction mode, as commented by Avise (2008), was the first documented instance in vertebrates [19], and the discovery firstly explained reproduction mechanism of high genetic diversity and an evolutionarily long-lived timescale of unisexual species [18]. In common carp (Cyprinus carpio), along with advances of polymorphic markers and genome sequencing, various density linkage maps were constructed [20–23], and used to locate cold tolerance [20] and muscle fiber-related [22] QTLs. A physical map was constructed from a BAC library containing a total of 92160 BAC clones with an average insert size of 141 kb [24,25]. In addition, a consensus linkage map was also constructed in grass carp [26]. As a required approach for positional gene cloning, physical mapping, and genome sequencing, several BAC libraries have been developed and characterized in other important aquaculture animals of China, such as gibel carp [27], zhikong scallop (Chlamys farreri)[28], pacific white shrimp (Litopenaeus vannamei) [29], grass carp [30], and halfsmooth tongue sole (Cynoglossus semilaevis)[31].

Table 1: Aquaculture animals that have been announced to be completely sequenced in China

Species	Genome size	Predicted gene number	Mainly announced institute	Announced time
Tongue sole (*Cynoglossus semilaevis*)	520 Mb	More than 20000	Yellow Sea Fisheries Research Institute, Chinese Academy of Fishery Sciences	August 2010
Pacific oyster (*Crassostrea gigas*)	8 Gb	More than 20000	Institute of Oceanology, Chinese Academy of Sciences	August 2010
Large yellow croaker (*Pseudosciaena crocea*)	750 Mb	Data not given	Zhejiang Ocean University	January 2011
Orange-spotted grouper (*Epinephelus coioides*)	1.1 Gb	More than 20000	Sun Yat-Sen University	March 2011
Common carp (*Cyprinus carpio*)	1.7 Gb	Data not given	The Centre for Applied Aquatic Genomics, Chinese Academy of Fishery Sciences	May 2011
Japanese flounder (*Paralichthys olivaceu*)	620 Mb	About 23000	Yellow Sea Fisheries Research Institute. Chinese Academy of Fishery Sciences	December 2011

Somatic Cell Nuclear Transfer and Stem Cell Technologies in Fish

Nuclear transfer has been successfully applied to fish genetic breeding for about half a century in China [32], and their developmental potentiality and reprogramming have been observed from the transplanted nuclei[33,34]. During this century, the cloned zebrafish were obtained by nuclear transfer from long-term-cultured cells [35], and the cytoplasmic impact was analyzed in cross-genus cloned fish derived from transgenic common carp (Cyprinus carpio) nuclei and goldfish (Carassius auratus) enucleated eggs [36]. Moreover, some differentially expressed genes were identified from the cloned zebrafish embryos at the dome stage [37], and the critical developmental stages for high integration efficiency were determined in zebrafish [38,39].

Stem cell culture and application have been rapidly developed in model fishes of medaka and zebrafish [40,41], and some stem cell technologies, such as gene targeting, germ cell transplantation and semi-cloning by nuclear transfer, have been successfully formed because of specific potentials of fish germ stem cells and embryonic stem cells [42,43]. Especially, Yi et al. [44] used the semi-cloning technology to create the semi-cloned female medaka which shows normal fertility and germline transmission over three generations. This approach represents a significant development for germline transmission of cultured cells, and will be a potential reproduction technology to combat infertility in vertebrates including human. Additionally, primordial germ cells (PGCs) and spermatogonial stem cells could be transplanted between two different salmonid species, and resulted in production of donor-derived offspring [45–47]. In combination with natural and artificial polyploid fishes with special reproduction modes [18,48], the surrogate approach might have significant values for producing polyploidy stocks in aquaculture or for restoring an endangered species in wild life conservation.

MOLECULAR BASES FOR ECONOMIC TRAITS OF AQUACULTURE ANIMALS

Reproduction Trait and Candidate Reproduction related Genes

In comparison with mammals, fish reproduction modes and strategies exhibit abundant diversities and extreme particularities[49,50]. Owing to the diversities and particularities, much attention has begun to be paid to some important fish with special reproduction modes. Gibel carp, because of the polyploid background and dual reproduction modes of gynogenesis and sexuality, has become a promising fish for screening and isolating reproduction-related genes that are involved in primordial germ cell formation, oocyte maturation and egg fertilization. Using suppression subtractive hybridization and protein purification [51,52], some important genes, such as germ plasm markers vasa [53] and Dazl [54], oocyte maturation-related factors cyclin A2 [55], C-type lectin [52], C1q-like [56–58], histone H2A variant h2af1o [59], and maternal-effect factor spindlin [60], have been identified and functionally characterized from gibel carp. For example, Sun et al. [61] have observed oocytespecific expression pattern and dynamic distribution of Spindlin during oocyte maturation and egg fertilization, demonstrated its association and interaction with β-tubulin and spindle, and found that its elimination destroys spindle assembly, and thereby disturbs the oocyte-to-embryo transition and first cleavage, which provided the first direct evidence for the critical oocyte-to-embryo transition function of Spindlin in vertebrates.

It is well known that fish reproduction is regulated and controlled by a cascade of hormones along the hypothalamus-pituitary-gonad (HPG) axis. In this cascade, gonadotropin-releasing hormones (GnRHs) produced by hypothalamus are found to be the primary neuropeptides, and gonadotropic hormones (GtHs) synthesized and released by pituitary are found to lead ovaries or testicles to produce mature eggs or sperms, whereas dopamine (DA) produced by pituitary has been found to exhibit an inhibiting effect on the functions of GnRHs[62–64]. Multiple GtHs have been reported by Chinese scientists in groupers[65] and Chinese sturgeon (Acipenser sinensis) [66]. In common carp, two GnRH variants have been isolated [67]. Significantly, when common carp sGnRH expression was inhibited by sGnRH antisense, the plasma GtH-II level was significantly reduced, and the gonad development was severely diminished or hampered [68]. Moreover, Xu et al. [69] used the gonad hampered transgenic fish as a model for defining global gene expression changes of the hypothalamic-pituitarygonadal axis, and revealed 9, 28, and 212 differentially expressed genes respectively in hypothalamus, pituitary, and ovary, of which 87 genes were novel. Obviously, this study not only indicated the progressive scaling-up effect of hypothalamic sGnRH antisense acting on the pituitary and ovary, but also provided new insights with respect to the identification of reproduction-related genes and the understanding of the molecular mechanisms and regulative pathways in fish reproduction system. Recently, Zhou et al. (unpublished data) have also conducted a detailed exploration and functional annotation of hypothalamus and pituitary transcriptomes of mature gibel carp by Roche 454 sequencing. By this way, 878430 high quality sequences (average sequence length of (480±50) bp) of hypothalamus and pituitary were assembled into 51111 contigs and 113046 singletons, resulting in 23012 unigenes. Moreover, RNA-Seq approach was applied to analyze the expression level difference of the 23012 unigenes between mature female and male hypothalamus and pituitary. The detailed report will be soon coming.

Sex Determination and Candidate Sex Determination Genes

Fishes are an attractive group of organisms for studies on vertebrate sex determination, because they include various types of sexuality from hermaphroditism to gonochorism and from environmental to genetic sex determination. Fishes can provide important insight into the plasticity of the sex determination process in vertebrates, and can be used to explore the cascade regulatory pathways and their evolutionary conservation of upstream or downstream function of sex determination genes [70,71]. In 2002, the DMY gene, a recent duplicate of DMRT1 (Drosophila doublesex and Caenorhabditis mab-3-related transcription factor 1) in the Japanese medaka, was identified as the first sex-determining gene among non-mammalian vertebrates [72], but it was quickly demonstrated that DMY is not the universal primary sex determining gene in fish [73]. Significantly, DMRT1, the first conserved zinc finger-like DNA binding domain (DM domain) gene identified in vertebrates, has also been shown to be intimately linked to male gonadogenesis and further differentiation among almost all studied fishes including gonochoristic and hermaphroditic species [74]. As a group of protogynous hermaphrodite fishes that the gonad development undergoes transition from ovary to intersexual gonad and then to testis, and cells at different stages during oogenesis and spermatogenesis can be synchronously observed in the transitional gonads, groupers have been used as a good model for mining sex determination or sex differentiation-related genes. Xia et al. [75] have found that grouper DMRT1 is specifically expressed in spermatogenic cells during female-to-male sex reversal, and the spermatogenic cell-specific and dynamic expression pattern was revealed to be basically similar in several sex reversal fishes including protogynous wrasse (Halichoeres tenuispinis)[76] and rice field eel (Monopterus albus) [77] as well as protandrous black porgy (Acanthopagrus schlegeli)[78,79] and gilthead seabream (Sparus aurata)[80].

SOX3 has been suggested to play significant roles in sex determination and differentiation in vertebrates, but the evidence is insufficient and controversial [81]. In grouper gonads, some special cysts synchronously contain primordial germ cells, oogonia and spermatogonia, and SOX3 is expressed very early and can be detected in the nuclei of some primordial germ cells along the germinal epithelium, which provide an excellent chance to examine the functional roles of SOX3. The detailed observation and analysis have indicated that when the SOX3-positive primordial germ cells develop toward oogonia and then oocytes, SOX3 expression continues to increase greatly, whereas, as the SOX3-positive primordial germ cells develop toward spermatogonia, SOX3 expression is stopped and expelled from the nuclei of spermatogonia. The data confirm that SOX3 is involved in oogonium differentiation and oocyte development [82].

In hermaphrodite red-spotted grouper and rice field eel [83,84], the gonad-specific aromatase (cyp19a1a) had been also demonstrated to be expressed by follicular cells of follicular layer around oocytes. The promoter activity analysis showed a number of potential binding sites for some transcriptional factors, such as SOX5, GATA gene family, CREB, AP1, FOXL1, C/EBP, ARE and SF-1, and most of the motifs are conserved among the studied fish species. Recently, a hypothesis proposes that the gonadal aromatase upregulation is necessary not only for triggering but also for maintaining ovarian differentiation, and the down-regulation is the only necessary step for inducing a testicular differentiation pathway [85].

Actually, several important genes, such as Fushi Tarazu factor-1 (FTZ-F1), Sry-related HMG-box gene 9 (SOX9), Wilms tumor gene 1 (WT1), anti-mullerian hormone (AMH), DMRT1 and the zinc finger transcription factor GATA binding proteins 4 (GATA4), have been supposed as candidate sex determination genes, and a hypothetical regulation hierarchy for sex determination and differentiation has been suggested in the hermaphroditic groupers[86]. In mammals, primary sex determination genes have been divided into 4 classes, in which class I comprises Y genes that have a male-specific effect required for spermatogenesis, class II are X genes that are expressed at a higher level in females than in males by virtue of the 2:1 ratio, class III are X genes that receive a parental imprint, and class IV are sex-specific regions of heterochromatin that exert effects on the epigenetic status of the rest of the genome [87]. Of course, sex determination mechanisms are also intricate in fishes, because fishes, like other aquatic animals of amphibians and reptiles, include male or female heterogametic genetic sex determination with or without strongly differentiated sex chromosomes [88,89]. A recent report indicates that sex-chromosome homomorphy in European tree frogs does not result from a recent turnover but is maintained over evolutionary timescales by occasional X-Y recombination, implicating that seemingly ever-young sex chromosomes may thus carry old-established sex-determining genes [90]. Moreover, B chromosomes in addition to standard chromosomes are also found to have a functional effect on sex determination in one cichlid species (Lithochromis rubripinnis) from Lake Victoria [91], which implies that a portion of sex chromosomes has been derived from B chromosomes. Along with the successive revelation of complete genome sequences in some main aquaculture animals and the quick development of next-generation sequencing techniques, more sex determination genes and the underlying mechanisms will be identified and clarified as performed previously in human and chimpanzee [92].

Growth Trait and Candidate Growth-Related Genes

Growth trait is one of the most valuable traits of interest in genetic improvement of aquaculture animals, because growth rate increase can directly reduce aquaculture costs and lead to the benefit increase. Since growth hormone (GH) gene was identified, many important growth-related genes have been demonstrated to play critical roles in growth regulation of vertebrates. Numerous candidate genes are correlated with growth, including main growth-related genes within the somatotropic axis, such as GH, growth hormone receptor (GHR), insulin-like growth factors (IGF-I and -II), growth hormone-releasing hormone (GHRH), Leptin, and growth hormone inhibiting hormone (GHIH or somatostatin), important transforming growth factor genes expressed in muscle tissues, such as myostatin (MSTN) and myogenic regulatory factors (MRFs), and other possible candidate genes have been recommended in fish through reviewing the livestock and fish literatures [93], and the regulative mechanisms of muscle growth have been explored by introducing different growth models including compensatory growth, genetic manipulation and in vitro tissue culture [94].

Recently, several important growth-related genes have been investigated by Chinese fish endocrinologists. For example, multiple somatostatin and somatostatin receptor genes were characterized in the Chinese sturgeon (Acipenser sinensis) [95] and in groupers [96–99]. Orangespotted grouper orexin was found to be involved in feeding and orexin-A was a stimulator of NPY mRNA expression in vivo and in vitro [100]. Zebrafish ghrelin was demonstrated to play an essential role in GH expression during adenohypophysis development [101]. Follistatin 1 (Fst1) overexpression was revealed to promote muscle growth in the transgenic zebrafish [102].

Disease Resistance Trait and Candidate Disease Resistance-Related Genes

Along with rapidly growing and highly intensifying in the past 30 years, many aquaculture species have been threatened by some serious diseases caused by viruses, bacteria, fungi, parasites or other undiagnosed and emerging pathogens. Especially in infectious virus diseases, a lot of pathogenic viruses were identified from freshwater and marine culture animals[103], and the complete genomes of several viruses, such as grass carp hemorrhage virus (GCHV)[104], shrimp white spot syndrome virus (WSSV) [105], mandarin fish infectious spleen and kidney necrosis iridovirus (ISKNV) [106], and flounder lymphocystis disease virus isolated in China (LCDV-C) [107], were sequenced in China. Moreover, their structure proteins were characterized [108–111], and some of them were functionally analyzed [112,113].

In order to identify innate antiviral genes that protect fish from some virus infections, suppressive subtractive hybridization technology has been used to isolate virally induced genes from UV-inactivated GCHV-infected CAB cells [114]. Through this model system, a series of interferon (IFN) induced genes have been cloned and characterized, and great progress has been made in identifying fish antidisease genes involved in IFN antiviral response [115–120]. Significantly, some novel interferon-inducible genes have been found [121], and their functional roles and signaling pathways have been well understood in fish [122–124]. Additionally, several candidate genes, such as PKR [125], PKR-like (or called PKZ) [126], β-Defensin [127], TRBP [128], C-type lectin [129], Enzyme E2 [130], MDA5 and LGP2 [131], have been revealed to play crucial roles in the antiviral defense response of fish and shrimp.

Cold Tolerance Trait and Candidate Cold Tolerance related Genes

Cold tolerance is an especially attractive trait in fish, and a recent experiment evidence suggests that some fishes might carry sufficient genetic variation to adapt to temperature and the cold tolerance should be under strong selection [132]. To identify candidate cold tolerance-related and cold change-related genes, a large-scale EST sequencing of multi-tissues including brain, liver, head kidney, and ovary had been performed in Antarctic notothenioid (Dissostichus mawsoni), and some important cold-specific expressed genes had been identified by comparative transcriptome analyses between D. mawsoni and non-notothenioid warm-water teleost fishes [133,134]. Moreover, the molecular origination of antifreeze protein III (AFP III) was demonstrated to come from a duplicate copy of sialic acid synthase (SAS) gene, supporting the evolutionary mechanism by neofunctionalization under escape from adaptive conflict [135].

Hypoxia Tolerance Trait and Candidate Hypoxia Tolerance-Related Genes

Variable hypoxia tolerance has been observed in aquatic animals including reptiles, amphibians, and fishes [136,137], but its molecular mechanisms remain largely unknown. Since Carassius species have been found to survive many days or even several months of anoxia at low temperature, Carassius auratus blastulae embryonic cells have been used to screen gene expression changes responding to hypoxia stress by suppression subtractive hybridization (SSH). Through this approach, a total of 211 differentially expressed genes are identified, in which most of them are reported for the first time from hypoxia-treated teleosts[138].

Moreover, several potential hypoxia tolerance-related genes have been characterized and functionally analyzed in fish. For example, heme oxygenase-1 (HO1) has been demonstrated to have protective role in fish under hypoxic stress [139]. Grass carp insulin-like growth factor-binding protein 1 (IGFBP1) has been revealed to play a key role in mediating the hypoxia-induced embryonic growth retardation and developmental delay [140]. A novel function of p53 that directly suppresses BNIP3 expression to protect against hypoxia-induced cell death has been found in model fish zebrafish [141]. Also in zebrafish, a stable transgenic fish has been created by expressing the Vitreoscilla hemoglobin (vhb) gene driven by common carp beta-actin promoter, and the expressed VHb can partially alleviate the hypoxia stress response to improve the survival rate has been confirmed by comparing the expression profiles of the vhb transgenic zebrafish [142].

APPLICATION OF GENETIC IMPROVEMENT IN AQUACULTURE ANIMALS

Genome technologies, especially molecular markers, have been designed to apply to breeding program and genetic improvement of main aquaculture animals including rainbow trout, Atlantic salmon, tilapia, channel catfish and ornamental fishes in the world, and their potential values have been evaluated by aquaculture geneticists[49,143–145].

In China, several nuclear genome markers, such as microsatellite patterns, AFLP profiles and transferrin alleles, and one cytoplasmic marker, the whole mitochondrial genome sequence, have been used to identify genetic organization of a novel clone A^+ created by sexual mating between clone D female and clone A male in polyploid gibel carp [18], which have demonstrated that the novel clone is a nucleo-cytoplasmic hybrid between the known clones A and D, because it contains an entire nuclear genome from the paternal clone A and a mtDNA genome (cytoplasm) from the maternal clone D [146]. Therefore, it is suggested to be formed via androgenesis of clone A sperm in clone D ooplasm. Significantly, the selected nucleo-cytoplasmic hybrid female still maintains its gynogenetic ability, and thereby produces a novel clone variety "CAS Ⅲ" that has been approved by National Certification Committee for Aquatic Varieties. Owing to the significant growth superiority, the novel clone variety has been propagated by gynogenesis in more than 10 hatcheries, and their offspring have been cultured throughout China [146]. Moreover, the finding will be of great significance for exploiting the genetic breeding approaches in gibel carp [18], because the nucleo-cytoplasmic hybrid clone escapes the genetic and developmental destruction caused by drastic treatments of irradiation and physical shocks in induced androgenesis and nuclear transplantation [146].

Additionally, sex-specific molecule markers have been developed and applied in China. Based on artificial propagation families, and gynogenetic XX, XY and YY individuals obtained previously [147], two pairs of Y chromosome- and X chromosome-specific markers were isolated from yellow catfish (Pelteobagrus fulvidraco) by AFLP screening, which thereby develops a Y- and X-specific allele marker-assisted sex control technique for mass production of all-male populations [148]. Using this technique, all-male yellow catfish has been approved as a novel variety "yellow catfish all-male No. 1" by National Certification Committee for Aquatic Varieties. Because yellow catfish males grow faster than females, and result in an ultimate size difference of about two to threefold, the all-male yellow catfish has been widely used for commercial production. Recently, all-female flounder has been also recognized as a novel variety for aquaculture in China. According to the developments, we have proposed an integration sex control technique for mass production of all-male populations in aquaculture fish [148]. As shown in Figure 1, the XY physiological females are selected by the Y chromosome-specific marker (YSM) and X chromosome-specific marker (XSM) from sex reversal progeny through 17α-ethinyloestradiol

(EE2) treatment, and about 25% YY super-males can be obtained from the mating progeny of the reversed XY female and normal XY male by YSM and XSM identification. Then, the YY physiological females can be induced from the EE2 treated YY fingerlings, and YY super-males can be produced continually from the mating of the YY physiological females and YY super-males. Subsequently, the YY super-males can be used to mate with XX females for commercial mass production of all-XY males.

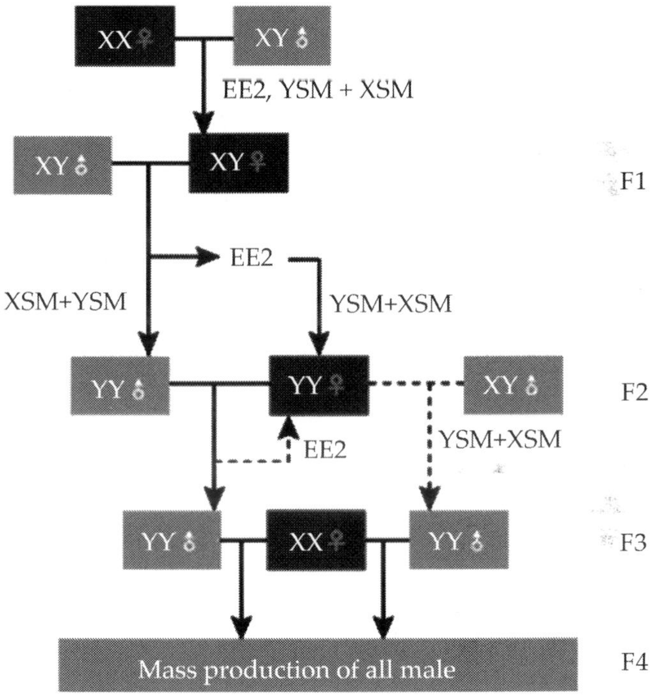

Figure 1: A schematic diagram of mass production of all-male populations through an integration sex control technique of female hormone 17α-ethinyloestradiol (EE2) treatment and genetic identification of Y chromosome-specific marker (YSM) and X chromosome-specific marker (XSM) in aquaculture fish.

CONCLUSION AND PERSPECTIVE

Along with complete genome sequence anatomy of some main aquaculture animals, ongoing development of genome technologies, somatic cell nuclear transfer and stem cell technologies, as well as identification and function analysis of candidate genes relative to economically important traits including reproduction, sex, growth, disease resistance, cold tolerance and hypoxia tolerance, genetic improvement research in aquaculture animals has begun to enter a new era for molecular design breeding. The future issues are mainly aimed at understanding gene regulation and control networks that are related to reproduction, growth, disease resistance, cold tolerance and hypoxia tolerance, at identifying the relative functional genes and molecular markers, and at creating novel breeding materials and technologies, which can be used to develop the theoretical strategies and feasible approaches for molecular design breeding in aquaculture animals. It will establish the theoretical and technical foundation for breeding new varieties with high yield and quality, and will make greater contribution for sustainable fisheries and biotechnological innovation.

REFERENCES

1. Wu H W, Zhong L. Progress and achievement on artificial proliferation of grass carp, black carp, silver carp and big head carp in China. Chin Sci Bull, 1964, 9: 900–907
2. Wu C J, Gui J F. Fish Genetics and Breeding Engineering (in Chinese). Shanghai: Shanghai Scientific and Technical Publishers, 1999
3. Naylor R L, Goldburg R J, Primavera J H, et al. Effect of aquaculture on world fish supplies. Nature, 2000, 405: 1017–1024
4. Pauly D, Christensen V, Guénette S, et al. Towards sustainability in world fisheries. Nature, 2002, 418: 689–695
5. James H T, Geoff L A. Fishes as food: aquaculture's contribution. EMBO Rep, 2001, 21: 958–963
6. Gui J F. The present and future of studies on genetic and developmental biological basis for improvement of important aquaculture fishes. Life Sci, 2005, 17: 112–118
7. Sarropoulou E, Nousdili D, Magoulas A, et al. Linking the genomes of nonmodel teleosts through comparative genomics. Mar Biotechnol, 10: 227–233
8. Star B, Nederbragt A J, Jentoft S, et al. The genome sequence of

Atlantic cod reveals a unique immune system. Nature, 2011, 477: 207–210

9. Chi W, Tong C, Gan X, et al. Characterization and comparative profiling of MiRNA transcriptomes in bighead carp and silver carp. PLoS ONE, 2011, 6: e23549
10. Liu Z J. Aquaculture Genome Technologies. Oxford, UK: Blackwell Publishing, Ames, IA, 2007
11. Yang L, Gui J F. Positive selection on multiple antique allelic lineages of transferrin in the polyploid *Carassius auratus*. Mol Biol Evol, 2004, 21: 1264–1277
12. Zhou L, Wang Y, Gui J F. Molecular analysis of silver crucian carp (*Carassius auratus gibelio* Bloch) clones by SCAR markers. Aquaculture, 2001, 201: 219–228
13. Guo W, Gui J F. Microsatellite marker isolation and cultured strain identification in *Carassius auratus gibelio*. Aquaculture Int, 2008, 16: 497–510
14. Wang D, Mao H L, Peng J X, et al. Discovery of a male-biased mutant family and identification of a male-specific SCAR marker in gynogenetic gibel carp *Carassius auratus gibelio*. Prog Nat Sci, 2009, 19: 1537–1544
15. Li F B, Gui J F. Clonal diversity and genealogical relationships of gibel carp in four hatcheries. Anim Genet, 2008, 39: 28–33
16. Jakovlic I, Gui J F. Recent invasion and low level of divergence between diploid and triploid forms of *Carassius auratus* complex in Croatia. Genetica, 2011, 139: 789–804
17. Zhou L, Wang Y, Gui J F. Genetic evidence for gonochoristic reproduction in gynogenetic silver crucian carp (*Carassius auratus gibelio* Bloch) as revealed by RAPD assays. J Mol Evol, 2000, 51: 498–506
18. Gui J F, Zhou L. Genetic basis and breeding application on clonal diversity and dual reproduction modes in polyploid *Carassius auratus gibelio*. Sci China Life Sci, 2010, 53: 409–415
19. Avise J C. Clonality: The Genetics, Ecology, and Evolution of Sexual Abstinence in Vertebrates. USA: Oxford University Press, 2008

20. Sun X W, Liang L Q. A genetic linkage map of common carp (*Cyprinus carpio* L.) and mapping of a locus associated with cold tolerance. Aquaculture, 2004, 238: 165–172
21. Cheng L, Liu L, Yu X, et al. A linkage map of common carp (*Cyprinus carpio*) based on AFLP and microsatellite markers. Anim Genet, 2009, 41: 191–198
22. Zhang Y, Xu P, Lu C, et al. Genetic linkage mapping and analysis of muscle fiber-related QTLs in common carp (*Cyprinus carpio* L.). Mar Biotechnol, 2010, 13: 376–392
23. Zheng X, Kuang Y, Zhang X, et al. A genetic linkage map and comparative genome analysis of common carp (*Cyprinus carpio* L.) using microsatellites and SNPs. Mol Genet Genomics, 2011, 286: 261–277
24. Li Y, Xu P, Zhao Z X, et al. Construction and characterization of the BAC library for common carp *Cyprinus Carpio* L. and establishment of microsynteny with Zebrafish *Danio Rerio*. Mar Biotechnol, 2010, 13: 706–712
25. Xu P, Li J, Li Y, et al. Genomic insight into the common carp (*Cyprinus carpio*) genome by sequencing analysis of BAC-end sequences. BMC Genomics, 2011, 12: 188
26. Xia J H, Liu F, Zhu Z Y, et al, A consensus linkage map of the grass carp(*Ctenopharyngodon idella*) based on microsatellites and SNPs. BMC Genomics, 2010, 11: 135
27. Geng F S, Zhou L, Gui J F. Construction and characterization of a BAC library for *Carassius auratus gibelio*, a gynogenetic polyploid fish. Anim Genet, 2005, 36: 535–536
28. Zhang Y, Zhang X, Scheuring C F, et al. Construction and characterization of two bacterial artificial chromosome libraries of Zhikong scallop, *Chlamys farreri* Jones et Preston, and identification of BAC clones containing the genes involved in its innate immune system. Mar Biotechnol, 2008, 10: 358–365
29. Zhang X, Zhang Y, Scheuring C, et al. Construction and characterization of a bacterial artificial chromosome (BAC) library of Pacific white shrimp, *Litopenaeus vannamei*. Mar Biotechnol, 2010, 12: 141–149
30. Jang S H, Liu H, Su J G, et al. Construction and characterization

of two bacterial artificial chromosome libraries of grass carp. Mar Biotechnol, 2010, 12: 261–266

31. Shao C W, Chen S L, Scheuring C F, et al. Construction of two BAC libraries from half-smooth tongue sole *Cynoglossus semilaevis* and identification of clones containing candidate sex-determination genes. Mar Biotechnol, 2010, 12: 558–568
32. Yan S Y, Lu D Y, Du M, et al. Nuclear transplantation in teleosts. Hybrid fish from the nucleus of crucian and the cytoplasm of carp. Sci Sin B. 1984, 27: 1029–1034
33. Chen H, Yi Y, Chen M, et al. Studies on the developmental potentiality of cultured cell nuclei of fish. Int J Biol Sci, 2010, 6: 192–198
34. Deng C, Liu H. An unknown piece of early work of nuclear reprogramming in fish eggs. Int J Biol Sci, 2010, 6: 190–191
35. Lee K Y, Huang H, Ju B, et al. Cloned zebrafish by nuclear transfer from long-term-cultured cells. Nat Biotechnol, 2002, 20: 795–799
36. Sun Y H, Chen S P, Wang Y P, et al. Cytoplasmic impact on cross-genus cloned fish derived from transgenic common carp (*Cyprinus carpio*) nuclei and goldfish (*Carassius auratus*) enucleated eggs. Biol Reprod, 2005, 72: 510–515
37. Luo D, Hu W, Chen S, et al. Identification of differentially expressed genes between cloned and zygote-developing zebrafish (*Danio rerio*) embryos at the dome stage using suppression subtractive hybridization. Biol Reprod, 2009, 80: 674–684
38. Hu W, Zhu Z Y. Integration mechanisms of transgenes and population fitness of GH transgenic fish. Sci China Life Sci, 2010, 53: 401–408
39. Luo D J, Hu W, Chen S P, et al. Critical developmental stages for the efficiency of somatic cell nuclear transfer in zebrafish. Int J Biol Sci, 2011, 7: 476–486
40. Hong Y, Liu T, Zhao H, et al. Establishment of a normal medakafish spermatogonial cell line capable of sperm production *in vitro*. Proc Natl Acad Sci USA, 2004, 101: 8011–8016
41. Hong N, Li Z, Hong Y. Fish Stem Cell Cultures. Int J Biol Sci, 2011, 7: 392–402
42. Xu H, Li M, Gui J, et al. Fish germ cells. Sci China Life Sci, 2010,

53: 435–446

43. Yi M, Hong N, Li Z, et al. Medaka fish stem cells and their applications. Sci China Life Sci, 2010, 53: 426–434
44. Yi M, Hong N, Hong Y. Generation of medaka fish haploid embryonic stem cells. Science, 2009, 326: 430–433
45. Takeuchi Y, Yoshizaki G, Takeuchi T. Biotechnology: Surrogate broodstock produces salmonids. Nature, 2004, 430: 629–630
46. Okutsu T, Suzuki K, Takeuchi Y, et al. Testicular germ cells can colonize sexually undifferentiated embryonic gonad and produce functional eggs in fish. Proc Natl Acad Sci USA, 2006, 103: 2725–2729
47. Okutsu T, Shikina S, Kanno M, et al. Production of trout offspring from triploid salmon parents. Science, 2007, 317: 1517
48. Liu S. Distant hybridization leads to different ploidy fishes. Sci China Life Sci, 2010, 53: 416–425
49. Gui J F. Genetic Basis and Artificial Control of Sexuality and Reproduction in Fish (in Chinese). Beijing: Science Press, 2007
50. Jalabert B. Particularities of reproduction and oogenesis in teleost fish compared to mammals. Reprod Nutr Dev, 2005, 45: 261–279
51. Xie J, Wen J J, Chen B, et al. Differential gene expression in fully grown oocytes between gynogenetic and gonochoristic crucian carps. Gene, 2001, 271: 109–116
52. Dong C H, Yang S T, Yang Z A, et al. A C-type lectin associated and translocated with cortical granules during oocyte maturation and egg fertilization in fish. Dev Biol, 2004, 265: 341–354
53. Xu H Y, Gui J F, Hong Y H. Differential expression of *vasa* RNA and protein during spermatogenesis and oogenesis in the gibel carp (*Carassius auratus gibelio*), a bisexually and gynogenetically reproducing vertebrate. Dev Dyn, 2005, 233: 872–882
54. Peng J X, Xie J L, Zhou L, et al. Evolutionary conservation of Dazl genomic organization and its continuous and dynamic distribution throughout germline development in gynogenetic gibel carp. J Exp Zool Part B, 2009, 312B: 855–871
55. Xie J, Wen J J, Yang Z A, et al. Cyclin A2 is differentially expressed during oocyte maturation between gynogenetic silver crucian carp

and gonochoristic color crucian carp. J Exp Zool, 2003, 295: 1–16

56. Chen B, Gui J F. Identification of a novel C1q family member in color crucian carp (*Carassius auratus*) ovary. Comp Biochem Phys B, 2004, 138: 285–293
57. Mei J, Chen B, Yue H M, et al. Identification of a C1q family member associated with cortical granules and follicular cell apoptosis in *Carassius auratus gibelio*. Mol Cell Endocrinol, 2008, 289: 67–76
58. Mei J, Zhang Q Y, Li Z, et al. *C1q-like* inhibits *p53*-mediated apoptosis and controls normal hematopoiesis during zebrafish embryogenesis. Dev Biol, 2008, 319: 273–284
59. Wu N, Yue H M, Chen B, et al. Histone H2A has a novel variant in fish oocytes. Biol Reprod, 2009, 81: 275–283
60. Wang X L, Sun M, Mei J, et al. Identification of a Spindlin homolog in gibel carp (*Carassius auratus gibelio*). Comp Biochem Physiol B Biochem Mol Biol, 2005, 141: 159–167
61. Sun M, Li Z, Gui J F. Dynamic distribution of spindlin in nucleoli, nucleoplasm and spindle from primary oocytes to mature eggs and its critical function for oocyte-to-embryo transition in gibel carp. J Exp Zool, 2010, 313A: 461–473
62. Mylonas C C, Fostier A, Zanuy S. Broodstock management and hormonal manipulations of fish reproduction. Gen Comp Endo, 2010, 165: 516–534
63. Zohar Y, Muñoz-Cueto J A, Elizur A, et al. Neuroendocrinology of reproduction in teleost fish. Gen Comp Endocrinol, 2010, 165: 438–455
64. Li W S, Lin H R. The endocrine regulation network of growth hormone synthesis and secretion in fish: Emphasis on the signal integration in somatotropes. Sci China Life Sci, 2010, 53: 462–470
65. Li C J, Zhou L, Wang Y, et al. Molecular and expression characterization of three gonadotropin subunits commonα, FSHβ and LHβ in groupers. Mol Cell Endocrinol, 2005, 233: 33–46
66. Cao H, Zhou L, Zhang Y Z, et al. Molecular characterization of Chinese sturgeon gonadotropins and cellular distribution in pituitaries of mature and immature individuals. Mol Cell Endocrinol, 2009, 303: 34–42

67. Li S, Hu W, Wang Y, et al. Cloning and expression analysis in mature individuals of two chicken type-II GnRH (cGnRH-II) genes in common carp (*Cyprinus carpio*). Sci China Ser C-Life Sci, 2004, 47: 349–358

68. Hu W, Li S, Tang B, et al. Antisense for gonadotropinreleasing hormone reduces gonadotropin synthesis and gonadal development in transgenic common carp (*Cyprinus carpio*). Aquaculture, 2007, 271: 498–506

69. Xu J, Huang W, Zhong C, et al. Defining global gene expression changes of the hypothalamic-pituitary-gonadal axis in female sgnrh-antisense transgenic common carp (*Cyprinus carpio*). PLoS ONE, 2011, 6: e21057

70. Brunner B, Hornung U, Shan Z, et al. Genomic organization and expression of the double-sex related gene cluster in vertebrates and detection of putative regulatory regions for DMRT1. Genomics, 2001, 77: 8–17

71. Delvin R H, Nagahama Y. Sex determination and sex differentiation in fish. Aquaculture, 2002, 208: 191–364

72. Matsuda M, Nagahama Y, Shinomiya A, et al. DMY is a Y-specific DM-domain gene required for male development in the medaka fish. Nature, 2002, 417: 559–663

73. Volff J N, Kondo M, Schartl M. Medaka dmY/dmrt1Y is not the universal primary sex-determining gene in fish. Trends Genet, 2003, 19: 196–199

74. Herpin A, Schartl M. *Dmrt1* genes at the crossroads: a widespread and central class of sexual development factors in fish. FEBS J, 2011, 278: 1010–1019

75. Xia W, Zhou L, Yao B, et al. Differential and spermatogenic cell-specific expression of DMRT1 during sex reversal in protogynous hermaphroditic groupers. Mol Cell Endocrinol, 2007, 263: 156–172

76. Jeong H B, Park J G, Park Y J, et al. Isolation and characterization of DMRT1 and its putative regulatory region in the protogynous wrasse, *Halichoeres tenuispinis*. Gene, 2009, 438: 8–16

77. Huang X, Guo Y, Shui Y, et al. Multiple alternative splicing and

differential expression of dmrt1 during gonad transformation of the rice field eel. Biol Reprod, 2005, 73: 1017–1024

78. He C L, Du J L, Wu G C, et al. Differential Dmrt1 transcripts in gonads of theprotandrous black porgy, *Acanthopagrus schlegeli.* Cytogenet Genome Res, 2003, 101: 309–313

79. Wu G C, Chiu P C, Lin C J, et al. Testicular dmrt1 is involved in the sexual fate of the ovotestis in the protandrous black porgy. Biol Reprod, 2012, 86: 41

80. Liarte S, Chaves-Pozo E, Garcia-Alcazar A, et al. Testicular involution prior to sex change in gilthead seabream is characterized by a decrease in DMRT1 gene expression and by massive leukocyte infiltration. Reprod Biol Endocrinol, 2007, 5: 20

81. Graves J A M. Weird animal genomes and the evolution of vertebrate sex and sex chromosomes. Annu Rev Genet, 2008, 42: 565–586

82. Yao B, Zhou L, Wang Y, et al. Differential expression and dynamic changes of SOX3 during gametogenesis and sex inversion in protogynous hermaphroditic fish. J Exp Zool, 2007, 307A: 207–219

83. Huang W, Zhou L, Li Z, et al. Expression pattern, cellular localization and promoter activity analysis of ovarian aromatase (Cyp19a1a) in protogynous hermaphrodite red-spotted grouper. Mol Cell Endocrinol, 2009, 307: 224–236

84. Zhang Y, Zhang W, Yang H, et al. Two cytochrome P450 aromatase genes in the hermaphrodite rice field eel *Monopterus albus*: mRNA expression during ovarian development and sex change. J Endocrinol, 2010, 199: 317–331

85. Guiguen Y, Fostier A, Piferrer F, et al. Ovarian aromatase and estrogens: a pivotal role for gonadal sex differentiation and sex change in fish. Gen Comp Endocrinol, 2010, 165: 352–366

86. Zhou L, Gui J F. Molecular mechanisms underlying sex change in hermaphroditic groupers. Fish Physiol Biochem, 2010, 36: 181–193

87. Arnold A P. The end of gonad-centric sex determination in mammals. Trends Genet, 2012, 28: 55–61

88. Quinn A E, Sarre S D, Ezaz T, et al. Evolutionary transitions between mechanisms of sex determination in vertebrates. Biol Lett, 2011, 7: 443–448

89. Mank J E. Small but mighty: the evolutionary dynamics of W and Y sex chromosomes. Chromosome Res, 2012, 20: 21–33
90. Stöck M, Horn A, Grossen C, et al. Ever-young sex chromosomes in European tree frogs. PLoS Biology, 2011, 9: e1001062
91. Yoshida K, Terai Y, Mizoiri S, et al. B chromosomes have a functional effect on female sex determination in lake victoria cichlid fishes. PLoS Genet, 2011, 7: e1002203
92. Hughes J F, Skaletsky H, Pyntikova T, et al. Chimpanzee and human Y chromosomes are remarkably divergent in structure and gene content. Nature, 2010, 463: 536–539
93. De-santis C, Jerry D R. candidate growth genes in finfish-Where should we be looking? Aquaculture, 2007, 272: 22–38
94. Johnston I A, Bower N I, Macqueen D J. Growth and the regulation of myotomal muscle mass in teleost fish. J Exp Biol, 2011, 214: 1617–1628
95. Li C J, Wei, Q W, et al. Molecular and expression characterization of two somatostatin genes in the Chinese sturgeon, *Acipenser sinensis*. Comp Biochem Physiol A Mol Integr Physiol, 2009, 154: 127–134
96. Ye X, Li W S, Lin H R. Polygenic expression of somatostatin in orange-spotted grouper (*Epinephelus coioides*): molecular cloning and distribution of the mRNAs encoding three somatostatin. Mol Cell Endocrinol, 2005, 241: 62–72
97. Zhang L, Li W S, Hong X, et al. Regulation of preprosomatostatin I (PSSI) gene expression by 17β-estradiol and identification of the PSSI promoter region in orange-spotted grouper (*Epinephelus coioides*). Mol Cell Endocrinol, 2009, 311: 87–93
98. Liu Y, Lu D, Zhang Y, et al. The evolution of somatostatin in vertebrates. Gene, 2010, 463: 21–28
99. Dong H, Li W, Lin H. Comparative analyses of sequence structure, evolution, and expression of four somatostatin receptors in orange-spotted grouper (*Epinephelus coioides*). Mol Cell Endocrinol, 2010, 323: 125–136
100. Yan A F, Zhang L J, Tang Z G, et al. Orange-spotted grouper (*Epinephelus coioides*) orexin: Molecular cloning, tissue expression, ontogeny, daily rhythm and regulation of *NPY* gene expression.

Peptides, 2011, 32: 1363–1370

101. Li X, He J, Hu W, et al. The essential role of endogenous ghrelin in growth hormone expression during zebrafish adenohypophysis development. Endocrinol, 2009, 150: 2767–2774

102. Li X, He J, Hu W, et al. Enhanced hyperplasia in muscles of transgenic zebrafish expressing *Follistatin 1*. Sci China Life Sci, 2011, 54: 159–165

103. Zhang Q Y. A review of viral diseases of aquatic in China. Acta Hydrobiol Sin, 2002, 26: 93–101

104. Qiu T, Lu R H, Zhang J, et al. Complete nucleotide sequence of the S10 genome segment of grass carp reovirus (GCRV). Dis Aqua Org, 2001, 44: 69–74

105. Yang F, He J, Lin X H, et al. Complete genome sequence of the shrimp white spot bacilliform virus. J Virol, 2001, 75: 11811–11820

106. He J G, Deng M, Weng S P, et al. Complete genome analysis of the mandarin fish infectious spleen and kidney necrosis iridovirus. Virology, 2001, 291: 126–139

107. Zhang Q Y, Xiao F, Xie J, et al. Complete genome sequence of lymphocystis disease virus (LCDV-C) isolated from China. J Virol, 2004, 78: 6982–6994

108. Xie X X, Xu L M, Yang F. Proteomic analysis of the major envelope and nucleocapsid proteins of white spot syndrome virus. J Virol, 2006, 80: 10615–10623

109. Zhou Q, Xu L, Li H, et al. Four major envelope proteins of white spot syndrome virus bind to form a complex. J Virol, 2009, 83: 4709–4712

110. Zhao Z, Ke F, Huang Y H, et al. Identification and characterization of a novel envelope protein in *Rana grylio* virus. J Gen Virol, 2008, 89: 1866–1872

111. Dong C F, Xiong X P, Shuang F, et al. Global landscape of structural proteins of infectious spleen and kidney necrosis virus. J Virol, 2011, 85: 2869–2877

112. Zhao Z, Shi Y, Ke F, et al. Constitutive expression of thymidylate synthase from LCDV-C induces a transformed phenotype in fish cells. Virology, 2008, 372: 118–126

113. Wang Z L, Xu X P, He B L, et al. Infectious spleen and kidney necrosis virus ORF48R functions as a new viral vascular endothelial growth factor. J Virol, 2008, 82: 4371–4383

114. Zhang Y B, Zhang Q Y, Xu D Q, et al. Identification of antiviral-relevant genes in the cultured fish cells induced by UV-inactivated virus. Chin Sci Bull, 2003, 48: 581–588

115. Zhang Y B, Gui J F. Molecular characterization and IFN signal pathway analysis of Carassius auratus *CaSTAT*1 identified from the cultured cells in response to virus infection. Dev Comp Immunol, 2004, 28: 211–227

116. Zhang Y B, Gui J F. Identification and expression analysis of two IFN-inducible genes in crucian carp (*Carassius auratus* L.). Gene, 2004, 325: 43–51

117. Hu C Y, Zhang Y B, Huang G P, et al. Molecular cloning and characterization of a fish PKR-like gene from the cultured CAB cells induced by UV-inactivated virus. Fish Shellfish Immunol, 2004, 17: 353–366

118. Zhang Y B, Jiang J, Chen Y D, et al. The Innate Immune Response to Grass Carp Haemorrhagic Virus (GCHV) in Cultured *Carassius auratus* Blastulae (CAB) Cells. Dev Comp Immunol, 2007, 31: 232–243

119. Zhang Y B, Wang Y L, Gui J F. Identification and characterisation of two homologs of interferon stimulated gene ISG15 in crucian carp. Fish Shellfish Immunol, 2007, 23: 52–61

120. Shi Y, Zhang Y B, Zhao Z, et al. Molecular characterization and subcellular localization of *Carassius auratus* interferon regulatory factor-1. Dev Comp Immunol, 2008, 32: 134–146

121. Jiang J, Zhang Y B, Li S, et al. Expression regulation and functional characterization of a novel interferon-inducible gene Gig2 and its promoter. Mol Immunol, 2009, 46: 3131–3140

122. Yu F F, Zhang Y B, Liu T K, et al. Fish virus-induced interferon exerts antiviral function through Stat1 pathway. Mol Immunol, 2010, 47: 2330–2341

123. Sun F, Zhang Y B, Liu T K, et al. Characterization of fish IRF3 as an IFN-inducible protein reveals evolving regulation of IFN response in vertebrates. J Immunol, 2010, 185: 7573–7582

124. Sun F, Zhang Y B, Liu T K, et al. Fish MITA activation serves as a mediator for distinct fish IFN gene activation dependent on IRF3 or IRF7. J Immunol, 2011, 187: 2531–2539

125. Zhu R, Zhang Y B, Zhang Q Y, et al. Functional domains and the antiviral effect of the double-stranded RNA-dependent protein kinase PKR from *Paralichthys olivaceus*. J Virol, 2008, 82: 6889–6901

126. Liu T K, Zhang Y B, Liu Y, et al. Cooperative roles of fish PKZ and PKR in IFN-mediated antiviral response. J Virol, 2011, 85: 12769–12780

127. Jin J Y, Zhou L, Wang Y, et al. Antibacterial and antiviral roles of a fish β-defensin expressed both in pituitary and testis. PLoS ONE, 2010, 5: e12883

128. Wang S, Liu N, Chen A J, et al. TRBP homolog interacts with eukaryotic initiation factor 6 (eIF6) in *Fenneropenaeus chinensis*. J Immunol. 2009, 182: 5250–5258

129. Zhao Z Y, Yin Z X, Xu X P, et al. A novel C-type lectin from the shrimp *Litopenaeus vannamei* possesses anti-white spot syndrome virus activity. J Virol, 2009, 83: 347–356

130. Chen A J, Wang S, Zhao X F, et al. Enzyme E2 from Chinese white shrimp inhibits replication of white spot syndrome virus and ubiquitinates its RING domain proteins. J Virol, 2011, 85: 8069–8079

131. Chang M, Collet B, Nie P, et al. Expression and functional characterization of the RIG-I-like receptors MDA5 and LGP2 in Rainbow trout (*Oncorhynchus mykiss*). J Virol, 2011, 85: 8403–8412

132. Barrett R D, Paccard A, Healy T M, et al. Rapid evolution of cold tolerance in stickleback. Proc Biol Sci, 2011, 278: 233–238

133. Chen Z, Cheng C H, Zhang J, et al. Transcriptomic and genomic evolution under constant cold in Antarctic notothenioid fish. Proc Natl Acad Sci USA, 2008, 105: 12944–12949

134. Xu Q, Cheng C H, Hu P, et al. Adaptive evolution of hepcidin genes in antarctic notothenioid fishes. Mol Biol Evol, 2008, 25: 1099–1112

135. Deng C, Cheng C H, Ye H, et al. Evolution of an antifreeze protein

by neofunctionalization under escape from adaptive conflict. Proc Natl Acad Sci USA, 2010, 107: 21593–21598

136. Bickler E, Buck T. Hypoxia tolerance in reptiles, amphibians, and fishes: life with variable oxygen availability. Annu Rev Physiol, 2007, 69: 145–170

137. Lushchak V I. Adaptive response to oxidative stress: Bacteria, fungi, plants and animals. Comp Biochem Physiol C Toxicol Pharmacol, 2011, 153: 175–190

138. Zhong X P, Wang D, Zhang Y B, et al. Identification and characterization of hypoxia-induced genes in *Carassius auratus* blastulae embryonic cells using suppression subtractive hybridization. Comp Biochem Physiol B, 2009, 152: 161–170

139. Wang D, Zhong X P, Qiao Z X, et al. Inductive transcription and protective role of fish heme oxygenase-1 under hypoxic stress. J Exp Biol, 2008, 211: 2700–2706

140. Sun C F, Tao Y, Jiang X Y, et al. IGF binding protein 1 is correlated with hypoxia-induced growth reduce and developmental defects in grass carp (*Ctenopharyngodon idellus*) embryos. Gen Comp Endocrinol, 2011, 172: 409–415

141. Feng X, Liu X, Zhang W, et al. p53 directly suppresses BNIP3 expression to protect against hypoxia-induced cell death. EMBO J, 2011, 30: 3397–3415

142. Guan B, Ma H, Wang Y, et al. Vitreoscilla hemoglobin (VHb) overexpression increases hypoxia tolerance in zebrafish (*Danio rerio*). Mar Biotechnol (NY), 2011, 13: 336–344

143. Liu Z J. Fish genomics and analytical genetic technologies, with examples of their potential applications in management of fish genetic resources. In: Bartley D M, Harvey B J, Pullin R S V, eds. Workshop on Status and Trends in Aquatic Genetic Resources: a basis for international policy. Food and Agriculture Organization of the United Nations, Rome, 2007. 145–178

144. McAndrew B, Napier J. Application of genetics and genomics to aquaculture development: current and future directions. J Agric Sci, 2010, 149: 143–151

145. Abernathy J W, Peatman E, Liu Z J. Basic aquaculture genetics. SRAC Publication No. 5001. 2010, 1–16

146. Wang Z W, Zhu H P, Wang D, et al. A novel nucleo-cytoplasmic hybrid clone formed via androgenesis in polyploid gibel carp. BMC Res Notes, 2011, 4: 82

147. Liu H Q, Cui S Q, Hou C C, et al. YY supermale generated gynogenetically from XY female in *Pelteobagrus fulvidraco* (Richardson). Acta Hydrobiol Sin, 2007, 31: 718–725

148. Wang D, Mao H L, Chen H X, et al. Isolation of Y- and X-linked SCAR markers in yellow catfish and application in the production of all-male populations. Anim Genet, 2009, 40: 978–981

CHAPTER

11

GENETIC PARAMETERS FOR UNIFORMITY OF HARVEST WEIGHT AND BODY SIZE TRAITS IN THE GIFT STRAIN OF NILE TILAPIA

Jovana Marjanovic[1,2], Han A. Mulder[1] , Hooi L. Khaw[3] and Piter Bijma[1]

[1] Animal Breeding and Genomics Centre, Wageningen University and Research, 6700 AH Wageningen, The Netherlands.

[2] Department of Animal Breeding and Genetics, Swedish University of Agricultural Sciences, 75007 Uppsala, Sweden.

[3] WorldFish, Jalan Batu Maung, 11960 Bayan Lepas, Penang, Malaysia.

ABSTRACT

Background

Animal breeding programs have been very successful in improving the mean levels of traits through selection. However, in recent decades, reducing the variability of trait levels between individuals has become a highly desirable objective. Reaching this objective through genetic selection requires that there is genetic variation in the variability of trait levels, a phenomenon known as genetic heterogeneity of environmental (residual) variance. The aim of our study was to investigate the potential for genetic improvement of uniformity of harvest weight and body size traits (length, depth, and width) in the genetically improved farmed tilapia (GIFT) strain. In order to quantify the genetic variation in uniformity of traits and estimate the genetic correlations between level and variance of the traits, double hierarchical generalized linear models were applied to individual trait values.

Results

Our results showed substantial genetic variation in uniformity of all analyzed traits, with genetic coefficients of variation for residual variance ranging from 39 to 58 %. Genetic correlation between trait level and variance was strongly positive for harvest weight (0.60 ± 0.09), moderate and positive for body depth (0.37 ± 0.13), but not significantly different from 0 for body length and width.

Conclusions

Our results on the genetic variation in uniformity of harvest weight and body size traits show good prospects for the genetic improvement of uniformity in the GIFT strain. A high and positive genetic correlation was estimated between level and variance of harvest weight, which suggests that selection for heavier fish will also result in more variation in harvest weight. Simultaneous improvement of harvest weight and its uniformity will thus require index selection.

BACKGROUND

In animal breeding, particular attention is paid to improving the mean level of traits through selection and this has been successful for many breeding programs. One such successful example is the genetically improved farmed tilapia (GIFT) project, which was led at WorldFish [1] and resulted in a line of tilapia known as the GIFT-strain. For this strain, a substantial realized genetic gain (>100 %) was achieved through 12 generations of genetic improvement for body weight at harvest [2, 3]. However, it is often desirable not only to improve the level of a trait, but also to reduce its variability [4, 5], because significant variation around the optimal value of a trait can have a negative impact on production performance, both in livestock and aquaculture [5, 6, 7]. In fish farming, differences in size among individuals are generally associated with competition for food within a group and the resulting feeding hierarchy [6, 8, 9]. The phenotypic coefficient of variation (CV) for body weight, apart from indicating variation of the trait is also an indicator of competitive interactions within a population [8]. For the GIFT strain, the CV ranges from 40 to 60 %, which is considered a high value [10].

Although good management during the grow-out phase can help reduce the CV, as noted by Ponzoni et al. [2], its average value across eight generations of GIFT remained at around 40 %. A common approach in fish farming to decrease phenotypic variation in body size and weight is to grade or sort fish into groups, according to size. If fish are not graded, the large variation in weight and size at harvest reduces their market value and has animal welfare consequences [11, 12]. From the point of view of fish farmers, uniformity of growth and body size is one of the key traits to be improved [11]. From the consumer's point of view, weight but also body size and appearance traits, play an important role in buying decisions [13, 14, 15].

An alternative approach to management procedures for reducing the variability of a trait is selective breeding. Selection for more uniform individuals requires that the variability of the trait itself has a genetic component i.e. that there is genetic variation, which is also known as genetic heterogeneity of environmental (residual) variance [16, 17]. In this case, within a population, some animals will be less prone than others to phenotypic changes in response to small environmental fluctuations, and thus will have a more stable performance. Several studies on livestock and laboratory animals have demonstrated the existence of genetic differences in residual variance among genotypes and have quantified their magnitude [7, 16, 18, 19, 20, 21, 22, 23, 24, 25, 26, 27, 28]. In aquaculture species, evidence for substantial genetic heterogeneity of residual variance comes from three studies on body weight in salmonids [29, 30, 31]. A previous study on uniformity in Nile tilapia that analyzed the standard deviation of harvest weight using a traditional linear mixed model indicated a genetic basis for variability of harvest weight [12]. However, to date, variability of harvest weight in Nile tilapia has not been analyzed at the variance level using double hierarchical generalized linear models (DHGLM). The DHGLM is a novel approach that can be used to study uniformity of individual trait values. The advantage of

DHGLM compared to analyzing variance or the standard deviation of a group is that it can take into account systematic effects on the variance of the individual record level such as sex of the fish. The genetic basis of the variability of body size traits has not been explored in any species, except in humans for height [32].

The main objective of our study was to investigate the potential for genetic improvement of uniformity of harvest weight and body size traits in the GIFT strain. For this purpose, we analyzed within-family variance of harvest weight, body length, depth, and width, by applying a DHGLM to individual trait values [33]. To quantify the genetic relationship between the level and the variance of these traits, we also estimated the genetic correlation between these two components.

METHODS

Environment

We used data that were obtained from an experiment that was specifically designed to estimate indirect genetic effects (IGE) for growth rate in the GIFT strain [34]. This experiment was carried out between 2009 and 2012 at the Jitra Aquaculture Extension Centre of the Department of Fisheries, which is managed by WorldFish and located at Kedah State of Malaysia. WorldFish complies with the Malaysian laws on animal experiments. During this experiment, four batches of fish were produced, i.e. one batch each year (batch named per year). However, for the last batch (2012), a high level of mortality occurred due to extreme weather conditions, which resulted in an insufficient number of records, and thus it was excluded from the analysis.

Experimental Design

To produce families, the GIFT breeding program uses a nested-mating design, where one male is mated to two females. For this work, we used the same mating scheme to produce the experimental fish, and thus two full-sib families were obtained from each father. Each full-sib family contributed 80 offspring to the experiment. Fry that belonged to the same full-sib family were nursed together and separately from other families. During the grow-out phase, fish were kept in groups. Before placing each fish in a group, they were individually identified with a PIT (Passive Integrated Transporter) tag. Following the optimal design for the estimation of IGE [35], families were assigned to groups so that each group consisted of members of two distinct, unrelated families. Both families contributed eight randomly selected individuals to each group to form groups of 16 members. Therefore, each family of 80 offspring contributed to 10 distinct groups (i.e. 80/10 members per group). Unique combinations of families in groups were created using a block design, with 11 families per block, where each family was combined only once with the other ten families in the same block. Hence, there were 55 family combinations i.e. groups, per block. If the number of available families for the last block was less than 11, an incomplete block was used with all the remaining families. An outline of the various steps that were carried out for each batch is in Fig. 1.

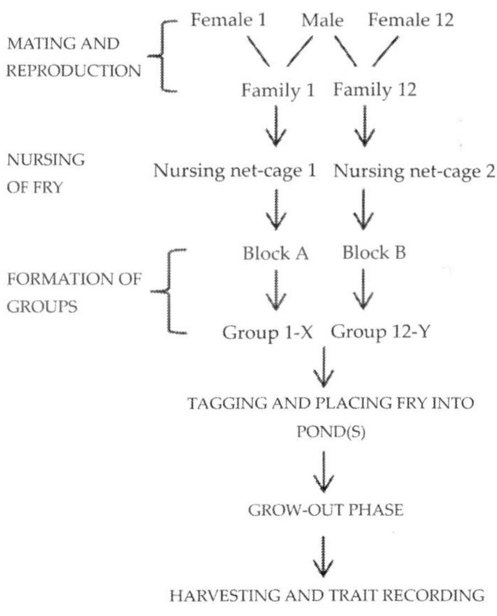

Figure 1: Outline of the experimental design for two paternal families. X represents any family from Block A, other than family 1; Y represents any family from Block B, other than family 12.

The groups were kept in net-cages that were placed in earthen ponds in rows and columns. For each batch, two ponds were available. Due to the small number of fish available for batch 2010, only one pond was used. The groups for each block were distributed randomly and as evenly as possible over both ponds. Thus, the 55 groups of a block were split into 27 groups for pond 1, and 28 groups for pond 2.

During the grow-out phase, fish were fed with commercial dry pellets containing 32 % of protein; the amount of pellets (3 to 5 % of average live weight) and feeding frequency (twice a day) were the same as for the GIFT selective breeding population. However, because the fish were kept in net-cages rather than in communal rearing, the feeding strategy differed from that in the standard GIFT program. Rather than spreading the food over the entire surface of a pond, it was placed in the corner of each net-cage so that the fish could express their competitive tendency (see Discussion). More details on the experiment are in Khaw et al. [12, 34]. The GIFT technology manual provides a description of key husbandry procedures [36].

Records

Fish were harvested 5 to 8 months after the grow-out period, when the average weight ranged from 200 to 250 g. At harvest, the following traits and parameters were recorded: live body weight (g), body measurements (length, depth, and width, in cm), tag number, sex, pond, and net-cage label. The age at harvest of each fish was computed from the recorded spawning and harvesting dates [34]. Over three batches, phenotypic observations on body weight and body measurements at harvest were available for 6330 fish from 493 groups.

Ideally, each group should contain 16 individuals at harvest. However, due to mortality, some groups contained very few individuals, and a threshold was set for group and family size. Thus, groups that contained less than seven individuals in total or less than three fish per family were discarded, which reduced the number of groups to 446. With two families in each group, 892 family-by-group combinations and 6090 individual records were available for each trait. Table 1 shows the number of observations at harvest (full dataset) and number of observations used in the analysis (edited or reduced dataset). The pedigree consisted of 34,517 records that traced the GIFT population back seven generations.

Table 1: Number of groups, families per group, and individuals at harvest (C-complete dataset) and after editing (R-reduced dataset)

Batch	Families		Groups		Families per group		Individuals	
	C	R	C	R	C	R	C	R
2009	66	66	209	188	418	376	2565	2461
2010	33	31	45	37	90	74	509	464
2011	68	68	239	221	478	442	3256	3165
Total	167	165	493	446	986	892	6330	6090

Statistical Analysis

The environmental component in the phenotypic variation of a trait can be measured either on the same individual for which repeated observations are available or on the individuals belonging to the same family [37]. In our dataset, body weight and body measurements were recorded at harvest. Hence, only one record for each trait was available for each individual, but eight observations were recorded per family per group. To analyze the genetic heterogeneity of the environmental variance, different approaches have been proposed [37] and we chose a DHGLM that models the residual variance of individual observations on the exponential scale, and can be interpreted as a multiplicative model [17]. On the level of the natural logarithm, the multiplicative model becomes additive.

Sire and dam, group, kin, and social maternal effect were included as random effects. A group effect was included to account for non-heritable indirect effects, which create a non-genetic covariance among individuals within the same group [38]. If this covariance is present but not accounted for, it can cause bias in the estimated genetic parameters [39]. According to the kin selection theory, relatives can cooperate with each other [40, 41], thus a non-genetic covariance between group mates belonging to the same family can arise. Therefore, we included a kin effect to account for this source of non-genetic covariance i.e. between group mates of the same family compared to group mates of the other family within a group [34]. Finally, a social maternal effect was included that accounts for the non-genetic effect of the common maternal environment of one full-sib family on the performance of the other full-sib family in the group [12]. In other words, we fitted a non-genetic effect of the mother of a full-sib family on the trait values of the other full sib family kept in the same group. Hence, we termed this effect "social", because it is expressed in the trait values of the social partners of the offspring of a mother, rather than in the offspring themselves.

Double Hierarchical Generalized Linear Models (DHGLM)

Lee and Nelder [42] developed a framework for the DHGLM, where level and residual variance of a trait can be modeled jointly with specified random effects. This approach has been applied in animal breeding by Rönnegård et al. [33] who implemented the DHGLM in the statistical software SAS and ASReml 2.0 [33]. The DHGLM algorithm iterates between two sets of mixed model equations i.e. a linear mixed model for the phenotypic records and a generalized linear mixed model for the response variable φ_i. φ_i is defined as $\phi_i = E(\hat{e}_i^2/(1-h_i))$, where $\hat{e}_i^2$ is the squared residual for the ith observation and h_i is the diagonal element of the hat matrix of y, corresponding to the same individual [33, 43]. As φ follows a χ^2 distribution, $\hat{e}_i^2/(1-h_i)$ can be linearized using a log link function so that $\log(\phi) = \log[\hat{e}_i^2/(1-h_i)]$ [33]. Instead of using a log link function, $\log[\hat{e}_i^2/(1-h_i)]$ can be linearized using a first order Taylor-series expansion as shown by Felleki et al. [44], which results in the response variable $\psi_i = \log(\hat{\sigma}_{e_i}^2) + (\{[e_i^2/(1-h_i)] - \hat{\sigma}_{e_i}^2\}/\hat{\sigma}_{e_i}^2)$, where $\hat{\sigma}_{e_i}^2$ denotes the predicted residual variance for observation i, and ei is the residual for individual i. Due to linearization, a bivariate DHGLM can then be used:

$$\begin{bmatrix} \mathbf{y} \\ \boldsymbol{\psi} \end{bmatrix} = \begin{bmatrix} \mathbf{X} & \mathbf{0} \\ \mathbf{0} & \mathbf{X_V} \end{bmatrix}\begin{bmatrix} \mathbf{b} \\ \mathbf{b_v} \end{bmatrix} + \begin{bmatrix} \mathbf{Z_{Par}} & \mathbf{0} \\ \mathbf{0} & \mathbf{Z_{Par_v}} \end{bmatrix}\begin{bmatrix} \mathbf{a} \\ \mathbf{a_v} \end{bmatrix}$$
$$+ \begin{bmatrix} \mathbf{V} & \mathbf{0} \\ \mathbf{0} & \mathbf{V_v} \end{bmatrix}\begin{bmatrix} \mathbf{g} \\ \mathbf{g_v} \end{bmatrix} + \begin{bmatrix} \mathbf{S} & \mathbf{0} \\ \mathbf{0} & \mathbf{S_v} \end{bmatrix}\begin{bmatrix} \mathbf{k} \\ \mathbf{k_v} \end{bmatrix}$$
$$+ \begin{bmatrix} \mathbf{U} & \mathbf{0} \\ \mathbf{0} & \mathbf{U_v} \end{bmatrix}\begin{bmatrix} \mathbf{m} \\ \mathbf{m_v} \end{bmatrix} + \begin{bmatrix} \mathbf{e} \\ \mathbf{e_v} \end{bmatrix},$$

where y is the vector of individual trait records (harvest weight, body length, depth, and width) and ψ is the vector of response variables for the variance part of the model, expressed per individual (ψ_i as defined above). b and b_v are the vectors of fixed effects, while a and a_v are the vectors of additive genetic effects of the sire and dam of each individual, with

$$\begin{pmatrix} \mathbf{a} \\ \mathbf{a_v} \end{pmatrix} \sim N\left(\mathbf{0}, \begin{bmatrix} \sigma_a^2 & \sigma_{a,a_v} \\ \sigma_{a,a_v} & \sigma_{a_v}^2 \end{bmatrix} \otimes \mathbf{A}\right),$$

where sire and dam variances are equal to a quarter of the additive genetic variance: $\sigma_{a(v)}^2 = \frac{1}{4}\sigma_{A(v)}^2$, $\sigma_{A(v)}^2$ denoting the ordinary additive genetic variance. Note that we assume equal additive genetic variances for the sire and dam, i.e. $\sigma_{sire(v)}^2 = \sigma_{dam(v)}^2 = \sigma_{a(v)}^2$. g and g_v are the vectors of random group effects, with $\begin{pmatrix} \mathbf{g} \\ \mathbf{g_v} \end{pmatrix} \sim N\left(\mathbf{0}, \begin{bmatrix} \sigma_g^2 & \sigma_{g,g_v} \\ \sigma_{g,g_v} & \sigma_{g_v}^2 \end{bmatrix} \otimes \mathbf{I}\right)$; k and k_v are the vectors of random kin effects, with $\begin{pmatrix} \mathbf{k} \\ \mathbf{k_v} \end{pmatrix} \sim N\left(\mathbf{0}, \begin{bmatrix} \sigma_k^2 & \sigma_{k,k_v} \\ \sigma_{k,k_v} & \sigma_{k_v}^2 \end{bmatrix} \otimes \mathbf{I}\right)$; m and m_v are the vectors of social maternal effects, with $\begin{pmatrix} \mathbf{m} \\ \mathbf{m_v} \end{pmatrix} \sim N\left(\mathbf{0}, \begin{bmatrix} \sigma_m^2 & \sigma_{m,m_v} \\ \sigma_{m,m_v} & \sigma_{m_v}^2 \end{bmatrix} \otimes \mathbf{I}\right)$; and e and e_v are the vectors of random residuals that are assumed to be independent and normally distributed $\begin{pmatrix} \mathbf{e} \\ \mathbf{e_v} \end{pmatrix} \sim N\left(\mathbf{0}, \begin{bmatrix} \mathbf{W}^{-1}\sigma_e^2 & 0 \\ 0 & \mathbf{W}_v^{-1}\sigma_{e_v}^2 \end{bmatrix} \otimes \mathbf{I}\right)$ with scaling variances σ_e^2 and $\sigma_{e_v}^2$. The expectations for the scaling variances σ_e^2 and $\sigma_{e_v}^2$ are equal to 1, because W and W_v already contain the reciprocals of the estimated residual variances per record. The $X(X_v)$, $Z(Z_v)$, $V(V_v)$, $S(S_v)$ and $U(U_v)$ are known design matrices assigning observations to the level of fixed, sire and dam, group, kin, and social maternal effects for y(ψ), respectively. The weights, $\mathbf{W} = \text{diag}\left(\hat{\psi}\right)^{-1}$ and $W_v = \text{diag}((1 - h)/2)$, are, together with vector ψ, updated at each iteration until convergence [43]. The social maternal effect was excluded for body width because the model did not converge, and for body length because it was not significant $\left(\chi_{1DF}^2 = 2.66, p = 0.264\right)$. The fixed effects included for trait level and the variance part of the model were interaction of batch (2009, 2010, and 2011), sex (male and female), pond (1 and 2) and the linear covariate 'age at harvest'.

To facilitate interpretation in the Results section, the group effect for trait level is presented as $g^2 = \hat{\sigma}_g^2/\hat{\sigma}_P^2$, where σ_P^2 is the phenotypic variance, and the kin effect as $k^2 = \hat{\sigma}_k^2/\hat{\sigma}_P^2$. Moreover, for the genetic estimates, the genetic coefficient of variation (GCV) for trait level and its residual variance (GCV_{Ve}) are provided. These are defined as, $GCV = \sigma_A / \mu$, where σ_A is the genetic standard deviation in trait level while μ is the population mean level of the trait [45], and, $GCV_{Ve} = \sigma_{A_V}/\sigma^2_E$, where σ_{AV} is the genetic standard deviation in the residual variance and σ^2_E is the mean residual variance from the additive model [37, 46]. When $\sigma^2_{A_V}$ is on the exponential scale, as is the case for the residual variance in our analysis, GCVVe is close to $\sqrt{\sigma^2_{A_V}}$ [37, 46].

RESULTS

Genetic Parameters for Trait Levels

Estimated genetic parameters for levels of harvest weight, body length, depth, and width are in Table 2. The estimated heritability for individual harvest weight (estimated by using the average residual variance across all observations) was equal to 0.25 (0.04) and the same value was obtained with a univariate model assuming a homogeneous residual variance (results not shown). The log-likelihood ratio tests indicated that both group and kin effects were highly significant ($p < 0.0001$). The group effect explained 13 % of the phenotypic variance, which shows that individuals within the same group are more similar to each other than to members of other groups. The kin effect explained 10 % of the phenotypic variance, which indicates that individuals within the same family are more alike compared to individuals of the other family in the group, in addition to their genetic similarity. We tested the model for harvest weight when group and kin effects were not included and found that removing one or both effects created an upward bias in the estimated variances for both the level and variance of the trait (results not shown). The social maternal effect was significant ($p < 0.001$) but small and explained 2 % of the phenotypic variance.

Table 2: Genetic parameters for level of harvest weight, length, depth, and width

Parameter	Harvest weight	Length	Depth	Width
$^{a}\sigma_A^2$	573.46 (115.80)	0.732 (0.136)	0.202 (0.037)	0.034 (0.007)
σ_e^2	1426.3 (27.99)	1.443 (0.028)	0.365 (0.007)	0.067 (0.001)
σ_g^2	300.26 (42.81)	0.354 (0.047)	0.104 (0.012)	0.037 (0.004)
σ_k^2	240.29 (35.45)	0.235 (0.035)	0.047 (0.008)	0.013 (0.002)
σ_m^2	43.64 (20.83)	–	0.013 (0.006)	–
σ_P^2	2297.2 (70.78)	2.418 (0.081)	0.631 (0.022)	0.136 (0.005)
h^2	0.25 (0.04)	0.30 (0.05)	0.32 (0.05)	0.25 (0.05)
$^{b}g^2$	0.13 (0.02)	0.15 (0.02)	0.16 (0.02)	0.27 (0.02)
$^{c}k^2$	0.10 (0.02)	0.10 (0.01)	0.08 (0.01)	0.10 (0.01)
$^{d}m^2$	0.02 (0.01)	–	0.02 (0.01)	–
^{e}GCV	0.14	0.05	0.06	0.06

Standard errors are indicated between brackets

[a] Additive genetic variance was calculated as four times the sire-dam variance

[b] Group effect, calculated as $g^2 = \sigma_g^2/\sigma_P^2$

[c] Kin effect, calculated as $k^2 = \sigma_k^2/\sigma_P^2$

[d] Social maternal effect, calculated as $m^2 = \sigma_m^2/\sigma_P^2$

[e] Genetic coefficient of variation

Heritabilities of harvest weight and body width were similar (0.25 ± 0.05), while heritabilities of body length and body depth were a little higher (~0.30 ± 0.05). The group effect explained ~15 % of the phenotypic variance for length and depth, and 27 % for width. The kin effect explained ~10 % of the phenotypic variance for all three body size traits.

Genetic Parameters for the Variance of Traits

Estimated genetic parameters for the variance of harvest weight, body length, depth, and width are in Table 3. For all traits, the contribution of genetic effects to their variance was highly significant ($p < 0.0001$). Estimated GCV_{Ve} for harvest weight was high and equal to 0.58, whereas for body size traits, GCV_{Ve} were lower i.e. 0.39, 0.42, and 0.45 for length, depth and width, respectively. These estimates indicate that there is substantial genetic variation in the residual variance compared to the average value of the residual variance, for all analyzed traits.

Table 3: Genetic parameters for the variance of harvest weight, length, depth, and width

Parameter	Harvest weight	Length	Depth	Width
[a] σ^2_A	0.343 (0.068)	0.156 (0.041)	0.184 (0.042)	0.203 (0.048)
σ^2_e	1.747 (0.034)	1.924 (0.038)	1.862 (0.036)	1.696 (0.033)
[b] σ^2_g	0.040 (0.021)	0.031 (0.021)	0.031 (0.018)	0.073 (0.020)
[c] σ^2_k	0.078 (0.027)	0.098 (0.029)	0.022 (0.023)	0.062 (0.023)
[d] σ^2_m	0.009 (0.009)	–	0.023 (0.011)	-
[e] GCV_{V_e}	0.58	0.39	0.42	0.45

Standard errors are indicated between brackets

[a] Additive genetic variance was calculated as four times the sire-dam variance

[b] Group variance

[c] Kin variance

[d] Social maternal variance

[e] Genetic coefficient of variation at variance level

Genetic Correlations between Level and Variance of Traits

Estimated genetic correlations between level and variance for harvest weight and body size traits are in Table 4. The genetic correlation between level and variance for harvest weight was high and positive (0.60 ± 0.09), which implies that selection for increased harvest weight will also yield more variation in the level of this trait. For body size traits, genetic correlations between level and variance were lower than for harvest weight, and were not significantly different from 0 for length and width, but moderate and positive for depth (0.37 ± 0.13).

Table 4: Genetic correlations between level and the variance for harvest weight, length, depth, and width

Harvest weight	Length	Depth	Width
0.60 (0.09)	0.11 (0.16)	0.37 (0.13)	0.20 (0.15)

Standard errors are indicated between brackets

DISCUSSION

In this study, we used a DHGLM to estimate genetic variation in uniformity of harvest weight and three body size traits, i.e. length, depth, and width. Our results showed substantial genetic variation in uniformity of all analyzed traits, with GCVVeGCVVe ranging from 39 to 58 %, while GCVGCVfor trait levels ranged from 5 to 15 %. A strong genetic correlation of 0.60 was found between trait level and variance, which suggests that selection for increased body weight at harvest will also result in more variation in the level of this trait.

Heritability of Individual Harvest Weight and Body Size Traits

Estimated heritability for individual harvest weight was moderate (0.25 ± 0.04), which is similar to results from previous studies on Nile tilapia [2, 47, 48]. To date, the GIFT strain has undergone 14 generations of selection for harvest weight. Our findings, together with the small average inbreeding coefficient of 3.1 % in the analyzed GIFT population, suggest that there is still a considerable amount of genetic diversity available for further selection, which is also in agreement with the positive genetic trend observed in the GIFT strain [3].

Heritabilities for individual body size traits were also moderate (0.25 to 0.32), which provide opportunities to improve body size traits in Nile tilapia. Body size traits could become traits of interest in future breeding programs since selection for heavier fish may lead to body shapes that deviate from the natural shape, the latter being favored by consumers [13, 49, 50].

Genetic Variance in Uniformity of Harvest Weight

Variance components that are estimated using the exponential model, as in this study, are independent of the scale of the trait, and thus, are comparable across traits and species [24, 30]. We found a substantial additive genetic variance for uniformity of harvest weight (0.34 ± 0.07; Table 3), which is larger than that in a similar study on Atlantic salmon by Sonesson et al. [30], who reported an additive genetic variance in the residual variance of 0.17 on the exponential scale. Our estimates are also higher than those reported for livestock traits [23, 24, 37, 51, 52]. These findings suggest that the observed phenotypic variability of harvest weight in the GIFT strain has a substantial genetic component.

Regardless of the underlying model, comparison of additive genetic parameters for uniformity across different studies can also be done by using the genetic coefficient of variation for residual variance (GCV_{Ve}) [37, 46]. GCV_{Ve} describes the change in residual variance when a genetic standard deviation of 1 is achieved in response to selection, relative to the mean of the residual variance. In our study, GCV_{Ve} for harvest weight was large i.e. 0.58. The proportional change in phenotypic variance can be calculated as $GCV_{Ve}(\sigma^2_E/\sigma^2_P)$, which in the case of harvest weight would be equal to 0.36. In the literature, GCV_{Ve} for variability of traits in livestock and laboratory animals usually ranges from 0.2 to 0.6 [37]. For uniformity of body weight in rainbow trout, GCV_{Ve} of 0.37 and ~0.2 were reported by Janhunen et al. [29] and Sae-Lim et al. [31], respectively, which are lower than the values found in our study. The estimated GCV_{Ve} for harvest weight suggests that there is sufficient genetic variation to allow a substantial change in the residual variance of this trait compared to its average value within a single generation of selection, which would be much larger than that for harvest weight level (Table 2). However, it should be noted that the accuracy of selection for uniformity tends to be lower than for trait levels [19], and that expressions for response to selection on environmental variability do not depend on GCV_{Ve} only [7, 17, 27, 46].

Effect of Data Distribution

The estimated level and variance for harvest weight could be influenced by the non-normal distribution of harvest weight. In data on aquaculture species, skewness is not unusual [6, 53]. A skewed distribution can result from inter-individual competition and subsequent feeding hierarchy, with a few individuals dominating the rest of the group. However, in many statistical inferences, normality is assumed and this is especially important in the analysis of the genetic heterogeneity of environmental variance [54]. To test whether genetic variation in residual variance is merely an artifact of a non-normal distribution, we applied a Box-Cox transformation to harvest weight. The transformation resulted in a normally distributed trait, which was then analyzed with the DHGLM. Results of the analysis showed that this transformation had only a minor effect on the estimated genetic parameters for trait level, but decreased the variance of the residual variance. Similar results were found in other studies that analyzed transformed traits [30, 31, 54]. Although the additive genetic variance of uniformity decreased somewhat after the Box–Cox transformation, this difference was not significant (p = 0.22). Thus, our results indicate that there is genetic variation in uniformity of harvest weight, irrespective of the scale of measurement of the trait. Unlike harvest weight, body length, depth, and width were normally distributed.

Genetic Correlation between Level and Variance of Harvest Weight and Body Size Traits

Our results imply that the observed variation in harvest weight in the GIFT strain could be reduced by selective breeding. However, selection for more uniform fish may result in a trade-off on improvement of harvest weight. The genetic correlation between level and variance of harvest weight is high and positive, 0.60 (Table 4), which means that single-trait selection for heavier fish will increase the variation in harvest weight among individuals. Similar correlations were obtained by Sae-Lim et al. [31]. Simultaneous improvement of harvest weight and its uniformity will therefore require index selection.

To maximize profit, not only uniformity of weight but also uniformity of size, may play an important role in fish farming, especially for markets where fish are sold as whole. The magnitudes of GCVGCV for uniformity of body size traits and harvest weight were similar but improvement of body size traits based on the estimated correlations (Table 4) is expected to have a limited effect on the level of these traits.

Factors Affecting Magnitude of Variability and Genetic Variance in Variability

In our analyses we used a sire and dam model, which fits the additive genetic mid-parent mean, while the Mendelian sampling deviation is part of the residual. This can potentially inflate the size of the estimated genetic variation in residual variance in case of heterogeneous Mendelian sampling variation, which is then confounded with the genetic part of the residual variance [30]. A Mendelian sampling variance that is heterogeneous among families can result from differential inbreeding coefficients of parents, or from the presence of a major gene that is segregating in some families but not in others [20].

In aquaculture species, maternal common environmental effects can have an important role in explaining differences among families. These effects can be included in the estimation of genetic parameters as non-genetic effects that account for covariances between full-sibs due to a shared environment. In this study, maternal common environmental effects were excluded from the models because of convergence problems, which arose when those effects were included. The same issue was observed in other studies that used the same dataset and for which the results showed confounding of maternal common environmental effects and direct genetic effects [12, 34]. The main difficulty that occurs when disentangling the two effects is due to the mating of one male with two females. Moreover, in our experiment, mating was often partly unsuccessful and resulted in 1 × 1 mating instead. However, even a perfect 2 × 2 mating design results in limited power to separate genetic and maternal common environmental effects, at least at the variance level, as reported by Sonesson et al. [30]. Previous studies on a larger GIFT population for which 1 × 2 mating was more successful, detected significant maternal common environmental effects (0.34) for individual harvest weight [2]. Thus, our estimates of the genetic variance of uniformity may be inflated by the inability to fit maternal common environmental effects.

A recent study on birth weight of mice treated environmental variability as a maternal trait, and found a positive response to selection [55]. In an earlier study, the same authors found evidence that environmental variability of birth weight was more likely to be a maternal genetic trait than a trait due to direct genetic effects [4]. In the study by Rutten et al. [56], the variance of body weight due to common environmental effects, which include maternal genetic and non-additive genetic effects, decreased with age. Since in our study, traits were measured at harvest, maternal genetic or common environmental effects probably explain only a small proportion of the heterogeneity of residual variance.

Mulder et al. [19] showed that the estimated genetic correlation between residual variances for body weight of both sexes was only 0.11, which suggests that they are different traits. A similar analysis could be conducted on our data, to investigate whether the large effect of sex is associated with a genetic correlation for variability between sexes that is less than 1. Ponzoni et al. [2] recorded the CV of body weight in the GIFT strain across eight generations and observed that good breeding management contributed to reduce the CV, although its average value remained at around 40 %. Thus, reducing uniformity will require both genetic and management interventions.

CV for Harvest Weight

In our experiment, the feeding strategy differed from that in the ordinary GIFT breeding program. Instead of spreading food on the surface of the pond as in the GIFT breeding population, we placed it in the corner of the net-cages so that the fish showed their competitive tendency. The CV for harvest weight in our study (35 %) was lower than the values found in previous studies on the GIFT strain where fish were communally reared (48 to 59 %) [10, 50]. Thus, there is no evidence that the level of competition between individuals was higher in our conditions than in the communal rearing conditions of these studies. In communal rearing, the feed is not spread over the entire pond's surface because auto-feeders are not available, which may cause some competition. In addition, the fish in our experiment were kept in small net-cages and stocked at low density, while in commercial ponds all fish are kept together at high density. Because of the differences in rearing conditions, the question of whether our results can be extended to commercial situations remains open. A selection experiment, in which parents are kept in many small groups and selected for uniformity while offspring are evaluated under commercial conditions, would constitute the ideal proof.

Future Prospects

Although studies on the genetic heterogeneity of environmental variance date as far back as 1942 [57], selection experiments to improve uniformity in livestock are scarce. Still, some experiments [58, 59, 60, 61] that were based on divergent selection for phenotypic variance, provided evidence for a genetic component in the phenotypic variability and suggested the possibility that this variability could be reduced by selective breeding. To our knowledge, selection for uniformity has never been performed in aquaculture species. Nevertheless, the high GCV_{Ve} found in our and other studies on aquaculture species suggest that aquaculture populations are suitable to validate the estimated genetic parameters by a selection experiment. Selection for uniformity of body weight or size could lead to increased profit by producing more fish in the size range that is favored by the consumers. Moreover, from the point of view of animal welfare, uniformity of fish body weight and size could reduce competition, and thus possible stress, injuries, and even mortality.

We studied the genetic variance of the residual variability. However, the total phenotypic variability also depends on other factors [62], as shown by the significant fixed effects on variability, for example sex effect (see above). Hence, decreasing the total phenotypic variability even more would require reducing the magnitude of these fixed effects. When the genetic correlation between growth rate in males and females differs from 1, it is possible, in principle, to remove the variability due to a difference in mean body weight between sexes. The magnitude of environmental effects, such as group and batch effects, is related to environmental sensitivity (and thus to genotype by environment interactions). Evaluating the prospects of reducing these components by genetic selection will require further research.

An interesting property of the specific design of our experiment is that it allows the simultaneous study of uniformity and social effects such as group and kin effects in our study and indirect genetic effects, which were analyzed in other studies on the same data [12, 34]. However, the experimental setting and feeding strategy that we applied differed from those in a commercial setting. Thus, genotype by environment interactions may be present and our results may not represent uniformity in the case of commercial tilapia farms. The DHGLM approach could be used to test whether the genetic background of uniformity differs between both environments. Results from such a study would be a useful addition to our findings.

CONCLUSIONS

Our study revealed substantial genetic variation in uniformity of harvest weight and body size traits, which opens promising prospects for the genetic improvement of uniformity by selective breeding of the GIFT strain. The genetic correlation between level and variance of harvest weight was high and positive, which indicates that selection for heavier fish may also result in more variation in harvest weight. Simultaneous improvement of harvest weight and its uniformity will thus require index selection.

NOTES

Authors' Contributions

JM performed the statistical analysis with the help from PB and HAM, and drafted the manuscript. PB and HAM contributed in designing the study, interpretation of the results and the writing of the manuscript. HLK helped in interpreting the results and drafting the manuscript. All authors read and approved the final manuscript.

Acknowledgements

The authors would like to thank DJ de Koning and L Rönnegård for their valuable comments on the draft. We would also like to thank the editor and two anonymous reviewers for their constructive comments and suggestions that helped us improve the manuscript. WorldFish is kindly acknowledged for providing the data for this research. JM benefited from a joint Grant from the European Commission and Wageningen University, within the framework of the Erasmus-Mundus joint doctorate programme "EGS-ABG".

REFERENCES

1. World Fish. http://www.worldfishcenter.org (2015). Accessed 8 July 2015.
2. Ponzoni RW, Nguyen NH, Khaw HL, Hamzah A, Bakar KRA, Yee HY. Genetic improvement of Nile tilapia (*Oreochromis niloticus*) with special reference to the work conducted by the WorldFish Center with the GIFT strain. Rev Aquac. 2011;3:27–41.
3. Khaw HL. Social interactions and variability in live body weight in the GIFT strain (Nile tilapia, *Oreochromis niloticus*) in Malaysia. PhD thesis, Wageningen University. 2015.
4. Pun A, Cervantes I, Nieto B, Salgado C, Pérez-Cabal MA, Ibáñez-Escriche N, et al. Genetic parameters for birthweight environmental variability in mice. J Anim Breed Genet. 2013;130:404–14.
5. Mulder HA, Bijma P, Hill WG. Selection for uniformity in livestock by exploiting genetic heterogeneity of residual variance. Genet Sel Evol. 2008;40:37–59.

6. Cutts CJ, Betcalfe NB, Caylor AC. Aggression and growth depression in juvenile Atlantic salmon: the consequences of individual variation in standard metabolic rate. J Fish Biol. 1998;52:1026–37.
7. SanCristobal-Gaudy M, Bodin L, Elsen JM, Chevalet C. Genetic components of litter size variability in sheep. Genet Sel Evol. 2001;33:249–71.
8. Jobling M. Simple indices for the assessment of the influences of social environment on growth performance, exemplified by studies on Arctic charr. Aquac Int. 1995;3:60–5.
9. Hart PJB, Salvanes AGV. Individual variation in competitive performance of juvenile cod and its consequences for growth. J Mar Biol Assoc UK. 2000;80:569–70.
10. Ponzoni RW, Hamzah A, Tan S, Kamaruzzaman N. Genetic parameters and response to selection for live weight in the GIFT strain of Nile Tilapia (*Oreochromis niloticus*). Aquaculture. 2005;247:203–10.
11. Sae-Lim P, Komen H, Kause A, van Arendonk JAM, Barfoot AJ, Martin KE, et al. Defining desired genetic gains for rainbow trout breeding objective using analytic hierarchy process. J Anim Sci. 2012;90:1766–76.
12. Khaw HL, Ponzoni RW, Yee HY, bin Aziz MA, Mulder HA, Marjanovic J, et al. Genetic variance for uniformity of harvest weight in Nile tilapia (*Oreochromis niloticus*). Aquaculture. 2016;451:113–20.
13. Colihueque N, Araneda C. Appearance traits in fish farming: progress from classical genetics to genomics, providing insight into current and potential genetic improvement. Front Genet. 2014;5:251.
14. Kause A, Ritola O, Paananen T, Eskelinen U, Mantysaari E. Big and beautiful? Quantitative genetic parameters for appearance of large rainbow trout. J Fish Biol. 2003;62:610–22.
15. Blonk RJW, Komen J, Tenghe A, Kamstra A, van Arendonk JAM. Heritability of shape in common sole, *Solea solea*, estimated from image analysis data. Aquaculture. 2010;307:6–11.
16. Sorensen D, Waagepetersen R. Normal linear models with genetically structured residual variance heterogeneity: a case study. Genet Res. 2003;82:207–22.

17. SanCristobal-Gaudy M, Elsen J-M, Bodin L, Chevalet C. Prediction of the response to a selection for canalisation of a continuous trait in animal breeding. Genet Sel Evol. 1998;30:423–51.Central
18. Waddington CH. Experiments on canalizing selection. Genet Res. 2009;1:140–50.
19. Mulder HA, Hill WG, Vereijken A, Veerkamp RF. Estimation of genetic variation in residual variance in female and male broiler chickens. Animal. 2009;3:1673–80.
20. Rowe SJ, White IMS, Avendaño S, Hill WG. Genetic heterogeneity of residual variance in broiler chickens. Genet Sel Evol. 2006;38:617–35.
21. Damgaard LH, Rydhmer L, Løvendahl P, Grandinson K. Genetic parameters for within-litter variation in piglet birth weight and change in within-litter variation during suckling. J Anim Sci. 2002;81:604–10.
22. Garreau H, Bolet G, Larzul C, Robert-Granié C, Saleil G, SanCristobal M, et al. Results of four generations of a canalising selection for rabbit birth weight. Livest Sci. 2008;119:55–62.
23. Vandenplas J, Bastin C, Gengler N, Mulder HA. Genetic variance in micro-environmental sensitivity for milk and milk quality in Walloon Holstein cattle. J Dairy Sci. 2013;96:5977–90.
24. Rönnegård L, Felleki M, Fikse WF, Mulder HA, Strandberg E. Variance component and breeding value estimation for genetic heterogeneity of residual variance in Swedish Holstein dairy cattle. J Dairy Sci. 2013;96:2627–36.
25. Ibáñez-Escriche N, Sorensen D, Waagepetersen R, Blasco A. Selection for environmental variation: a statistical analysis and power calculations to detect response. Genetics. 2008;180:2209–26.
26. Ibáñez-Escriche N, Moreno A, Nieto B, Piqueras P, Salgado C, Gutiérrez JP. Genetic parameters related to environmental variability of weight traits in a selection experiment for weight gain in mice; signs of correlated canalised response. Genet Sel Evol. 2008;40:279–93.
27. Ibáñez-Escriche N, Varona L, Sorensen D, Noguera JL. A study of heterogeneity of environmental variance for slaughter weight in pigs. Animal. 2008;2:19–26.

28. Gutiérrez J, Nieto B, Piqueras P, Ibáñez N, Salgado C. Genetic parameters for canalisation analysis of litter size and litter weight traits at birth in mice. Genet Sel Evol. 2006;38:445–62.
29. Janhunen M, Kause A, Vehviläinen H, Järvisalo O. Genetics of microenvironmental sensitivity of body weight in rainbow trout (*Oncorhynchus mykiss*) selected for improved growth. PLoS One. 2012;7:e38766.
30. Sonesson AK, Odegård J, Rönnegård L. Genetic heterogeneity of within-family variance of body weight in Atlantic salmon (*Salmo salar*). Genet Sel Evol. 2013;45:41.
31. Sae-Lim P, Kause A, Janhunen M, Vehviläinen H, Koskinen H, Gjerde B, et al. Genetic (co)variance of rainbow trout (*Oncorhynchus mykiss*) body weight and its uniformity across production environments. Genet Sel Evol. 2015;47:46.
32. Yang J, Loos RJF, Powell JE, Medland SE, Speliotes EK, Chasman DI, et al. FTO genotype is associated with phenotypic variability of body mass index. Nature. 2012;490:267–72.
33. Rönnegård L, Felleki M, Fikse F, Mulder HA, Strandberg E. Genetic heterogeneity of residual variance-estimation of variance components using double hierarchical generalized linear models. Genet Sel Evol. 2010;42:8.
34. Khaw HL, Ponzoni RW, Yee HY, Aziz M, Bijma P. Genetic and non-genetic indirect effects for harvest weight in the GIFT strain of Nile tilapia (*Oreochromis niloticus*). Aquaculture. 2016;450:154–61.
35. Bijma P. Estimating indirect genetic effects: precision of estimates and optimum designs. Genetics. 2010;186:1013–28.
36. Thodesen J, Ponzoni RW. GIFT technology manual: an aid to tilapia selective breeding. Bayan Lepas: WorldFish; 2004.
37. Hill WG, Mulder HA. Genetic analysis of environmental variation. Genet Res (Camb). 2010;92:381–95.
38. Bergsma R, Kanis E, Knol EF, Bijma P. The contribution of social effects to heritable variation in finishing traits of domestic pigs (*Sus scrofa*). Genetics. 2008;178:1559–70.
39. Bijma P, Muir WM, Ellen ED, Wolf JB, Van Arendonk JAM. Multilevel selection 2: estimating the genetic parameters determining inheritance and response to selection. Genetics. 2007;175:289–99.

40. Hamilton WD. The genetical evolution of social behaviour I. J Theor Biol. 1964;7:1–16.

41. File AL, Murphy GP, Dudley SA. Fitness consequences of plants growing with siblings: reconciling kin selection, niche partitioning and competitive ability. Proc Biol Sci. 2012;279:209–18.

42. Lee Y, Nelder JA. Double hierarchical generalized linear models (with discussion). J R Stat Soc Ser C Appl Stat. 2006;55:139–85.

43. Mulder HA, Rönnegård L, Fikse WF, Veerkamp RF, Strandberg E. Estimation of genetic variance for macro- and micro-environmental sensitivity using double hierarchical generalized linear models. Genet Sel Evol. 2013;45:23.

44. Felleki M, Lee D, Lee Y, Gilmour AR, Rönnegård L. Estimation of breeding values for mean and dispersion, their variance and correlation using double hierarchical generalized linear models. Genet Res (Camb). 2012;94:307–17.

45. Houle D. Comparing evolvability and variability of quantitative traits. Genetics. 1992;130:195–204.

46. Mulder HA, Bijma P, Hill WG. Prediction of breeding values and selection responses with genetic heterogeneity of environmental variance. Genetics. 2007;175:1895–910.

47. Khaw HL, Bovenhuis H, Ponzoni RW, Rezk MA, Charo-Karisa H, Komen H. Genetic analysis of Nile tilapia (*Oreochromis niloticus*) selection line reared in two input environments. Aquaculture. 2009;294:37–42.

48. Khaw HL, Ponzoni RW, Hamzah A, Abu-Bakar KR, Bijma P. Genotype by production environment interaction in the GIFT strain of Nile tilapia (*Oreochromis niloticus*). Aquaculture. 2012;326–329:53–60.

49. Trọng TQ, Mulder HA, van Arendonk JAM, Komen H. Heritability and genotype by environment interaction estimates for harvest weight, growth rate, and shape of Nile tilapia (*Oreochromis niloticus*) grown in river cage and VAC in Vietnam. Aquaculture. 2013;384–387:119–27.

50. Nguyen NH, Khaw HL, Ponzoni RW, Hamzah A, Kamaruzzaman N. Can sexual dimorphism and body shape be altered in Nile

tilapia (*Oreochromis niloticus*) by genetic means? Aquaculture. 2007;272:S38–46.

51. Mulder HA, Hill WG, Knol EF. Heritable environmental variance causes nonlinear relationships between traits: application to birth weight and stillbirth of pigs. Genetics. 2015;199:1255–69.
52. Sell-Kubiak E, Bijma P, Knol EF, Mulder HA. Comparison of methods to study uniformity of traits: application to birth weight in pigs. J Anim Sci. 2015;93:900–11.
53. Jobling M, Reinsnes TG. Physiological and social constraints on growth of Arctic charr, *Salvefinus alpinus L.*: an investigation of factors leading to stunting. J Fish Biol. 1986;28:379–84.
54. Yang Y, Christensen OF, Sorensen D. Analysis of a genetically structured variance heterogeneity model using the Box–Cox transformation. Genet Res (Camb). 2011;93:33–46.
55. Formoso-Rafferty N, Cervantes I, Ibáñez-Escriche N, Gutiérrez JP. Genetic control of the environmental variance for birth weight in seven generations of a divergent selection experiment in mice. J Anim Breed Genet. 2015. doi:10.1111/jbg.12174.
56. Rutten MJM, Komen H, Bovenhuis H. Longitudinal genetic analysis of Nile tilapia (*Oreochromis niloticus* L.) body weight using a random regression mode. Aquaculture. 2005;246:101–13.
57. Waddington CH. Canalization of development and the inheritance of aquired characters. Nature. 1942;150:563–5.
58. Kaufman PK, Enfield FD, Comstock RE. Stabilizing selection for pupa weight in *Tribolium castaneum*. Genetics. 1977;87:327–41.
59. Ibanez-Escriche N, Argente MJ, García ML, Muelas R, Santacreu MA, Blasco A. Preliminary results in a divergent selection experiment on variance for litter size in rabbits. I. Geznetic parameters. In: Proceedings of the 9th World Rabbit Congress: 10–13 June 2008, Verona; 2014. p. 121–124.
60. Bodin L, Bolet G, Garcia M, Garreau H, Larzul C, David I. Robustesse et canalisation: vision de généticiens. Inra Prod Anim. 2012;23:11–22.
61. Rendel JM, Sheldon BL, Finlay DE. Selection for canalization of the scute phenotype. II. Am Nat. 1966;100:13–31.

62. Moreno A, Ibáñez-Escriche N, García-Ballesteros S, Salgado C, Nieto B, Gutiérrez JP. Correlated genetic trend in the environmental variability of weight traits in mice. Livest Sci. 2012;148:189–95.

CHAPTER

12

BREEDING SALMONIDS FOR FEED EFFICIENCY IN CURRENT FISHMEAL AND FUTURE PLANT-BASED DIET ENVIRONMENTS

Cheryl D. Quinton[1], Antti Kause[1], Juha Koskela[2] and Ossi Ritola[3]

[a] MTT Agrifood Research Finland, Biotechnology and Food Research, Biometrical Genetics, 31600 Jokioinen, Finland

[b] Finnish Game and Fisheries Research Institute Jyväskylä, Survontie 9, 40500 Jyväskylä, Finland

[c] Finnish Game and Fisheries Research Institute, Tervo Fisheries Research and Aquaculture, 72210 Tervo, Finland

ABSTRACT

The aquaculture industry is increasingly replacing fishmeal in feeds for carnivorous fish with soybean meal (SBM). This diet change presents a potential for genotype-environment (G × E) interactions. We tested whether current salmonid breeding programmes that evaluate and select within fishmeal diets also improve growth and efficiency on potential future SBM diets. A total of 1680 European whitefish from 70 families were reared with either fishmeal- or SBM-based diets in a split-family design. Individual daily gain (DG), daily feed intake (DFI) and feed efficiency (FE) were recorded. Traits displayed only weak G × E interactions as variances and heritabilities did not differ substantially between the diets, and cross-diet genetic correlations were near unity. In both diets, DFI exhibited moderate heritability and had very high genetic correlation with DG whereas FE had low heritability. Predicted genetic responses demonstrated that selection to increase DG and FE on the fishmeal diet lead to favourable responses on the SBM diet. Selection for FE based on an index including DG and DFI achieved at least double FE gain *versus* selection on DG alone. Therefore, current breeding programmes are improving the biological ability of salmonids to use novel plant-based diets, and aiding the aquaculture industry to reduce fishmeal use.

Keywords: Feed efficiency genotype-environment interaction selection aquaculture *Coregonus lavaretus*

INTRODUCTION

The use of fishmeal in aquaculture feeds has become a major issue as the global industry continues to grow [10]. Fishmeal is mostly produced from wild-caught small pelagic fish species, and is considered to be the superior protein source for teleost fishes [18]. It is currently the major component of diets for farmed carnivorous salmonids such as rainbow trout (Oncorhynchus mykiss) and Atlantic salmon (Salmo salar) whose grower diets typically contain 30–50% fishmeal (2000 statistics [35,45]). However, several environmental and economic reasons exist for reducing fishmeal use. These include the decline and fluctuation of wild stocks harvested for fishmeal production due to overfishing and natural environmental factors, instability and predicted increases in fishmeal prices, and negative consumer perception of the safety of fishmeal-fed products [10,34]. Accordingly, feed manufacturers aim to replace fishmeal with alternatives such as plant products. Soybean meal is one source of protein that can be successfully substituted in part for fishmeal in carnivorous fishes' feeds [18,23,36,41,43]. Some major feed companies have pledged to reduce fishmeal use by at least 50% by 2010, and it has been predicted that within 10 years, fish diets will be close to 100% vegetarian [39].

Aquaculture genetic improvement programmes aim to complement such production system changes. Breeding programmes exist for all major farmed carnivorous fish species, and have achieved improvements in growth, feed efficiency, disease resistance and product quality traits [12]. Improved feed efficiency in particular is a goal of most fish breeding programmes, but genetic parameter estimates for feed consumption and efficiency are rare [21,22,24,42]. This is because recording methods for individual fish feed intake have only recently been implemented in large-scale genetics research [21, 22, 42]; thus no current breeding programmes select directly for feed efficiency. Furthermore, all breeding programmes evaluate and select broodstock based on performance with current fishmeal-based diets. Future plant-based diets may impact fish breeding programmes if genotype-environment (G × E) interactions occur. G × E interactions may occur in the form of genotype re-ranking across environments, or scaling effects as indicated by environmental differences in trait genetic variation [9, 30]. In a worst-case scenario, superior genotypes on a fishmeal diet would actually be inferior on a plant-based diet. This would be shown by a negative genetic correlation between diets [8]. In this case, current selection on fishmeal diets would compromise performance on future plant based diets.

In this study, the impact of a novel soybean protein-based diet on selective improvement of growth, feed intake and feed efficiency was investigated in European whitefish (Coregonus lavaretus L.), a salmonid now farmed commercially in Finland. A breeding programme has recently been established to improve growth and feed efficiency in this species. To assess the degree to which these traits are affected by G × E interactions when reared with either a traditional fishmeal diet or a potential future soybean meal-based diet, we tested whether these traits express diet-specific phenotypic and genetic variation, and estimated trait genetic correlations between the diets. To quantify the impact of G × E interaction on selection response, we predicted genetic changes occurring on both diets in response to alternative strategies of selection for growth or feed efficiency on either fishmeal or soybean meal-based diets.

MATERIALS AND METHODS

Diet Formulations

Two practical isonitrogenous and isocaloric diets were formulated (Tab. I). In the fishmeal (FM) diet, fishmeal supplied 100% of the dietary protein. This diet represents a typical commercial diet used in whitefish farming. In the soybean meal (SBM) diet, 50% of the dietary protein was replaced with SBMderived protein. This diet represents a realistic future diet. Methionine, lysine and phosphorus supplements were added to the SBM diet to balance these levels with the FM diet. Ingredients were mixed in a Hobart-type mixer, extruded (Clextral BC 45, FR) to 3.5 mm pellets, re-dried at 40–45 °C, top dressed with fish oil and stored in a freezer until use.

Population and Experimental Design

Whitefish in the experiment originated from the breeding programme based at the Tervo station of the Finnish Game and Fisheries Research Institute (FGFRI). The original broodstock was established in 1998 by mating 50 wild males and 150 wild females originating from the Kokemäki River, Finland. In October 2003, the generation for the current experiment was established: 45 sires were mated with 52 dams in a partial factorial design to create 70 fulland half-sib families. Each sire was mated to an average of 1.6 dams and each dam to an average of 1.3 sires (both ranges 1–2). Genetic relationships among sires and dams are unknown.

Table I: Formulation and analyzed nutrient composition of the experimental fishmeal and soybean meal diets

	Fishmeal ($g \cdot kg^{-1}$)	Soybean meal ($g \cdot kg^{-1}$)
Ingredient		
Fishmeal, capelin[a]	480	255
Soybean meal, defatted, hulls removed by sieve[a]	0	411
Fish oil, sandeel[a]	160	180
Wheat gluten[a]	20	20
Wheat meal[a]	267.5	34.0
Wheat starch, pregelatinized[a]	52.5	52.5
Mono calcium phosphate[b]	0	16
L-lysine[c]	0	7
DL-methionine[d]	0.0	4.5
Vitamin and mineral mix[e]	20	20
Analysed composition		
Water	23	27
Crude protein	398	404
Crude lipid	210	210
Crude fibre	10	22
N-free extracts[f]	359	337
Gross energy ($MJ \cdot kg^{-1}$)	22.6	22.3

Sources: [a] Raisio, FI; [b] Kemira, FI; [c] BASF, DE; [d] Welding GmbH, DE.
[e] Added to supply (per kg diet): retinol acetate 8000 IU, cholecalciferol 3000 IU, *all-race-α*-tocopheryl acetate 300 IU, menadione sodium bisulfite 10 mg, thiamine HCl 21 mg, riboflavin 30 mg, calcium *d*-pantothenate 92 mg, biotin 0.3 mg, folic acid 6 mg, vitamin B_{12} 0.04 mg, niacin 120 mg, pyridoxineHCl 20 mg, ascorbic acid (35% Stay C) 900 mg, inositol 200 mg, manganese oxide (62% Mn) 100 mg, zinc oxide (74% Zn) 200 mg, potassium iodide (76% I) 6 mg.
[f] Calculated as 1000 – (water + protein + lipid + fibre).

Families were kept separate during incubation and early rearing. At the eyed-egg stage (January 2004), the families were transported to the FGFRI Laukaa Research Station and incubated (water 4–6 °C). After hatching (February 2004), 100 to 150 fish per family were held in indoor 150 L fibreglass tanks (water 13–15 °C) and fed commercial dry diets (larvae: AgloNorse, EWOS Ltd., NO; juveniles: Nutra Parr and Royal Silver, Raisio Ltd., FI). In June 2004, approximately 40 random individuals from each family were tagged by injecting a passive integrated transponder into the body cavity. To give all the fish the same initial nutritional environment and to identify if tagging harmed any fish, all fish were fed with a 1:1 mixture of the two experimental diets for four weeks prior to the trial.

Twenty-four tagged fish per family were randomly sampled for the diet trial. To construct a split-family design, each family was first randomly split into two groups to be reared with the alternative diets. Each group was evenly distributed over 6 round 0.6 m^3 replicate tanks (12 experimental tanks in total). Consequently, the diet trial began with a total of 1680 fish, each tank containing 140 fish (two fish from each family). The trial was conducted from July 29 to October 21, 2004, during which fish tripled in weight. Fish were fed 6 $h \cdot d^{-1}$ using belt feeders. Fish were counted and bulk-weighed biweekly and feeding was adjusted according to average weight and tank biomass. To ensure excess feeding, the feed amounts supplied were adjusted to be 1.3 times higher than predicted by Koskela [27]. Tanks were supplied with fresh water (14.8–15.1 °C; flow rate 8–16 $L \cdot min^{-1}$; outlet water O_2 level > 80% saturation) and 24 h light was provided with ceiling fluorescent tubes. The experimental conditions were standardized to permit accurate comparison of families across diets. This test environment deviates from commercial circumstances where fish are reared in large outdoor net cages under naturally varying environmental conditions.

Traits Recorded

Individual body weights were recorded to the nearest g at the beginning and end of the trial (Tab. II). Daily feed intake was measured by X-radiography [17] 5 times per individual, with 2-week intervals between measurements. On feed intake measurement days, fish were fed with the same methods and quantity as during normal days, but feed pellets included lead glass beads (Ballotini size 8.5; Jencons Ltd., UK) visible in X-ray. Afterward, fish were anaesthetized, identified, and X-rayed (Bennett HFQ 3000P X-ray machine, US).

To transform the number of glass beads fed to the amount of feed ingested, predictive regression models were established in a separate study. For each diet, 16 samples of known weights were taken from the bead-labelled pellets and X-rayed. The number of beads present in each sample was counted, and diet-specific regression equations were obtained ($R^2 = 0.97–0.98$). Feed intake (g) for each individual was predicted using these equations from the number of beads observed under X-ray [17].

Traits analyzed were individual daily weight gain (DG), average daily feed intake (DFI) and feed efficiency (FE). Daily gain was calculated as the difference between the initial and the final body weights, divided by the number of days in the trial (77–80 d, depending on the tank). Individual DFI was calculated by fitting repeated measures analysis of variance with measurement time (1–5) as the random repeated factor, and then calculating least squares means for each individual (MIXED procedure, SAS® 9.1; SAS® Inst. Inc., US). This was done separately within each experimental tank. Feed efficiency was calculated as the ratio of DG to DFI.

Statistical Analysis of Diet Effects

Diet effects on the means of body weights, DG, DFI and FE were tested with analysis of variance (MIXED procedure). Statistical models included diet as a fixed effect, and replicate tank nested within diet, family, and diet-family interaction as random effects. For all traits, variance due to random experimental tank-family interaction was zero and thus was excluded. Standard errors and degrees of freedom for the F-tests of the fixed effects were calculated using the Kenward and Roger option. Additional analyses were performed to standardize DG and DFI to a common body weight by adding initial weight as a regression covariate to the above model.

Genetic Analysis

In order to examine G × E interactions, observations recorded under each diet treatment were treated as separate traits. For instance, DG recorded on FM (DG_{FM}) and SBM diets (DG_{SBM}) were defined as different traits.

Phenotypic and genetic parameters of DG, DFI and FE were estimated using multiple-trait animal models with DMU software, applying restricted maximum likelihood and average information methods [31]. Models contained experimental tank as a fixed effect, and full-sib family, individual genetic and residual error as random effects. The individual genetic effect included additive genetic effects and parts of potential dominance effects. The random full-sib family effect contained (co)variance due to common incubation and early rearing of full sibs, as well as parts of potential dominance (co)variances. Residual covariances between traits measured in different diets were set to zero. Standard errors of (co)variances were obtained by a first-order Taylor series expansion of the average information matrix of the estimated (co) variances. Heritability (h^2) was calculated as the ratio of genetic variance to total phenotypic variance. Full-sib family effect (c^2) was calculated as the ratio of full-sib family variance to total phenotypic variance. When calculating trait correlations within diets, we did not estimate correlations between FE and its component traits because this practice can be considered statistically vague due to autocorrelation effects. To aid the reader in perceiving the results, we do however present diet means, variances and heritabilities for FE.

Daily gain and feed intake are commonly expressed relative to body weight. Thus, the analysis for DG and DFI was also carried out using a model that included initial body weight as a covariate. For this study, we use the terms "absolute" and "relative" to refer to traits analyzed without and with initial body weight as a covariate, respectively.

Prediction of Genetic Responses to Selection

Selection index calculations were used to predict responses in DG, DFI and FE on both FM and SBM diets to phenotypic selection [15]. Selection was not practiced directly for FE because when selecting on a ratio, genetic changes in the individual component traits are very difficult to control [14]. Four selection strategies were compared: (a) selection for DG_{FM}; (b) selection for maximum FEFM where increased DG_{FM} and decreased DFI_{FM} were selected for simultaneously and selection index weights were set to obtain maximum genetic change in FE_{FM}; (c) selection for DG_{SBM}; and (d) selection for maximum FE_{SBM} where increased DG_{SBM} and decreased DFI_{SBM} were selected to obtain maximum genetic change in FE_{SBM}. Strategy (a) is comparable to current aquaculture breeding programmes that only select for growth rate. Strategies (c) and (d) predict the effects of selection performed on the potential future soybean diets.

Direct and correlated genetic responses to one generation of selection were calculated by $R = i(b' G) (b' Pb)^{-1/2}$, where R is the vector of responses, i is intensity of selection (set to 1), b is the vector of relative index weights which sum to 1, G is the genetic covariance matrix and P is the phenotypic covariance matrix. G and P were results from the 4-trait genetic parameter estimation model for DG_{FM}, DFI_{FM}, DG_{SBM} and DFI_{SBM} described above. The parameters for FE were not needed because all results for this ratio can be predicted from its component traits DG and DFI [14].

To generate the alternative selection strategies, relative index weights were modified as follows. To obtain strategy (a), the index weight for DG_{FM} was set to 1, while the weights for the three remaining traits were zero. Similarly, for strategy (c), the index weight for DG_{SBM} was set to one. For the selection strategies to maximize FE (b and d), the maximum genetic response in FE on both diets was obtained when half of the index weight was on DG and half against DFI.

Table II: Least squares means (± s.e.), and statistical tests for the diet effect for traits recorded on fishmeal and soybean meal diets. Denominator degrees of freedom (ddf) shown; all numerator df = 1

Trait	Fishmeal		Soybean meal		Test statistics		
	n	mean ± s.e.	*n*	mean ± s.e.	ddf	*F*	*P*
Initial body weight (g)	818	40.9 ± 0.85	829	40.6 ± 0.87	10.0	0.78	0.3984
Final body weight (g)	765	131.2 ± 2.38	768	125.8 ± 2.24	9.7	9.36	0.0125
Daily gain ($g \cdot d^{-1}$)	765	1.145 ± 0.0212	768	1.091 ± 0.0201	9.5	20.0	0.0014
Daily feed intake ($g \cdot d^{-1}$)	817	0.948 ± 0.0236	829	1.027 ± 0.0285	9.6	7.23	0.0235
Feed efficiency	765	1.208 ± 0.0150	768	1.078 ± 0.0240	8.5	23.2	0.0011

To calculate genetic response in FE, mean FE was first calculated for each diet from the data, i.e., before selection. Then, genetic responses to selection were calculated for DG and DFI, and the new mean FE was calculated from these [22].

RESULTS

Diet Differences

As expected, whitefish performed better on the FM diet. Fish fed the FM diet had significantly higher final body weight and DG, lower DFI, and better FE than those fed the SBM diet (Tab. II). These diet differences remained (results not shown) when DG and DFI were standardized by including initial body weight as a covariate ($P < 0.0001$) in the statistical models.

Phenotypic and Genetic (co) Variation

There was no evidence for diet differences in trait phenotypic or genetic variation. In both diets, DG and DFI showed moderate heritability whereas FE showed very low heritability that did not differ from zero (Tab. III). For a given trait recorded on both diets, the differences in heritabilities between diets were relatively small (mean absolute difference = 0.07) and confidence intervals overlapped considerably. Because heritabilities may remain constant even if the underlying genetic and residual variations change, diet-specific coefficients of variation were also calculated for each trait. Diet differences between coefficients of phenotypic variation for pairs of traits were small (mean absolute difference = 1.1%) as well.

Coefficients of genetic variation for absolute and relative DG, DFI and FE were only slightly higher on the FM diet (mean difference = 2.9%).

Table III: Phenotypic variance (V_P), coefficients of genetic (CV_G) and phenotypic variation (CV_P), heritability (h^2 ±s.e.) and full-sib family effect (c^2 ±s.e.) for absolute and relative (Rel.) traits measured on fish reared with fishmeal and soybean meal diets

Diet, Trait	V_P	CV_G	CV_P	h^2 ± s.e.	c^2 ± s.e.
Fishmeal					
Daily gain	0.134	16.2	31.9	0.26 ± 0.18	0.13 ± 0.09
Daily feed intake	0.095	15.6	32.5	0.23 ± 0.15	0.08 ± 0.07
Feed efficiency	0.044	4.3	17.4	0.06 ± 0.10	0.04 ± 0.05
Rel. daily gain	0.090	17.0	26.1	0.42 ± 0.17	0.04 ± 0.07
Rel. daily feed intake	0.068	15.6	27.6	0.32 ± 0.14	0.03 ± 0.06
Soybean meal					
Daily gain	0.100	13.1	29.0	0.20 ± 0.15	0.08 ± 0.08
Daily feed intake	0.104	12.8	31.3	0.17 ± 0.15	0.09 ± 0.08
Feed efficiency	0.033	4.3	16.9	0.07 ± 0.11	0.07 ± 0.06
Rel. daily gain	0.076	14.3	25.2	0.32 ± 0.14	0.04 ± 0.06
Rel. daily feed intake	0.081	12.6	27.8	0.21 ± 0.15	0.10 ± 0.08

Genetic correlations between the same traits recorded in each diet indicated very little re-ranking of families across the diets. Genetic correlations (± s.e.) between the diets for DG (0.97±0.21), DFI (0.93±0.28) and FE (1.00±0.95) were all close to unity. The large standard error of the FE correlation was likely caused by the low heritability of this trait. Genetic correlations between the diets for relative DG (0.99 ± 0.13) and relative DFI (0.97 ± 0.22) were also very high.

Within-diet trait correlations were similar in both diets. Phenotypic (r_P) and genetic correlations (r_G ± s.e.) between DG and DFI were high and positive on both FM ($r_P = 0.88$, $r_G = 0.97 \pm 0.05$) and SBM diets ($r_P = 0.86$, $r_G = 0.93 \pm 0.10$). Correlations between relative DG and DFI (FM $r_P = 0.82$, $r_G = 0.97 \pm 0.05$; SBM $r_P = 0.82$, $r_G = 0.96 \pm 0.10$) were similar to those for absolute traits.

Prediction of Selection Responses

The selection index calculations showed that current selection on fishmeal diets will lead to strong favorable correlated genetic changes for performance on SBM diets (Fig. 1). Little difference was observed between the diets for genetic response to selection. We emphasize that although the genetic and phenotypic parameter point estimates used in the selection index differed slightly between diets, there were overlapping confidence intervals associated with these estimates. Therefore, the genetic responses predicted from the selection index represent average expected differences between the diets and also are subject to variation around the mean.

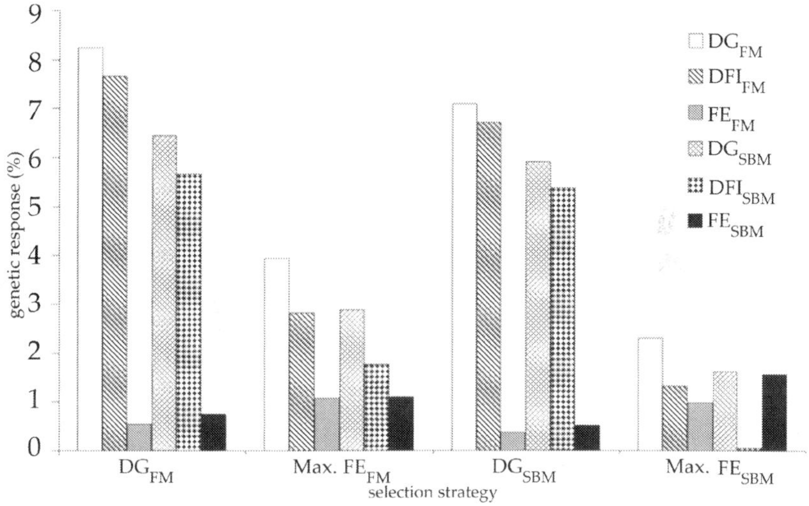

Figure 1: Genetic responses in daily gain (DG), daily feed intake (DFI) and feed efficiency (FE) in fishmeal (FM) and soybean meal (SBM) diets in response to four alternative selection strategies: selection on DG_{FM} alone, selection for maximum FE_{FM}, selection for DG_{SBM} alone, and selection for maximum FE_{SBM}. Genetic responses are given as percent change from the original means in the data.

Selection for DG_{FM} caused increased DG and DFI in both diets, and genetic responses on the FM diet were approximately 2.0% higher than on the SBM diet. In contrast, genetic response in FE was 0.2% higher on the SBM diet than on the FM diet. Selection for maximum FE_{FM} also resulted in genetic responses that differed only slightly between diets; responses for DG and DFI were approximately 1% higher on the FM diet than on the SBM diet, and FE responses were similar across diets (Fig. 1).

Direct selection for DG_{SBM} or maximum FE_{SBM} did not generally achieve better genetic responses in SBM traits than the correlated responses obtained by selection on the respective FM traits (Fig. 1).

Genetic responses in DG_{SBM} and FE_{SBM} were 0.5% and 0.2% higher, respectively, in the DG_{FM} selection strategy than in the DG_{SBM} strategy. However, genetic response in FE_{SBM} was 0.46% higher in the maximum FE_{SBM} strategy than in the FE_{FM} strategy. Compared with the DG selection strategies, the maximum FEFM and FE_{SBM} strategies doubled the genetic response in FE_{FM}, and tripled the FE_{SBM} response, respectively (Fig. 1). However, the FE strategies also caused approximately 4% lower responses in DG compared to DG strategies.

DISCUSSION

Impacts of G × E on Breeding Programmes

In the current study, only weak diet G × E interactions were detected for growth, feed intake and feed efficiency in European whitefish. Therefore, increased soybean meal (SBM) use in feeds does not compromise the genetic improvements achieved by breeding programmes. In effect, results from selection index calculations show that current fish breeding programmes that select on fishmeal diet (FM) growth or feed efficiency performance also improve their broodstock's ability to use novel soybean meal diets.

Furthermore, this study confirms that the current fishmeal diet is an appropriate environment for evaluating and selecting genotypes for improved growth and efficiency on future soybean meal-based diets. Direct selection for FM growth was as effective for improving SBM diet growth and feed efficiency as direct selection for SBM growth. The only situation in which the SBM diet was a superior selection environment was for direct selection for SBM diet feed efficiency. Breeding programmes operating within future diet environments should therefore achieve genetic gains comparable to those with current diets.

These results are a consequence of two factors. First, trait heritabilities and phenotypic variations did not differ substantially across diets. Although several studies have shown G × E interaction expressed by elevated genetic variation in stressful or novel conditions [6], this did not occur in our study. Modern SBM products are well processed and thus may only cause mild environmental stress in salmonids. Second, genetic correlations for DG, DFI and FE between diets were close to unity, indicating little re-ranking of families across the diets. These weak re-rankings agree with other farmed fish studies that have found little or no family interactions with diet nutrient levels [1, 2, 13, 22, 37] or production environments [20, 44].

Although novel soybean meal feeds should not be problematic for selective breeding of whitefish, novel high soy diets can cause reduced growth and feed efficiency in salmonids (e.g., [4, 40]). In the current study, the SBM diet supported slower growth but higher feed intake which indicate that it was digested or processed less well than the FM diet, probably due to anti-nutritional factors present in soybean meal [11] for which whitefish are not adapted. Major efforts around the world are refining plant-protein based fish feeds through ingredient selection, supplementation and processing techniques [11]. Our results show that selective breeding is an effective complementary method to improve biological utilization of novel feed ingredients by farmed fish.

Genetic Improvement of Feed Efficiency

Our results show that genetic improvement of feed efficiency for both fishmeal and soybean meal based diets can be achieved through selection in European whitefish. Simultaneous selection for fast growth and reduced feed intake achieved at least double the genetic response in feed efficiency versus selection on growth alone.

However, genetic progress in feed efficiency occurs slowly; the greatest predicted selection responses in FE were 7–8 times smaller than gains achieved in growth rate. Accordingly, the heritability for FE was low (0.06–0.07). This was unexpected because in terrestrial farm animals the average heritability of FE is moderate, 0.25, and has similar magnitude as for growth traits [38]. The only known previous estimates of efficiency heritability in fish (0.3 in rainbow trout [24]) are also higher than in our data, however those were based on family means and thus represent overestimates. In our data, feed intake exhibited moderate heritability (0.17–0.32) that is typical of feed consumption traits in terrestrial species (e.g., 0.25 [5], 0.23 [19]) and midway between previous estimates for rainbow trout (average 0.10 [22]) and catfish Ictalurus punctatus (0.41 [42]).

In terrestrial animals, improvement of feed efficiency has been attributed to lowered maintenance energy requirements through endocrine changes [28] and reduced thermogenesis, physical activity [29, 33] and protein turnover rates [32]. We hypothesize that cultured salmonids can display very low FE heritability for two reasons. First, maintenance or resting metabolism rates of poikilothermic fish are 10 to 30 times less than mammals and up to 100 times less than birds of the same weight [3]. Thus, potential variation in maintenance requirements generates very little genetic variation for FE in fish. Second, because FE studies in salmonids are usually conducted during an exponential growth phase, most feed consumed is directed to growth, and there is very little variation in nutrient diversion to other body functions. These contribute to an almost unity correlation between feed intake and growth that reduces variation in FE. Genetic correlations between daily gain and feed intake in both European whitefish (0.93, 0.97; this study) and rainbow trout (0.86–0.96 [22]) were very high. In contrast, in terrestrial animals this correlation is typically lower (0.6–0.7 [5, 7, 26]).

At this time, no known fish breeding programmes select directly for feed efficiency; all rely on a correlated genetic response obtained by sole selection for high body weight. Previous studies have proposed that feed efficiency in fish can be indirectly improved by selection for fast growth [16, 24, 25]. The selection index results in the current study also support the effectiveness of selection for growth as a strategy to indirectly improve efficiency. However, simultaneous selection for fast growth and reduced feed intake achieved at least double the genetic response in feed efficiency versus selection on growth alone. Although the increase in feed efficiency generated by including feed intake to the selection index was small (< 2%), on a production scale such improvements translate to considerable economic benefits [25] and therefore may merit the measurement expense in a selective breeding programme.

ACKNOWLEDGEMENTS

The authors thank M. Henryon and two anonymous referees for constructive comments on early versions of this manuscript. The study was funded by the Employment and Economic Development Centre of Central Finland.

REFERENCES

1. Austreng E., Refstie T., Effect of varying dietary protein level in different families of rainbow trout, Aquaculture 18 (1979) 145–156.
2. Austreng E., Risa S., Edwards D.J., Hvidsten H., Carbohydrate in rainbow trout diets. II. Influence of carbohydrate levels on chemical composition and feed utilization of fish from different families, Aquaculture 11 (1977) 39–50.
3. Brett J.R., The metabolic demand for oxygen in fish, particularly salmonids, and a comparison with other vertebrates, Respir. Physiol. 14 (1972) 151–170.
4. Bureau D.P., Harris A.M., Cho C.Y., The effects of purified alcohol extracts from soy products on feed intake and growth of chinook salmon (Oncorhynchus tshawytscha) and rainbow trout (Oncorhynchus mykiss), Aquaculture 161 (1998) 27–43.
5. Cammack K.M., Leymaster K.A., Jenkins T.G., Nielsen M.K., Estimates of genetic parameters for feed intake, feeding behavior, and daily gain in composite ram lambs, J. Anim. Sci. 83 (2005) 777–785.
6. Charmantier A., Garant D., Environmental quality and evolutionary potential: lessons from wild populations, Proc. R. Soc. B 272 (2005) 1415–1425.
7. Clutter A.C., Brascamp E.W., Genetics of performance traits, in: Rothschild M.F., Ruvinsky A. (Eds.), The Genetics of the Pig, CAB International, Wallingford, 1998, pp. 427–462.
8. Falconer D.S., The problem of environment and selection, Am. Nat. 86 (1952) 293–298.
9. Falconer D.S., Mackay T.F.C., Introduction to quantitative genetics, 4th edn., Longman, Essex, 1996.
10. FAO Fisheries Department, The state of world fisheries and aquaculture (SOFIA) 2004, Food and Agriculture Organization of the United Nations, Rome, 2004.
11. Francis G., Makkar H.P.S., Becker K., Antinutritional factors present in plantderived alternate fish feed ingredients and their effects in fish, Aquaculture 199 (2001) 197–227.
12. Gjedrem T., Thodesen J., Selection, in: Gjedrem T. (Ed.), Selection

and Breeding Programs in Aquaculture, Springer, Dordrecht, 2005, pp. 89–111.

13. Glover K.A., Taggart J.B., Skaala O., Teale A.J., A study of inadvertent domestication selection during start-feeding of brown trout families, J. Fish Biol. 64 (2004) 1168–1178.
14. Gunsett F.C., Linear index selection to improve traits defined as ratios, J. Anim. Sci. 59 (1984) 1185–1193.
15. Hazel L.N, The genetic basis for constructing selection indexes, Genetics 28 (1943) 476–490.
16. Henryon M., Jokumsen A., Berg P., Lund I., Pedersen P.B., Olesen N.J., Slierendrecht W.J., Genetic variation for growth rate, feed conversion effi- ciency, and disease resistance exists within a farmed population of rainbow trout, Aquaculture 209 (2002) 59–76.
17. Jobling M., Coves D., Damsgård B., Kristiansen H.R., Koskela J., Petursdottir T.E., Kadri S., Gudmundsson O., Techniques for measuring feed intake, in: Houlihan D., Boujard T., Jobling M. (Eds.), Food Intake in Fish, Blackwell Science Inc., Oxford, 2001, pp. 49–87.
18. Jobling M., Gomes E., Dias J., Feed types, manufacture and ingredients, in: Houlihan D., Boujard T., Jobling M. (Eds.), Food Intake in Fish, Blackwell Science Inc., Oxford, 2001, pp. 25–48.
19. Johnson Z.B., Chewning J.J., Nugent R.A. III, Genetic parameters for production traits and measures of residual feed intake in large white swine, Anim. Sci. 77 (1999) 1679–1685.
20. Kause A., Ritola O., Paananen T., Breeding for improved appearance of large rainbow trout in two production environments, Aquacult. Res. 35 (2004) 924–930.
21. Kause A., Tobin D., Dobly A., Houlihan D., Martin S., Mäntysaari E.A., Ritola O., Ruohonen K., Recording strategies and selection potential of feed intake measured using the X-ray method in rainbow trout, Genet. Sel. Evol. 38 (2006) 389–409.
22. Kause A., Tobin D., Houlihan D.F., Martin S.A.M., Mäntysaari E.A., Ritola O., Ruohonen K., Feed efficiency of rainbow trout can be improved through selection: Different genetic potential on alternative diets, J. Anim. Sci. 84 (2006) 807–817.

23. Kaushik S.J., Cravedi J.P., Lalles J.P., Sumpter J., Fauconneau B., Laroche M., Partial or total replacement of fish meal by soybean protein on growth, protein utilization, potential estrogenic or antigenic effects, cholesterolemia and flesh quality in rainbow trout, Oncorhynchus mykiss, Aquaculture 133 (1995) 257–274.
24. Kinghorn B., Genetic variation in food conversion efficiency and growth in rainbow trout, Aquaculture 32 (1983) 141–155.
25. Kolstad K., Grisdale-Helland B., Gjerde B., Family differences in feed efficiency in Atlantic salmon (Salmo salar), Aquaculture 241 (2004) 169–177.
26. Koots K.R., Gibson J.P., Smith C., Wilton J.W., Analyses of published genetic parameter estimates for beef production traits, 1. Heritability, Anim. Breed. Abstr. 62 (1994) 309–338.
27. Koskela J., Growth rate and feeding level of European whitefish (Coregonus lavaretus L. s.l.) under hatchery conditions, Pol. Arch. Hydrobiol. 39 (1992) 731–738.
28. Luiting P., The role of genetic variation in feed intake and its physiological aspects: results from selection experiments, in: van der Heide D., Huisman E.A., Kanis E., Osse J.W.M., Verstegen M.W.A. (Eds.), Regulation of Feed Intake, CAB International, Wallingford, 1999, pp. 75–87.
29. Luiting P., Schrama J.W., van der Hel W., Urff E.M., Metabolic differences between White Leghorns selected for high and low residual food consumption, Brit. Poult. Sci. 32 (1991) 763–782.
30. Lynch M., Walsh B., Genetics and analysis of quantitative traits, Sinauer Associates, Sunderland, 1998.
31. Madsen P., Jensen J., A user's guide to DMU, a package for analysing multivariate mixed models, Danish Institute of Agricultural Sciences, Tjele, 2006.
32. McDonagh M.B., Herd R.M., Richardson E.C., Oddy V.H., Archer J.A., Arthur P.F., Meat quality and the calpain system of feedlot steers following a single generation of divergent selection for residual feed intake, Aust. J. Exp. Agric. 41 (2001) 1013–1021.
33. McPhee C.P., Kerr J.C., Cameron N.D., Peri-partum posture and behaviour of gilts and the location of their piglets in lines selected

for components of efficient lean growth, Appl. Anim. Behav. Sci. 71 (2001) 1–12.

34. Naylor R.L., Goldburg R.J., Primavera J.H., Kautsky N., Beveridge M.C.M., Clay J., Folke C., Lubchenco J., Mooney H., Troel M., Effect of aquaculture on world fish supplies, Nature 405 (2000) 1017–1024.
35. New M.B., Wijkström U.N., Use of fishmeal and fish oil in aquafeeds: further thoughts on the fishmeal trap, FAO Fisheries Circular C975, Food and Agriculture Organization of the United Nations, Rome, 2002.
36. Olli I.J., Krogdahl Å., Nutritive value of four soybean products in diets for rainbow trout (Oncorhynchus mykiss, Walbaum) reared in fresh water, Acta Agric. Scand. A 44 (1994) 185–192.
37. Palti Y., Silverstein J.T., Wieman H., Phillips J.G., Barrows F.T., Parsons J. E., Evaluation of family growth response to fishmeal and gluten-based diets in rainbow trout (Oncorhynchus mykiss), Aquaculture 255 (2006) 548–556.
38. Pitchford W.S., Genetic improvement of feed efficiency of beef cattle: what lessons can be learnt from other species? Aust. J. Exp. Agricult. 44 (2004) 371–382.
39. Powell K., Fish farming: Eat your veg, Nature 426 (2003) 378–379.
40. Refstie S., Storebakken T., Roem A.J., Feed consumption and conversion in Atlantic salmon (Salmo salar) fed diets with fish meal, extracted soybean meal or soybean meal with reduced content of oligosaccharides, trypsin inhibitors, lectins and soya antigens, Aquaculture 162 (1998) 301–312.
41. Refstie S., Storebakken T., Baeverfjord G., Roem A.J., Long-term protein and lipid growth of Atlantic salmon (Salmo salar) fed diets with partial replacement of fish meal by soy protein products at medium or high lipid level, Aquaculture 193 (2001) 91–106.
42. Silverstein J.T., Bosworth B.G., Waldbieser G.C., Wolters W.R., Feed intake in channel catfish: is there a genetic component? Aquacult. Res. 32 s1 (2001) 199–205.
43. Storebakken T., Refstie S., Ruyter B., Soy products as fat and protein sources in fish feeds for intensive aquaculture, in: Drackley

J.K. (Ed.), Soy in Animal Nutrition, Federation of Animal Science Societies, Savoy, 2000, pp. 127–170.

44. Sylvén S., Rye M., Simianer H., Interaction of genotype with production system for slaughter weight in rainbow trout (Oncorhynchus mykiss), Livest. Prod. Sci. 28 (1991) 253–263.

45. Tacon A.G.J., Use of fish meal and fish oil in aquaculture: a global perspective, Aquat. Resour. Cult. Dev. 1 (2004) 3–14.

CHAPTER

13

INCREASING SALINITY TOLERANCE IN TILAPIAS: SELECTIVE BREEDING USING LOCALLY AVAILABLE STRAINS

[1]Cecilia J. Jaspe and Christopher M. A. Caipang[2]

[1] Brackishwater Aquaculture Center, Institute of Aquaculture, College of Fisheries and Ocean Sciences, University of the Philippines Visayas, Miag-ao 5023, Iloilo, Philippines,

[2] Faculty of Biosciences and Aquaculture, University of Nordland, Bodø 8049, Norway

ABSTRACT

The culture of tilapias is an important industry in most aquaculture producing countries in the world. Due to the many favorable attributes of this species of fish, including tolerance to wide range of environmental conditions and fast growth rates, interest in its culture is increasing at a rapid pace. There have been research initiatives on breeding, genetic improvement and stock manipulation in order to increase the resistance of this fish to high salinity levels, enabling them to thrive in brackish water and almost seawater conditions. Saline-tolerant tilapias have been co-cultured with shrimp in brackish water ponds as it was believed that tilapias are able to inhibit the occurrence of luminous vibriosis in ponds. This paper describes the selective breeding program that we developed in order to produce high-saline tolerant tilapias by using locally available strains.

Key Words: Tilapia, Oreochromis sp., saline-tolerant, breeding

INTRODUCTION

Tilapias are one of the most economically-important groups of aquaculture species because they serve as major sources of protein in most countries. They are versatile species of fish, which is found in almost any type of tropical aquaculture systems ranging from traditional to highly intensive production systems. They withstand wide range of environmental conditions and perform well regardless of the water salinity to which they are exposed to. Some species are even able to thrive and breed in full strength seawater (Cnaani & Hulata 2011).

Among the tilapia species, the Nile tilapia, Oreochromis niloticus is the preferred species for culture as this fish dominates production in freshwater ponds and cages (Kamal & Mair 2005). However, it has low tolerance to high salinity levels. On the other hand, the Mozambique tilapia, Oreochromis mossambicus is a euryhaline species and is one of the best studied tilapias in terms of elucidating the mechanisms involved in euryhalinity among fishes (Uchida et al 2000). However, Mozambique tilapia exhibits poor growth due to inbreeding of the founder stock. A tilapia hybrid, the Florida red tilapia also grows well in high saline waters and studies have been done in line with some biotechnical and socio-economic aspects of its culture in saline environments (Watanabe et al 2006).

The culture of tilapias in saline waters is well-documented based on numerous research studies done in the past (reviewed in Cnaani & Hulata 2011). Given the limited space for freshwater aquaculture and pressures on providing the food demands of the population, tilapias are now being cultured in brackishwater ponds and even in marine cages. This scenario will further intensify in the years to come in order to cope with food requirements of the increasing human population. One of the constraints in the culture of tilapias in high saline environments is its sensitivity to handling and susceptibility to secondary infections (Chang

& Plumb 1996). Hence, tilapias reared in high salinities experience higher disease outbreaks and mortalities than those that are reared in freshwater environment. Because of these problems, there have been intensive research efforts on improving the salinity tolerance of tilapias either through modifications in the culture techniques or stock improvement.

This paper discusses the breeding program that was developed at our research facility on selective breeding of tilapia using local strains in order to produce saline tolerant tilapia hybrids.

INCREASING TOLERANCE OF SALINITY THOUGH SELECTIVE BREEDING

The different approaches on the improvement of tolerance in tilapias to saline environment as well as the underlying mechanisms involved in their salinity tolerance have been discussed by Cnaani & Hulata (2011). The authors stated that salt tolerance of the fish is the overall fitness of that particular species of fish in a saline environment, and is a combination of traits including metabolism, growth, osmoregulation, immunocompetence and fecundity. Each trait is influenced by multiple genes leading to genetic variation. They further stressed that the underlying phenotypic differences in salt tolerance can result in intra- and inter-specific variation among tilapia populations.

The physiological mechanisms in tilapias during exposure to changes in salinity levels have been extensively reviewed (Prunet & Bornancin 1989). O. mossambicus is more adaptable to high salinity levels than O. niloticus because of the ability of the former to modify the structures of its esophageal epithelial cells during transfer to seawater (Cataldi et al 1988). Other modifications that tilapias undergo when transferred to high salinity levels include functional transformation in the kidney (Cataldi et al 1991), changes in the chloride cell characteristics (Fiess et al 2007), changes in the amino acid composition of the digestive enzymes (Fang & Chiou 1989), changes in the neuroendocrine hormones (Sakamoto & McCormick 2006) and regulation of the different osmosensors (Loretz 2008).

Transcriptomic approaches have also been used to analyze the mechanisms involved in salinity tolerance among tilapias. Fiol et al (2006) analyzed the transcriptional profile of the genes involved in hyperosmotic stress in the gill epithelial cells of O. mossambicus during seawater transfer. They found that most of the genes were immediately upregulated following transfer to seawater, and the genes involved in osmotic response can be grouped into six categories, namely, (1) stress response signal transduction, (2) compatible organic osmolyte accumulation, (3) energy metabolism, (4) lipid transport and cell membrane protection, (5) actin-based cytoskeleton dynamics, and (6) protein and mRNA stability.

Selective breeding is one of the approaches that is used to improve the ability of tilapias to withstand high salinity levels. A simple method to evaluate salinity tolerance in parental strains and hybrids of tilapia is imperative when conducting selective breeding programs for salinity tolerance. An example of this method is known as the median lethal salinity (MLS), which is a simple index based on the short-term capacity of the fish to adapt to daily increments of salinity (Lemarié et al 2004). MLS is the salinity at which 50% of the fish population dies at a given daily rate of increasing salinity levels. This method has been successfully used to evaluate salinity tolerance in O. niloticus and Sarotherodon melanotheron. This technique takes into account the capacity of fish to adapt to increasing salinity levels and is a simple, optimized and efficient criterion for assessing salinity tolerance of tilapias. Selective breeding programs for both purebreds and crossbreds of tilapias in terms of salinity tolerance and growth have been done in the Philippines (Rosario et al 2004; Tayamen et al 2004). Also, quantitative genetics through the use of diallele mating design was employed to evaluate genetics effects influencing salinity tolerance (Armas-Rosales 2006).

SELECTIVE BREEDING INITIATIVES USING LOCAL STRAINS

Tilapias are not endemic to the Philippines but several of the introduced species and strains have become well-adapted to the conditions of the country. We have developed a breeding program for the production of saline-tolerant tilapia hybrids using different strains that are locally available. We produced a saline-tolerant tilapia hybrid by crossing O. mossambicus with another tilapia hybrid, O. spilorus x O. niloticus GIFT x O. aureus. The parental strains, O. mossambicus were obtained from the holding ponds of the research center, while the tilapia hybrids were purchased from a commercial tilapia farm in Central Philippines. These two tilapia strains are tolerant to high salinity levels even at full seawater (Jaspe et al 2007); hence, they were good candidates for selective breeding to produce saline-tolerant hybrids. The broodstock were placed in 1 m x 1 m mesh nets, which were installed inside the concrete outdoor tanks, supplied with flow-through freshwater and ample aeration. The fish were fed with a commercial feed twice a day until satiation. During breeding, the male to female ratio is 1:3 in order to facilitate effective fertilization and to ensure that all females will be mouth-brooders. F_1 progenies from the following crosses: (1) O. mossambicus (female) crossed with the hybrid, O. spilorus x O. niloticus GIFT x O.aureus (male), and (2) O. mossambicus (male) crossed with O. spilorus x O. niloticus GIFT x O. aureus (female).

The F_1 progenies of these crosses have been subjected to a salinity test in order to determine their level of tolerance to abrupt changes in salinity and how they cope with the extreme changes. All the F_1 progenies of the parental strains as well as the crosses were able to tolerate abrupt changes in salinity and there were no mortalities in the fish even at a salinity level of 50 ppt over a monitoring period of 20 days (Jaspe et al 2007). During the course of the study, the fish did not develop lesions; thus, there is least likely a possibility of secondary infections. However, we have noticed that when there was an abrupt change in salinity to a maximum increase of 5 ppt per day, the fish were sluggish a few minutes after exposure but recovered a within one hour after and resumed their normal feeding activity. We have also observed that when the salinity level reached at least 30 ppt, the fish had dark pigmentation, showed erratic swimming behavior and stopped feeding but resumed to normal conditions within a few hours after they adjusted to the salinity conditions. In the course of the salinity tolerance test, we maintained the optimum water quality parameters in the containers. One important factor that we considered was water temperature, in which the range was 26-28°C. This optimum water temperature is crucial in maintaining a steady state plasma osmolality in the fish during exposure to salinity changes (Sardella et al 2004) and could also be partly responsible in preventing mortalities in the fish.

Next, we reared to maturity the F_1 progenies of the saline-tolerant tilapia hybrids and assessed their reproductive potential. An assessment of the reproductive capacity of the fish is needed in order to find out if the resulting hybrids are able to support the fry requirement during the grow-out phase. Also, this information is necessary to ensure that there is sufficient number of broodstock to be maintained in the hatcheries. Mature eggs were collected from the female spawners of the F_1 of the parental strains (O. mossambicus and the tilapia hybrid, Oreochromis spilorus x O. niloticus GIFT x O. aureus) and their crosses. We observed that among the reproductive parameters that were assessed, total fecundity and gonadosomatic varied among the spawners but not the relative fecundity. Total fecundity is positively correlated with the body weight of the spawner; thus, when choosing the spawner for selective breeding, it is important to take into consideration the size of the fish. Since we did not find any significant difference in the relative fecundity, that is, the number of eggs produced per unit weight of the fish, therefore any combination of the parental strains during selective breeding can be used to produce saline-tolerant progenies.

In terms of the performance of the F_1 progenies in the grow-out ponds, our initial observations showed that the growth, survival and production of the tilapia hybrids did not vary significantly, therefore combinations of the parental strains during selective breeding do not affect the grow-out performance of the resulting progenies. We are currently mass producing these saline-tolerant strains in order to establish an effective number of the founder stock to prevent depression in growth of future generations due to inbreeding. A flow-chart for this breeding program is shown in Figure 1.

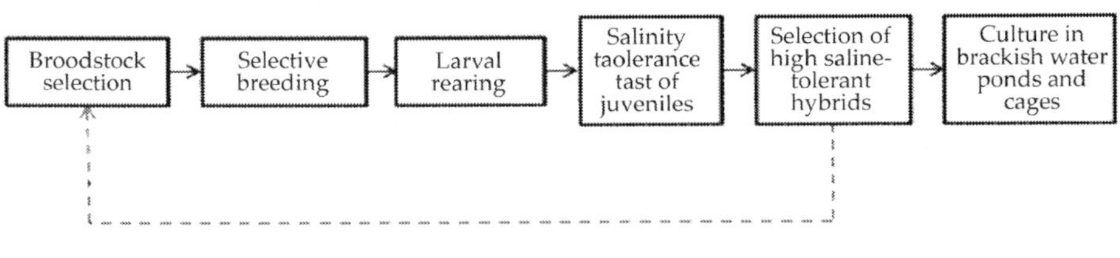

Figure 1: Schematic diagram of the breeding program for locally-available strains of tilapia.

Aside from the establishment of the selective breeding program for the production of saline-tolerant tilapia strains, the water obtained from the broodstock tanks of the salinetolerant tilapia strains is a good source of "green water". It is called "green water" because it contains considerable populations of green microalgae such as Chlorella sp. (Tendencia et al 2005). In a small-scale study, the "green water" that we obtained from the broodstock tanks of the saline-tolerant tilapias was able to inhibit the proliferation of Vibrio harveyi, the causative agent of luminous vibriosis in shrimp aquaculture (Huervana et al 2006). As such, the saline-tolerant tilapia strains that we produced in our selective breeding program can be cultured together with marine shrimp in ponds for the production of the "green water technology", thus, preventing future outbreaks of luminous Vibriosis (Corre et al 2000; Cruz et al 2008).

CONCLUSIONS

We have developed a selective breeding program for the production of saline-tolerant strains of tilapia using parental strains that are available locally. The salinity tolerance, growth performance and survival of the F_1 progenies from the different crosses were not significantly different; hence, any combination of the parental strains during spawning can be used without affecting the physiological characteristics of the offspring. In addition, the saline-tolerant hybrids that we produced can be a good source for the green water technology in shrimp aquaculture and is a good management strategy to control bacterial diseases such as luminous vibriosis during the shrimp growout phase.

ACKNOWLEDGEMENTS

The authors would like to thank their respective institutions (Brackishwater Aquaculture Center and University of Nordland) for the technical support given during the conduct of the studies and the logistics provided during the preparation of the manuscript. C. M. Caipang also gratefully acknowledges the support provided by the National Institute of Molecular Biology and Biotechnology (NIMBB), UP Visayas.

REFERENCES

1. Armas-Rosales A. M., 2006 Genetic Effects Influencing Salinity and Cold Tolerance in Tilapia. M.Sc. thesis, Louisiana State Univ., Baton Rouge, LA. 140 pp.

2. Cataldi E., Crosetti D., Leone C., Cataudella S., 1988 Oesophagus structure during adaptation to salinity in Oreochromis niloticus (Perciformes, Pisces) juveniles. Bull Zool 55:59-62.
3. Cataldi E., Garibaldi L., Crosetti D., Leoni C., Cataudella S., 1991 Variations in renal morphology during adaptation to salinities in tilapias. Environ Biol Fish 31:101- 106.
4. Chang P. H., Plumb J. A., 1996 Effects of salinity in Streptococcus infection of Nile tilapia, Oreochromis niloticus. J Appl Aquacult 6:39-45.
5. Cnaani A., Hulata G., 2011 Improving salinity tolerance in tilapias: past experience and future prospects. Isr J Aquacult-Bamidgeh, in press.
6. Corre V. L. Jr., Janeo R., Caipang C. M., Calpe A. T., 2000 Use of probiotics and reservoir with "green water" and other tips of a successful culture. Aquacult Asia 5:14-18.
7. Cruz P. S., Andalencio M. N., Bolivar R. B., Fitzsimmons K., 2008 Tilapia-shrimp polyculture in Negros Island, Philippines: a review. J World Aquac Soc 39:713- 725.
8. Fang L. S., Chiou S. F., 1989 Effect of salinity on the activities of digestive proteases from the tilapia fish, Oreochromis niloticus in different culture environments. Comp Biochem Physiol 93A:439-443.
9. Fiess J. C., Kunkel-Patterson A., Mathias L., Riley L. G., Yancey P. H., Hirano T., Grau E. G., 2007 Effects of environmental salinity and temperature on osmoregulatory ability, organic osmolytes, and plasma hormone profiles in the Mozambique tilapia (Oreochromis mossambicus). Comp Biochem Physiol A Mol Integr Physiol 146: 252-264.
10. Fiol D. F., Chan S. Y., Kültz D., 2006 Identification and pathway analysis of immediate hyperosmotic stress responsive molecular mechanisms in tilapia (Oreochromis mossambicus) gill. Comp Biochem Physiol D 1:344-356.
11. Huervana F. H., de la Cruz J. J. Y., Caipang C. M. A., 2006 Inhibition of luminous Vibrio harveyi by "green water" obtained from tank culture of tilapia, Oreochromis mossambicus. Acta Ichthyol et Piscat 36:17-23.

12. Jaspe C. J., Caipang C. M., Corre V. L. Jr., 2007 Salinity resistance of F1 progenies from crosses between Oreochromis mossambicus and a tilapia hybrid, Oreochromis spilorus x O. niloticus GIFT x O. aureus: implications to high-saline culture. UPV J Nat Sci 12:250-259.
13. Kamal A. H., Mair G. C., 2005 Salinity tolerance in superior genotypes of tilapia, Oreochromis niloticus, Oreochromis mossambicus and their hybrids. Aquaculture 247:189-201.
14. Lemarié G., Baroiller J. F., Clota F., Lazard J., Dosdat A., 2004 A simple test to estimate the salinity resistance of fish with specific application to O. niloticus and S. melanotheron. Aquaculture 240:575-587.
15. Loretz C. A., 2008 Extracellular calcium-sensing receptors in fishes. Comp Biochem Physiol 149A:225-245.
16. Prunet P., Bornancin M., 1989 Physiology of salinity tolerance in tilapia: an update of basic and applied aspects. Aquat Living Resour 2:91-97.
17. Rosario W. R., Georget C., Chevassus-Au-Louis B., Morissens P., Muyalde N. C., de la Cruz A. E., de Vera E., Poivey J.-P., 2004 Selection from an interspecific hybrid population of two strains of fast growing and salinity tolerant tilapia. In: Proc 6th Int Symp Tilapia in Aquaculture. Bolivar R.B., Mair G.C., Fitzsimmons K. (eds.), p. 73, ICLARM, Manila, Philippines.
18. Sakamoto T., McCormick S. D., 2006 Prolactin and growth hormone in fish osmoregulation. Gen Comp Endocrinol 147:24-30.
19. Sardella B. A., Cooper J., Gonzalez R. J., Brauner C. J., 2004 The effect of temperature on juvenile Mozambique tilapia hybrids (Oreochromis mossambicus x O. urolepis hornorum) exposed to full-strength and hypersaline seawater. Comp Biochem Physiol A 137:621-629.
20. Tayamen M. M, Abella T. A., Reyes R. A., Danting M. J. C., Mendoza A. M., Marquez E. B., Salguet A. C., Apaga M. M., Gonzales R. C., 2004 Development of tilapia for salinity waters in the Philippines. In: Proc 6th Int Symp Tilapia in Aquaculture. Bolivar R.B., Mair G.C., Fitzsimmons K. (eds.), pp.463-478, ICLARM, Manila, Philippines.

21. Tendencia E. A., de la Peña M. R., Choresca C. H. Jr., 2005 Efficiency of Chlorella sp. and Tilapia hornorum in controlling the growth of luminous bacteria in a simulated shrimp culture environment. Aquaculture 249:55-62.
22. Uchida K., Kaneko T., Miyazaki H., Hasegawa S., Hirano T., 2000 Excellent salinity tolerance of Mozambique tilapia (Oreochromis mossambicus): elevated chloride cell activity in the branchial epithelia of the fish adapted to concentrated seawater. Zool Sci 17:149-160.
23. Watanabe W. O., Fitzimmons K., Yi Y., 2006 Farming tilapia in saline waters. In: Tilapia: Biology, Culture, and Nutrition. C. Lim, C.D. Webster (eds.), pp.347-447, Food Products Press/Haworth Press, New York, London, Oxford (2006).

SECTION III

SELECTION STRATEGIES

CHAPTER

14

DERIVATION OF ECONOMIC VALUES FOR PRODUCTION TRAITS IN AQUACULTURE SPECIES

Kasper Janssen[1], Paul Berentsen[2], Mathieu Besson[1,3] and Hans Komen[1]

[1] Animal Breeding and Genomics, Wageningen University and Research, Droevendaalsesteeg 1, 6708 PB Wageningen, The Netherlands.

[2] Business Economics Group, Wageningen University and Research, Hollandseweg 1, 6706 KN Wageningen, The Netherlands.

[3] Génétique Animale Biologie Intégrative, INRA, AgroParisTech, Université Paris-Saclay, 78350 Jouy-en-Josas, France

ABSTRACT

Background

In breeding programs for aquaculture species, breeding goal traits are often weighted based on the desired gains but economic gain would be higher if economic values were used instead. The objectives of this study were: (1) to develop a bio-economic model to derive economic values for aquaculture species, (2) to apply the model to determine the economic importance and economic values of traits in a case-study on gilthead seabream, and (3) to validate the model by comparison with a profit equation for a simplified production system.

Methods

A bio-economic model was developed to simulate a grow-out farm for gilthead seabream, and then used to simulate gross margin at the current levels of the traits and after one genetic standard deviation change in each trait with the other traits remaining unchanged. Economic values were derived for the traits included in the breeding goal: thermal growth coefficient (*TGC*), thermal feed intake coefficient (*TFC*), mortality rate (*M*), and standard deviation of harvest weight (σ_{HW}). For a simplified production system, improvement in *TGC* was assumed to affect harvest weight instead of growing period. Using the bio-economic model and a profit equation, economic values were derived for harvest weight, cumulative feed intake at harvest, and overall survival.

Results

Changes in gross margin showed that the order of economic importance of the traits was: *TGC*, *TFC*, *M*, and σ_{HW}. Economic values in € (kg production)$^{-1}$ (trait unit)$^{-1}$ were: 0.40 for *TGC*, −0.45 for *TFC*, −7.7 for *M*, and −0.0011 to −0.0010 for σ_{HW}. For the simplified production system, similar economic values were obtained with the bio-economic model and the profit equation. The advantage of the profit equation is its simplicity, while that of the bio-economic model is that it can be applied to any aquaculture species, because it can include any limiting factor and/or environmental condition that affects production.

Conclusions

We confirmed the validity of the bio-economic model. *TGC* is the most important trait to improve, followed by *TFC* and *M*, and the effect of σ_{HW} on gross margin is small.

BACKGROUND

In Europe, over 80% of aquaculture production originates from breeding programs, which in most cases apply family selection with the aim of improving multiple traits simultaneously [1]. Breeding goal traits are often weighted based on desired gains rather than economic values [2], which compromises economic gain [3, 4].

In aquaculture species, economic values are available only for a few species, although their importance has repeatedly been underlined, e.g. [5, 6]. Profit equations have been used to derive economic values for Nile tilapia (*Oreochromis niloticus*) [7], common carp (*Cyprinus carpio*) [8], Australian abalones (*Haliotis rubra* and *H. laevigata*) [9], and crayfish (*Cherax tenuimanus*) [10]. Besson et al. [11, 12] used a bio-economic model to derive economic values for African catfish (*Clarias gariepinus*) that were produced in a land-based aquaculture system in which water is treated and recirculated and for growth rate in European seabass (*Dicentrarchus labrax*) under varying temperature conditions, respectively.

For livestock species with simple and highly controlled production systems, such as pig production, economic values can be derived from a profit equation, e.g. [13]. For production systems with a higher degree of complexity, partly due to seasonal variation, such as in dairy cattle and sheep farming, profit equations may fail to provide an adequate description of the farming system and bio-economic models are required [14, 15, 16]. In general, bio-economic models provide a more accurate description of farming systems than profit equations and are, therefore, increasingly used to estimate economic values [2].

Fish farms are complex production systems for two reasons. First, fish are kept outdoors in most farming systems and, thus, are exposed to fluctuating environmental conditions. Seasonal variation in temperature causes variation in growth rate of fish, because of their ectothermic nature. Fish are harvested at a constant weight rather than at a constant age, hence the length of a production cycle depends on the stocking date. Second, production output of a farm is determined by constraints such as oxygen availability [11] and stocking density. Stocking density constrains production output for many important aquaculture species, including Atlantic salmon (*Salmo salar*), European seabass, and gilthead seabream (*Sparus aurata*). Thus, bio-economic models could prove useful to derive economic values for aquaculture species.

The objectives of this study were: (1) to develop such a bio-economic model, (2) to apply the model to determine the economic importance and economic values for various traits in a case-study on gilthead seabream, and (3) to validate the model by comparison with a profit equation for a simplified production system.

METHODS

Traits

The breeding goal considered here includes growth rate, feed intake rate, mortality rate, and uniformity in harvest weight. Growth rate affects revenues, feed costs and juvenile costs; feed intake rate affects feed costs; mortality rate affects feed costs and juvenile costs. Feed and juveniles are major costs in production [17], thus including growth and feed intake in the breeding goal is common practice in livestock [18, 19]. Uniformity, i.e. size variation around the mean harvest weight, determines the distribution of fish over price categories at harvest, and thus affects revenues via the average sales price of fish.

Economic values are specific for the unit in which a trait is expressed [20]. Here, growth rate is expressed in units of thermal growth coefficient (*TGC*) [21]. *TGC* is a standardized measure of growth in fish that takes stocking weight and temperature variation over the lifespan of a fish into account. *TGC* is widely used [22], is more accurate than other measures of growth rate [23], and is relatively robust to differences in temperature regimes [24, 25]. Feed intake is assumed to be determined by the same variables as bodyweight and gain in bodyweight, because energy requirement is largely determined by bodyweight and gain in bodyweight [26]. In this study, feed intake rate is, therefore, expressed in units of thermal feed intake coefficient *TFC*, a *TGC* analogue. *TFC* takes stocking weight and temperature variation over the lifespan of a fish into account. The *TFC* model is independent of the *TGC* model, i.e. a modelled change in growth rate will not affect modelled feed intake rate and vice versa, which is a prerequisite to derive economic values. Mortality rate (*M*) is expressed as % of mortality per day. Uniformity is expressed as the standard deviation of harvest weight (σ_{HW}) in grams.

Bio-economic Model

The bio-economic model developed by Besson et al. [12] was adapted to simulate production systems with seasonal variation in temperature and in which density constrains production output, in this case a typical grow-out farm for gilthead seabream in Greece. The model is a deterministic simulation model that is programmed in R version 2.12.2. [27]. The model simulates operation of the farm during an average year. The farm consists of 20 cages of 2800 m^3 each and produces about 550 tons annually. As shown in Fig. 1, the model consists of three hierarchical parts: a fish model, a cage model, and a farm model. Inputs into the fish model are: stocking date, temperature coefficients, *TGC*, and *TFC*. Outputs of the fish model for each stocking date are: bodyweight per day per fish, feed consumption per day per fish, and harvest date. Inputs into the cage model are: outputs of the fish model, *M*, cage volume, and feed prices. Outputs of the cage model for each stocking date are per production cycle of a cage: fish production, number of juveniles stocked, feed consumption, and feed costs. A production cycle is the period between stocking and harvesting a cage. Inputs of the farm model are: outputs of the cage model, number of cages, price of juveniles, price of packing, and sales prices. Outputs of the farm model are total per year: fish production, number of juveniles stocked, feed consumption, feed costs, juvenile costs, packing costs, revenues from fish sales, and gross margin.

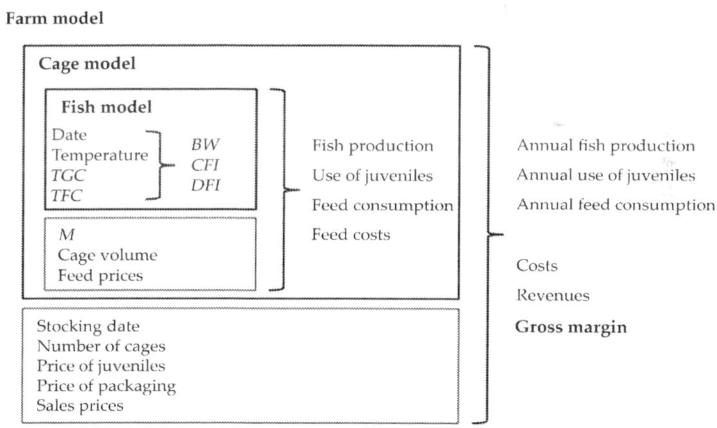

Figure 1: Schematic overview of the bio-economic model. *TGC* thermal growth coefficient, *TFC* thermal feed intake coefficient, *BW* bodyweight, *CFI* cumulative feed intake, *DFI* daily feed intake, *M* mortality rate.

The model was used to derive economic values of the traits mentioned above. Economic values give the expected change in profit from a small change in trait level, keeping the level of all other traits constant. Genetic change does not affect fixed costs, hence change in profit due to genetic change equals change in gross margin. When change in trait level equals the additive genetic standard deviation (σ_A) [15], the resulting change in gross margin indicates how important that trait is, because σ_A is indicative of the rate at which breeding values can be improved [28]. To determine the relative importance of the traits and to derive economic values, the model was run under two situations: first before genetic change and second, after a change of one σ_A in one trait with the other traits kept constant. For the trait 'uniformity', minimum and maximum values of the possible range of σ_A were used, because the actual value was unknown. Trait levels were changed in the desired direction of the genetic change. Economic values were expressed per kg of fish produced in the situation before genetic change [29] and were calculated as:

$$\textit{Economic value} = \left(\frac{\textit{gross margin}_A - \textit{gross margin}_B}{\textit{trait level}_A - \textit{trait level}_B}\right) \Big/ \textit{fish production}_B, \tag{1}$$

where subscripts indicate before (B) and after (A) genetic change.

Model Equations

The "Estimation of model coefficients" section describes the derivation of model coefficients from farm data. These coefficients (see Table 1) are used as the equations' coefficients of the bio-economic model. The three parts of the bio-economic model are described in the following three subsections.

Table 1: Estimated coefficients used in model equations

Symbol	Meaning	Value	Standard error	Unit
TM	Annual mean temperature	19.57	0.0119	°C
TA	Amplitude of temperature	−4.806	0.0167	°C
tA	Date at which temperature equals TM	−32.49	0.2057	Day
TGC	Thermal growth coefficient	12.6	0.0847	$g^{2/3}$/(day degrees · 1000)
TFC	Thermal feed intake coefficient	8.25	0.157	$g^{0.544}$/(day degrees · 1000)

p	Weight exponent to predict cumulative feed intake	0.544	0.00282	–
M	Mortality rate	0.0300	0.000164	%/day

Estimation of Model Coefficients

Coefficients in the equations to describe temperature, fish growth, feed intake, and number of fish per cage were derived from recent farm data of the company Andromeda S.A., which is hereafter referred to as 'data'. The data included daily records of temperature, feed provided, and mortality, and regular records of bodyweight from 15 cages of a farm located in Vonitsa, Northern Greece during the period 2013 through 2015.

Seasonal variation in daily water temperature throughout the year followed a sinusoidal pattern. Therefore, the equation to describe daily temperature ($T_{ts,a}$) in °C was [30]:

$$T_{t_{s,a}} = TM - TA \cdot sin\left(2\pi \cdot \left(t_{s,a} - t_A\right)/365\right), \tag{2}$$

where TM is the average annual temperature (°C), TA is the range of temperatures around TM (°C), t_A is the time of the year at which $T_{ts,a}$ equaled TM, and $t_{s,a}$ represents the date defined as:

$$t_{s,a} = s + a, \tag{3}$$

where stocking date s (s = 1, ... , n) equals 1 on January 1 2013 and a (a = 0, ... , n) is the age of the fish (days). To estimate TM, TA, and t_A, Eq. 2 was fitted to the data by means of non-linear least-squares regression in R. Table 1 shows the resulting coefficients.

Bodyweight in seabream can be predicted from the stocking weight and the sum of daily effective temperatures. Daily effective temperature is the daily temperature minus 12°C, where 12°C represents the minimum temperature for seabream growth [31]. Therefore, the equation to describe bodyweight (in g) at $t_{s,a}$ ($BW_{ts,a}$) was [32]:

$$BW_{t_{s,a}} = \left(BW_{t_{s,0}}^{2/3} + \frac{TGC}{1000} \cdot \sum_{i=t_{s,0}}^{(t_{s,a})-1} (T_i - 12)\right)^{3/2}, \tag{4}$$

where $BW_{ts,0}$ is bodyweight at stocking (in g), and $\sum_{i=t_{s,0}}^{(t_{s,a})-1} (T_i - 12)$ represents the sum of effective temperatures (day degrees) over the lifespan of a fish excluding $t_{s,a}$. To estimate TGC, Eq. 4 was fitted to the data by means of non-linear least-squares regression in R. Instead of fixing exponents 2/3 and 3/2, fitting these to the data resulted in values of 0.612 and 1/0.612 but this barely improved accuracy of the model. Values of 2/3 and 3/2 were preferred, because standardization of growth models allows for a better comparison of growth rate across studies. Analogous to $BW_{ts,a}$, the model to describe cumulative feed intake (in g) at $t_{s,a}$ ($CFI_{ts,a}$) was:

$$CFI_{t_{s,a}} = \left(BW_{t_{s,0}}^{p} + \frac{TFC}{1000} \cdot \sum_{i=t_{s,0}}^{t_{s,a}} (T_i - 12) \right)^{1/p} - BW_{t_{s,0}}, \tag{5}$$

where p is a weight exponent, and $\sum_{i=t_{s,0}}^{t_{s,a}} (T_i - 12)$ represents the sum of effective temperatures (day degrees) over the lifespan of a fish, including $t_{s,a}$. The term $BW_{ts,0}$ was subtracted from

$$\left(BW_{t_{s,0}}^{p} + \frac{TFC}{1000} \cdot \sum_{i=t_{s,0}}^{t_{s,a}} (T_i - 12) \right)^{1/p}$$

to force the model through the intercept. To estimate TFC and p, Eq. 5 was fitted to the data by means of non-linear least-squares regression in R. Parameter M (%/day) was assumed to be constant over time, hence the number of fish alive decreased exponentially in time. The model to describe the number of fish at $t_{s,a}$ ($N_{ts,a}$) was:

$$N_{t_{s,a}} = N_{t_{s,0}} \cdot \left(1 - \frac{M}{100\%} \right)^{(t_{s,a})-1}, \tag{6}$$

where $N_{ts,0}$ is the number of fish stocked. To estimate M, Eq. 6 was fitted to the data by means of non-linear least squares regression in R.

Fish Model

Date ($t_{s,a}$) was modelled as in Eq. 3. Temperature ($T_{ts,a}$) was modelled as in Eq. 2. Bodyweight ($BW_{ts,a}$) was modelled as in Eq. 4, where stocking weight ($BW_{ts,0}$) was 4.4 g, equal to the average in the data. Equation 4 was rewritten to calculate the harvest date ($t_{s,h}$) as:

$$\sum_{i=t_{s,0}}^{(t_{s,h})-1} (T_i - 12) = \frac{BW_{t_{s,h}}^{2/3} - BW_{t_{s,0}}^{2/3}}{TGC/1000}, \tag{7}$$

where $BW_{ts,h}$ is the average harvest weight, here set to the desired market weight of 400 g. Solving the right hand side of the equation, yields $\sum_{i=t_{s,0}}^{(t_{s,h})-1} (T_i - 12) = 4084$ day degrees. Cumulative feed intake $CFI_{ts,a}$ was modelled as in Eq. 5. $CFI_{ts,h}$ was set equal to $CFI_{(ts,h)}-1$, because fish are not fed on the day that they are harvested. Daily feed intake ($DFI_{ts,a}$) was modeled as:

$$DFI_{t_{s,a}} = CFI_{(t_{s,a})+1} - CFI_{t_{s,a}}. \tag{8}$$

Cage Model

To maximize production, standing stock in a cage reaches the maximum allowable density of 15 kg/m^3 at harvest. For the 2800 m^3 cages, per production cycle fish production is thus 42,000 kg or 100,500 fish. To compensate for mortality, a larger number of fish is stocked than harvested. The number of fish in a cage at $t_{s,a}$ ($N_{ts,a}$) was modeled as:

$$N_{t_{s,a}} = 100,500 \cdot \left(1 - \frac{M}{100\%}\right)^{(t_{s,h} - t_{s,a})}. \tag{9}$$

The number of juveniles stocked per cage ($N_{ts,0}$) was calculated by substituting $t_{s,a}$ for ts,0. Daily feed intake per cage (kg) at $t_{s,a}$ ($DFIcage_{ts,a}$) was modeled as:

$$DFIcage_{t_{s,a}} = N_{t_{s,a}} \cdot DFI_{t_{s,a}}/1000. \tag{10}$$

Total feed consumption per cage (kg) stocked at date $t_{s,0}$ ($TFIcage_{ts,0}$) was calculated as:

$$TFIcage_{t_{s,0}} = \sum_{i=t_{s,0}}^{t_{s,h}} DFIcage_i. \tag{11}$$

Depending on bodyweight, fish are fed different feed types. Daily feed costs per cage ($DFCcage_{ts,a}$) were calculated as the product of $DFIcage_{ts,a}$ and feed price per size category. Table 2 shows the price of feed types for the fish size categories. Total feed costs per cage (€) stocked at date $t_{s,0}$ ($TFCcage_{ts,0}$) were calculated as:

$$TFCcage_{t_{s,0}} = \sum_{i=t_{s,0}}^{t_{s,h}} DFCcage_i. \tag{12}$$

Farm Model

For the farm model, the cage model was run repeatedly over the whole range of stocking dates, from 1 (January 1th) to 365 days (December 31th) by one-day steps. Per stocking date, age at harvest ($t_{s,h} - t_{s,0}$), $TFIcage_{ts,0}$, $TFCcage_{ts,0}$, and $N_{ts,0}$ were calculated. These results were averaged over all stocking dates to compute the average production cycle of a cage. The period between two successive production cycles is three days [Personal communications Andromeda S.A., 2015]. For the 20 cages that are present on the farm, the number of production cycles per year was calculated as:

$$Production\ cycles\ per\ year = 20 \cdot \frac{365}{average(t_{s,h} - t_{s,0}) + 3}. \tag{13}$$

Average results per production cycle were multiplied by the number of production cycles per year to compute outputs at the farm level per year: fish production, number of juveniles stocked, feed consumption, and feed costs. Juvenile costs at the farm level were calculated by multiplying the number of juveniles stocked by the price of juveniles of €0.20 per piece. Packing costs at the farm level were calculated by multiplying fish production by the price of packing of €0.33 per kg fish. Revenues from fish sales at the farm level were calculated as the product of fish production and average sales price. Average sales price was computed as the proportion of fish of each category in Table 3 multiplied by its corresponding sales price. Three percent of the fish harvested are deformed. The size distribution of the remaining fish was calculated from a normal probability density function with $\mu = 400$ and $\sigma_{HW} = 60$ [Personal communications Andromeda S.A., 2015].

Table 2: Feed price per fish size category in 2014 [Personal communications Andromeda S.A., 2015]

Fish size (g)	Price (€/kg)
<7	2.21
7–13	1.97
13–30	1.65
30–80	1.26
80–300	1.12
>300	1.17

Additive Genetic Standard Deviation of Traits

The economic importance of each trait in the breeding goal depends on the change in gross margin, which itself depends on the change in trait level of one σ_A. The genetic coefficient of variation (CVA) can be used to estimate σ_A from the mean trait level (μ) [28]:

$$\sigma_A = \frac{CV_A}{100\%} \cdot \mu. \qquad (14)$$

For TFC, σ_A can be estimated from the genetic variation in $BW_{ts,h}$. For this study, $BW_{ts,h}$ was set equal to 400 g, and $\sum_{i=t_{s,0}}^{(t_{s,h})-1} (T_i - 12)$ was 4084 day degrees (Eq. 7). CVA of bodyweight was estimated to be 10.6% based on data from Navarro et al. [33]. For $BW_{ts,h}$, σ_A was thus 42.4 g. The distribution of $BW_{ts,h}$ was simulated in R as $BW_{ts,h,n} = \mu + z_n \cdot \sigma_A$, where $\mu = 400$ and $\sigma_A = 42.4$, and z_n is a standard normal distribution ($z_n \sim N(0, 1)$) with $n = 1, \ldots, 10^6$. From this simulation, σ_A of TGC was estimated as:

$$\sigma_A \text{of } TGC \approx \sqrt{Var(TGC_n)} \approx \sqrt{\left(\frac{1000}{4084}\right)^2 \cdot Var\left(BW_{t_{s,h},n}^{2/3}\right)}$$

$$= \frac{1000}{4084}\sqrt{\frac{1}{10^6 - 1} \cdot \sum_{i=1}^{n}\left(BW_i^{2/3} - 400^{2/3}\right)^2}$$

$$= 0.95\ g^{2/3}/\left(\text{day degrees} \cdot 1000\right). \qquad (15)$$

Table 3: Sales price per fish size category in 2014 [Personal communications Andromeda S.A., 2015]

Category (g)	Price (€/kg)
<100	0
100–200	1.65
200–300	4.15
300–400	4.52
400–600	4.63
>600 g	5.27
Deformed	2.52

For TFC, σ_A can be estimated from the genetic variation in both $BW_{ts,0}$ and $CFI_{ts,h}$. For $CFI_{ts,h}$, σ_A can be approximated by:

$$\sigma_A \text{ of } CFI_{t_{s,h}} \approx \frac{1.75}{r_A} \cdot \sigma_A \text{ of } BW_{t_{s,h}}, \tag{16}$$

where r_A is the genetic correlation between $BW_{ts,h}$ and $CFI_{ts,h}$, which was assumed to be 0.90 [34]. Solving Eq. (16), σ_A of $CFI_{ts,h}$ was equal to 82 g. Based on an average $CFI_{ts,h}$ of 713 g in our study, the CVA of $CFI_{ts,h}$ was 12%, which is close to values reported for other species [34, 35]. Genetic variances for $CFI_{ts,h}$ and BWts,0 were simulated to calculate σ_A of TFC. In our study, the average $BW_{ts,0}$ was 4.4 g. Based on a CVA of 10.6% for bodyweight, σ_A of $BW_{ts,0}$ was equal to 0.45 g. The distribution of $CFI_{ts,h}$ was simulated in R as $CFI_{ts,h,n} = \mu + z_n \cdot \sigma_A$, where $\mu = 713$, $\sigma_A = 82$, and z_n is a standard normal distribution ($z_n \sim N(0, 1)$) with $n = 1, \dots, 10^6$. The distribution of $BW_{ts,0}$ was simulated in R as $BW_{ts,0,n} = \mu + z_n \cdot \sigma_A$, where $\mu = 4.4$, $\sigma_A = 0.45$ and z_n is a standard normal distribution ($z_n \sim N(0, 1)$) with $n = 1, \dots, 10^6$. A covariance of zero was assumed between $CFI_{ts,h}$ and $BW_{ts,0}$. Based on the simulations, σ_A of TFC was estimated as:

$$\sigma_A \text{ of } TFC \approx \sqrt{Var(TFC_n)}$$

$$= \sqrt{Var\left(1000 \cdot \frac{(CFI_{t_{s,h},n} + BW_{t_{s,0},n})^p - BW^p_{t_{s,0},n}}{4084}\right)}$$

$$= \frac{1000}{4084} \cdot \sqrt{\frac{1}{10^6 - 1} \cdot \sum_{i=1}^{n} \left(\begin{array}{l} (CFI_i + BW_i)^{0.544} - BW_i^{0.544} \\ -((713 + 4.4)^{0.544} - 4.4^{0.544}) \end{array} \right)^2}$$

$$= 0.55\ g^{0.544} / (\text{degree days} \cdot 1000). \tag{17}$$

For M, σ_A can be estimated from the genetic variation in cumulative mortality at harvest ($CM_{ts,h}$). The average $CM_{ts,h}$ was 14.9%. In animal breeding, an underlying liability scale is commonly used to analyze mortality and survival [36]. Heritability of $CM_{ts,h}$ on the liability scale was assumed to be 0.17 [37] and by definition σ_P is equal to 1, hence $\sigma_A = \sqrt{h^2} = \sqrt{0.17}$. Before genetic change, the deviation of the threshold from the mean (x_B) was calculated from the quantile function of a normal distribution in R as $x_B = -\text{qnorm}(0.149) = 1.04$. After genetic change by one σ_A, the deviation from the threshold from the mean (x_A) becomes: $x_A = x_B + \sigma_A = 1.04 + \sqrt{0.17} = 1.45$. After genetic change, $CM_{ts,h}$ was calculated from the distribution function of a normal distribution in R as:

$$CM_{t_{s,h}} = (1 - pnorm(x_A)) \cdot 100\%$$
$$= (1 - pnorm(1.45)) \cdot 100\% = 7.34\%. \tag{18}$$

Average age at harvest was equal to 539 days (Table 4). M after genetic change was calculated as:

$$M = \left(1 - \left(1 - \frac{CM_{t_{s,h}}}{100}\right)^{\frac{1}{539}}\right) \cdot 100$$

$$\left(1 - \left(1 - \frac{7.34}{100}\right)^{\frac{1}{539}}\right) \cdot 100 = 0.014\%/\text{day}. \tag{19}$$

The difference in M before and after genetic change was 0.016, which was treated as the σ_A of M. Genetic improvement of uniformity reduces the environmental variance of bodyweight. For environmental variance of bodyweight, CV_A was calculated as [38]:

$$CV_A = \frac{SD(\sigma_E^2)}{\overline{\sigma_E^2}} \cdot 100\%, \tag{20}$$

where $SD(\sigma_E^2)$ is the genetic standard deviation of environmental variance and $\overline{\sigma_E^2}$ is the mean environmental variance. Environmental variance equals phenotypic variance minus genetic variance [39]. The CV_A of environmental variance of bodyweight is about 20% in rainbow trout [40] and 41.7% in Atlantic salmon (Salmo salar) [41]. For seabream, the actual value was unknown, hence a minimum of 20% and maximum of

40% were used to represent both extremes of the possible range of σ_A. In this study, the trait uniformity was expressed on the standard deviation scale instead of the variance scale. On the standard deviation scale, the CV_A is half as large as on the variance scale [42, 43]. For $BW_{ts,h}$, $\overline{\sigma_E} = \sqrt{\sigma_{HW}^2 - \sigma_A^2} = \sqrt{60^2 - 42.4^2} = 42.45$ g. For the minimum CV_A of the environmental standard deviation of bodyweight of 10%, σ_A equals 4.2 g, and for the maximum CV_A of the environmental standard deviation of bodyweight of 20%, σ_A equals 8.5 g.

Validation of the BioEconomic Model

To validate the bio-economic model, a simplified production system was assumed for which a profit equation can be developed. In this simplified production system, fish were harvested at a constant sum of effective temperatures instead of constant bodyweight. The sum of effective temperatures at harvest was assumed to be unaffected by genetic change. This allowed a profit equation to be set up as a function of the traits: harvest weight ($BW_{ts,h}$), cumulative feed intake at harvest ($CFI_{ts,h}$), and survival at harvest ($S_{ts,h}$). In the bio-economic model, $S_{t_{s,h}} = \frac{N_{t_{s,h}}}{N_{t_{s,0}}} \cdot 100\%$. The bio-economic model was adapted by changing the harvest criterion from a bodyweight of 400 g to a sum of effective temperatures of 4084 day degrees. Thus, an increase in TGC resulted in a greater harvest weight instead of a shorter growing period. One σ_A change in TGC led to change in BWts,h ; one σ_A change in TFC led to change in $CFI_{ts,h}$; one σ_A change in M led to change in $S_{ts,h}$. Economic values were derived from the bio-economic model using Eq. 1 and trait levels of $BW_{ts,h}$, $CFI_{ts,h}$, and $S_{ts,h}$. In the profit equation, profit at the farm level was described as:

$$Profit = \frac{1000 \cdot Q}{BW_{t_{s,h}}} \cdot \left(BW_{t_{s,h}} \cdot \frac{sales\ price - packing\ costs}{1000} \right.$$
$$\left. -CFI_{t_{s,h}} \cdot \frac{1}{0.5 + S_{t_{s,h}}/200} \cdot \left(\frac{feed\ price}{1000} \right) - \frac{juvenile\ price}{S_{t_{s,h}}/100} \right)$$
$$-\ fixed\ costs, \quad (21)$$

where Q represents production output of the farm (kg) and was calculated as the product of maximum stocking density, cage volume and number of production cycles per year. Q is not affected by genetic change when the harvest criterion is a sum of effective temperatures of 4084 day degrees. Economic values were calculated as partial derivatives of the profit equation and divided by Q to express them per kg fish production. For $BW_{ts,h}$, the economic value was calculated as:

$$Economic\ value_{BW_{t_{s,h}}} = \frac{\delta Profit}{\delta BW_{t_{s,h}}} \cdot \frac{1}{Q} = \frac{1000}{BW^2_{t_{s,h}}} \cdot \left(\frac{CFI_{t_{s,h}} \cdot (feed\ price/1000)}{0.5 + S_{t_{s,h}}/200} + \frac{juvenile\ price}{S_{t_{s,h}}/100} \right). \quad (22)$$

For $CFI_{ts,h}$, the economic value was calculated as:

$$Economic\ value_{CFI_{t_{s,h}}} = \frac{\delta Profit}{\delta CFI_{t_{s,h}}} \cdot \frac{1}{Q} = -\frac{feed\ price}{BW_{t_{s,h}} \cdot (0.5 + S_{t_{s,h}}/200)}. \quad (23)$$

For $S_{ts,h}$, the economic value was calculated as:

$$Economic\ value_S = \frac{\delta Profit}{\delta S} \cdot \frac{1}{Q} = \frac{CFI_{t_{s,h}} \cdot feed\ price}{BW_{t_{s,h}} \cdot \left(50 + S_{t_{s,h}} + S^2_{t_{s,h}}/200\right)} + \frac{100,000 \cdot juvenile\ price}{BW_{t_{s,h}} \cdot S^2_{t_{s,h}}}. \quad (24)$$

RESULTS

Production Results before Genetic Change

Tables 4 and 5 show the results of the model for key production variables and costs, respectively. The annual fish production was about 565 tons and gross margin about 759,000€. Average feed costs were €1.18/kg feed and average sales price was €4.49/kg fish.

Table 4: Key production variables of the gilthead seabream farm before genetic change

Item *FCR*	Value
Number of juveniles stocked ($year^{-1}$)	1,659,945
Feed consumption (kg/year)	1,070,177
Fish production (kg/year)	564,661
Cages stocked ($year^{-1}$)	13.4
Average age at harvest (day)	539
Survival (%)	85.1
Biological *FCR*[a] (kg feed/kg fish)	1.80
Economic *FCR*[b] (kg feed/kg fish)	1.92

[a]Biological *FCR* = feed consumption/(fish production + biomass mortality – biomass juveniles)

[b]Economic *FCR* = feed consumption/fish production

Table 5: Economic results for the gilthead seabream farm before genetic change

Item	Farm level (€)	Fish level (€/kg)
Feed costs	1,259,917	2.23
Juvenile costs	331,989	0.59
Packing costs	186,338	0.33
Total variable costs	1,778,244	3.15
Total revenues	2,537,166	4.49
Gross margin	758,922	1.34

Production Results after Genetic Change and Economic Values

The effect of the genetic change on production results is illustrated in Table 6. Changes in gross margin show that the order of economic importance of traits was: TGC, TFC, M, and σ_{HW}. The effect on gross margin of one σ_A change for each trait relative to the effect of a 8.5 g decrease in σ_{HW} was 43-fold for TGC, 28-fold for TFC, and 12-fold for M. Non-linearity was strongest for σ_{HW} (results not presented for TGC, TFC, and M), for which a doubling of change in trait level from −4.2 to −8.5 g led to 7.7% overestimation of the increase in gross margin.

The mechanisms by which changes in trait levels determined changes in gross margin were as follows. An increase in TGC resulted in a lower age at harvest (Eq. 7) and consequently, the number of production cycles per year increased (Eq. 13) and at the farm level, the annual number of juveniles stocked and annual fish production increased. An increase in TGC did not affect daily feed consumption (Eq. 4) and consequently, cumulative feed intake at harvest decreased because the sum of effective temperatures at harvest decreased (Eq. 5) and at the farm level, the annual feed consumption decreased. A decrease in TFC decreased total feed consumption per production cycle (Eqs. 5, 11) but the number of juveniles stocked per production cycle and the fish production per production cycle remained unaltered, thus at the farm level, only annual feed consumption decreased. A decrease in M reduced the number of juveniles stocked per production cycle (Eq. 9) and consequently, daily feed intake per cage decreased because the average number of fish per cage per day was smaller (Eq. 10), but fish production per production cycle was unaltered. Thus at the farm level, the annual number of juveniles stocked and annual feed consumption decreased. The effect of σ_{HW} on the average sales price is illustrated in Fig. 2: less variation led to more sales in size category 300 to 400 g (€4.52/kg) at the expense of sales in size category 200 to 300 g (€4.15/kg). Production results were unaltered by a change in σ_{HW}.

Table 6: Effect of genetic change on production results relative to the situation without genetic change

Trait	Genetic change (trait unit)	Δ Juveniles stocked ($year^{-1}$)	Δ Feed consumption (kg/year)	Δ Fish production (kg/year)	Δ Gross margin (€)
TGC [a]	+0.95	89,400	−68,409	36,309	213,131
TFC [b]	−0.55	0	−120,300	0	140,891
M [c]	−0.016	−137,294	−34,446	0	69,531
σ_{HW} [d]	−4.2	0	0	0	2636
σ_{HW}	−8.5	0	0	0	4952

[a] Thermal growth coefficient $[g^{2/3}/(\text{day degrees} \cdot 1000)]$

[b] Thermal feed intake coefficient $[g^{0.544}/(\text{day degrees} \cdot 1000)]$

[c] Mortality rate (%/day)

[d] Standard deviation of harvest weight (g)

Economic values are in Table 7, which shows that the economic value of σ_{HW} was similar for both levels of genetic change.

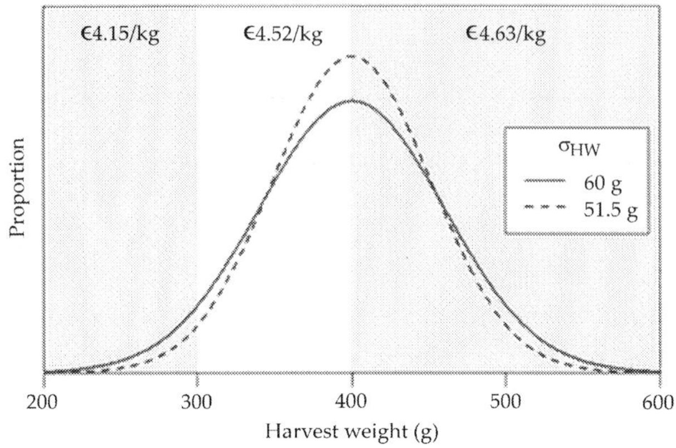

Figure 2: Distribution of harvest weight over sales price categories at different standard deviations of harvest weight (σ_{HW}).

Table 7: Economic values of traits for gilthead seabream

Trait	Baseline trait level (trait unit)	Genetic change (trait unit)	Economic value [€ (kg production)$^{-1}$ (trait unit)$^{-1}$]
TGC[a]	12.6	+0.95	0.40
TFC[b]	8.25	−0.55	−0.45
M[c]	0.0300	−0.016	−7.7
σ_{HW}[d]	60	−4.2	−0.0011
σ_{HW}	60	−8.5	−0.0010

[a] Thermal growth coefficient [$g^{2/3}$ /(day degrees · 1000)]

[b] Thermal feed intake coefficient [$g^{0.544}$ /(day degrees · 1000)]

[c] Mortality rate (%/day)

[d] Standard deviation of harvest weight (g)

Comparison of Economic Values from the Bio-Economic Model and the Profit Equation

Table 8 shows that, for the simplified production system, the economic values derived from the bio-economic model and the profit equation were similar.

Table 8: Economic values derived from the bio-economic model and profit equation

Trait	Baseline trait level (trait unit)	Genetic change (trait unit)	Economic value [€ (kg production)$^{-1}$ (trait unit)$^{-1}$]	
			Bio-economic model	Profit equation
$BW_{t_{s,h}}$ [a]	400	43.6	0.0074	0.0072
$CFI_{t_{s,h}}$ [b]	713	−80.0	−0.0031	−0.0032
$S_{t_{s,h}}$ [c]	85.1	7.66	0.016	0.019

[a] Harvest weight (g)

[b] Cumulative feed intake at harvest (g)

[c] Survival at harvest (%)

DISCUSSION

Validity of the Bio-Economic Model

For the simplified production system, the profit equation and the bio-economic model return similar economic values, which confirm the validity of the bio-economic model. To further validate the bio-economic model, production results were compared to those of other studies. FCR (Table 4) was within the range of 1.5 to 2 reported by Sola et al. [44] but considerably lower than the 2.3 value reported by EAS-EATiP [45]. Overall survival was 85%, which is within the range reported by EAS-EATiP [45]. A comparison with a cost-breakdown for large-scale production of gilthead seabream and European seabass (*Dicentrarchus labrax*) is in Table 9 [17]. In the FAO data, variable costs are higher, largely because labor, energy, and medicines and veterinary services were not considered to be variable costs in our study. Trends in the increase in productivity per person [46] support the assumption that labor should be treated more as a fixed than a variable cost. Medicine costs may vary, but veterinary costs are likely to be fixed per farm. Energy costs are to a larger extent determined by farm layout than by realized production and thus can be considered as fixed. Altogether, total variable costs may have been slightly underestimated in our study, but FCR and overall survival matched well to current industry standards.

Table 9: Cost-breakdown for gilthead seabream production

Item	Proportion of total costs (%)	
	Our study[a]	**Barazi-Yeroulanos [17]**
Feed	50	48

Juveniles	13	11
Marketing (incl. packing)	7	18
Labor	–	3
Energy	–	4
Medicines and veterinary services	–	2
Other	–	4
Total variable costs	70	89
Total fixed costs	–	11

[a]Relative to revenues

Breeding Goal

In the breeding goal, TGC, TFC, and M are equivalent to respectively $BW_{ts,h}$, $CFI_{ts,h}$, and $S_{ts,h}$, when BWts,0 is much smaller than $BW_{ts,h}$. When the sum of effective temperatures is the harvest criterion, one σ_A change in TGC, TFC, and M led to changes in $BW_{ts,h}$, $CFI_{ts,h}$, and $S_{ts,h}$ that were very similar to the σ_A of these traits (Table 8), which demonstrates their equivalence. If the economic values of TGC, TFC, and M were calculated for the sum of effective temperatures instead of harvest weight as the harvest criterion, they would be slightly lower for TGC [0.34€ (kg production)$^{-1}$ ($g^{0.544}$ /(day degrees · 1000))$^{-1}$] and unaltered for TFC and M. In agreement with Wilton and Goddard [47], economic values were similar for both harvest criteria. Although both sets of traits are equivalent in the breeding goal, there are pros and cons to each one. $BW_{ts,h}$ is commonly used as a selection criterion and thus its use in the breeding goal is straightforward. However, $BW_{ts,h}$ is a management parameter that is strongly influenced by the growing period and temperature regime. TGC corrects for heterogeneity in stocking weight, growing period, and temperature regime, and, therefore, allows for a better comparison of breeding values across conditions than $BW_{ts,h}$ [11, 24, 25].

FCR could be used as an alternative to TFC in the breeding goal. An advantage of feed intake compared to FCR is that it relates directly to feed costs [48]. An advantage of FCR is that it illustrates the effect of improvement in efficiency on, for example, environmental impacts, as in Besson et al. [49]. Feed intake is often considered more appropriate as a breeding goal trait than FCR [18, 19], with a common argument that traits expressed as ratio's are disadvantageous in animal breeding [19]. Selection for a ratio, e.g. FCR, results in a lower selection response than selection for both components of the ratio, e.g. feed intake and growth [50]. However, in the same way that FCR is a ratio, growth is the ratio of feed intake to FCR, thus a breeding goal that includes both growth and FCR is equivalent to a breeding goal that includes growth and feed intake. The economic value of growth depends on which other trait, feed intake or FCR, is included in the breeding goal [48].

Economic Values

Bio-economic models and profit equations are both suitable to derive economic values. An advantage of a profit equation compared to a bio-economic model is its simplicity. However, its applicability is limited to specific situations, because environmental conditions are ignored. For example, the profit equation cannot be used to derive economic values for a range of temperature regimes, as was done in Besson et al. [11] by using the bio-economic model. Such properties may be of particular interest for breeding programs that aim at supplying many farms. In addition, alternative constraints on production output such as oxygen availability cannot be dealt with by the profit equation but can be incorporated in the bio-economic model, as discussed later. Furthermore, the profit equation is rigid in terms of trait definition, which has led to the false assumption that harvest weight changes following genetic improvement, whereas in the bio-economic model genetic improvement of growth rate leads to a reduction in the growing period.

From a profit function, economic values can be computed from either its partial derivative with respect to trait level, or from an increase or decrease in trait level relative to the current mean. In this study, simulated changes in trait levels correspond to desired directions of genetic change of one genetic standard deviation. However, for a non-linear profit function, Goddard [51] demonstrated that economic values that maximize profit in the next generation may depend on selection responses. Dekkers et al. [52] showed that economic gain is slightly higher when economic values are derived as the partial derivative of a non-linear profit equation at the genetic level of the next generation than at the genetic level of the current generation. This implies two things:

- Economic values are closer to optimum when the simulated change in trait level resembles its expected rather than its desired direction.
- Economic values are closer to optimum when the simulated change in trait level equals the difference between trait levels in the current and next generation than when it is the partial derivative at current genetic levels.

In our study, expected and desired directions of change in trait levels were identical, except for TFC, which may increase in practice due to its genetic correlation with TGC [34]. A genetic standard deviation generally provides a better proxy for the difference between trait levels in the current and next generation than an infinitesimal change, and hence will result in an economic value that is closer to optimum than the conventional partial derivative at the current trait level.

This is the first time that economic values have been derived for uniformity in aquaculture, here expressed as σ_{HW}. In recent years, there has been increasing interest to improve uniformity [40, 53–55]. Improvement in uniformity affects average sales price and reduces the need for size-grading. However, for seabream production, reducing the need for size-grading would not result in major cost-savings, because seabream is size-graded only once during grow-out. Thus, a potential effect on grading frequency was excluded from the economic value of uniformity. Furthermore, uniformity has been suggested to affect feed intake, growth rate and mortality [56, 57].

Economic consequences of changes in other traits were accounted for in their respective economic values. By exploiting genetic correlations, selection for uniformity may be used to improve the other traits in the breeding goal.

Application to Other Aquaculture Species

In its current form, the bio-economic model can be easily applied for the derivation of economic values for other species produced in systems where stocking density limits production output, such as cages and flow-through tanks. This would require different values for the coefficients of Table 1, maximum stocking density, stocking and harvest weight, and input and output prices. Equations 4 and 5 require some species-specific modifications, such as alternative values for exponents 2/3 and 3/2 in Eq. 4 [58] or a different minimum temperature for growth.

Adaptations to the model are required for species that are reared in production systems for which constraints on production output are different, such as recirculating aquaculture systems and ponds. When the constraint on production output is different from stocking density, the number of fish stocked per production cycle (Eq. 9) is determined by other parameters. For recirculating aquaculture systems, treatment capacity of the biofilter can be a constraint on production output [12]. In this case, daily nitrogen excretion by fish is the parameter that determines the number of fish stocked per production cycle. Daily nitrogen excretion by fish can be predicted from the difference between daily feed consumption and daily gain in bodyweight, as described in Besson et al. [12]. In both cages and ponds, oxygen availability can be a constraint on production output. In this case, daily oxygen consumption per fish is the parameter that determines the number of fish stocked per production cycle. Daily oxygen consumption per fish can be predicted from daily feed consumption and daily gain in bodyweight, as described in Besson et al. [11]. With the above modifications, the same bio-economic model was applied for the derivation of economic values for African catfish produced in recirculating aquaculture systems [12], European seabass produced in cages [11], gilthead seabream produced in cages (this study), turbot produced in tanks (unpublished results), and Nile tilapia produced in ponds (unpublished results).

CONCLUSIONS

We developed a bio-economic model to derive economic values for a wide range of aquaculture species. Its validity was confirmed by the comparison to a profit equation for a simplified production system and by comparison of the production results to those of other studies. Application of the bio-economic model to gilthead seabream resulted in economic values for TGC, TFC, M, and σ_{HW}. TGC was the most important trait to improve, followed by TFC and M. The effect of σ_{HW} on gross margin was small.

NOTES

Authors' Contributions

KJ developed the bio-economic model, performed the analysis, and wrote the manuscript. HK and PB contributed to discussions and writing of the manuscript. MB helped in developing the bio-economic model and contributed to discussions. All authors read and approved the final manuscript.

Acknowledgements

We are grateful for the extensive production data provided by Andromeda S.A. and their feedback on model assumptions. We thank Piter Bijma for his help in the computation of genetic standard deviations.

Availability of Data and Material

Production data of Andromeda S.A. will not be shared, because confidentiality was agreed.

Funding

The research leading to these results has received funding from the European Union's Seventh Framework Programme (KBBE.2013.1.2-10) under grant agreement no. 613611.

APPENDICES

Appendix 1: Genetic Variation in *TGC*

Genetic variation in TGC depends on genetic variation both in $BW_{ts,0}$ and $BW_{ts,h}$:

$$Var(A_{TGC}) = Var\left(1000 \cdot \frac{A_{BW_{t_{s,h}}}^{2/3} - A_{BW_{t_{s,0}}}^{2/3}}{\sum_{i=t_{s,0}}^{(t_{s,h})-1} (T_i - 12)}\right), \quad (25)$$

where ATGC is the genotype for TGC, $A_{BW_{t_{s,h}}}$ is the genotype for harvest weight, and $A_{BW_{t_{s,0}}}$ is the genotype for stocking weight. Equation 25 can be rewritten as:

$$Var(A_{TGC}) = \left(\frac{1000}{\sum_{i=t_{s,0}}^{(t_{s,h})-1}(T_i - 12)}\right)^2 \cdot \left(Var\left(A_{BW_{t_{s,h}}}^{2/3}\right) + Var\left(A_{BW_{t_{s,0}}}^{2/3}\right) - 2 \cdot cov\left(A_{BW_{t_{s,h}}}^{2/3}, A_{BW_{t_{s,0}}}^{2/3}\right)\right). \tag{26}$$

Because BWts,0 is much smaller than $BW_{ts,h}$, $Var\left(A_{BW_{t_{s,0}}}^{2/3}\right) - 2 \cdot cov\left(A_{BW_{t_{s,h}}}^{2/3}, A_{BW_{t_{s,0}}}^{2/3}\right)$ is much smaller than $Var\left(A_{BW_{t_{s,h}}}^{2/3}\right)$ [59]. Thus, Eq. 26 can be reduced to Eq. 15.

Appendix 2: Genetic Variation in Cumulative Feed Intake at Harvest

The regression coefficient of the genotype for $CFI_{ts,h}$ ($A_{CFI_{t_{s,h}}}$) on the difference between $A_{BW_{t_{s,h}}}$ and $A_{BW_{t_{s,0}}}$, can be calculated as:

$$b\left(A_{CFI_{t_{s,h}}}, A_{BW_{t_{s,h}}} - A_{BW_{t_{s,0}}}\right) = r_A \cdot \sqrt{\frac{Var\left(A_{CFI_{t_{s,h}}}\right)}{Var\left(A_{BW_{t_{s,h}}} - A_{BW_{t_{s,0}}}\right)}}, \tag{27}$$

where $b\left(A_{CFI_{t_{s,h}}}, A_{BW_{t_{s,h}}} - A_{BW_{t_{s,0}}}\right)$ is the regression coefficient, and r_A is the genetic correlation coefficient. For the regression of $A_{CFI_{t_{s,h}}}$ on $A_{BW_{t_{s,h}}} - A_{BW_{t_{s,0}}}$, the intercept corresponds to feed consumption to meet maintenance energy requirements ($CFI_{maintenance}$) at zero growth. Assuming a digestible energy content of the diet of 17 kJ/g, $CFI_{maintenance}$ is calculated as [26]:

$$CFI_{maintenance} = \sum_{i=t_{s,0}}^{t_{s,h}} 47.89 \cdot (BW_i/1000)^{0.80}/17 = 20 \text{ g}. \tag{28}$$

For $A_{BW_{t_{s,h}}} - A_{BW_{t_{s,0}}} = 400 - 4.4 = 395.6$ g, $A_{CFI_{t_{s,h}}}$ equals $1.80 \cdot 395.6 = 713$ g,, where 1.80 is the biological FCR (Table 4). The regression coefficient can thus be approximated as (713 − 20)/395.6 = 1.75 and Eq. 27 can be rewritten as:

$$1.75 = r_A \cdot \sqrt{\frac{Var\left(A_{CFI_{t_{s,h}}}\right)}{Var\left(A_{BW_{t_{s,h}}}\right) + Var\left(A_{BW_{t_{s,0}}}\right) - 2 \cdot cov\left(A_{BW_{t_{s,h}}}, A_{BW_{t_{s,0}}}\right)}}. \quad (29)$$

Because $BW_{ts,0}$ is much smaller than $BW_{ts,h}$, $Var\left(A_{BW_{t_{s,0}}}\right) - 2 \cdot cov\left(A_{BW_{t_{s,h}}}, A_{BW_{t_{s,0}}}\right)$ is much smaller than $Var\left(A_{BW_{t_{s,h}}}\right)$ [59]. Thus Eq. 29 can be reduced to Eq. 16.

REFERENCES

1. Janssen K, Chavanne H, Berentsen P, Komen H. Impact of selective breeding on European aquaculture. Aquaculture. 2016;. doi:10.1016/j.aquaculture.2016.03.012.
2. Nielsen HM, Amer PR, Byrne TJ. Approaches to formulating practical breeding objectives for animal production systems. Acta Agric Scand A Anim. 2014;64:2–12.
3. Shook GE. Major advances in determining appropriate selection goals. J Dairy Sci. 2006;89:1349–61.
4. Gibson JP, Kennedy BW. The use of constrained selection indexes in breeding for economic merit. Theor Appl Genet. 1990;80:801–5.
5. Gjedrem T, Baranski M. Selective breeding in aquaculture: an introduction. Dordrecht: Springer; 2009.
6. Gjedrem T, Thodesen J. Selection. In: Gjedrem T, editor. Selection and breeding programs in aquaculture. Dordrecht: Springer; 2005. p. 89–111.
7. Ponzoni RW, Nguyen NH, Khaw HL. Investment appraisal of genetic improvement programs in Nile tilapia (*Oreochromis niloticus*). Aquaculture. 2007;269:187–99.
8. Ponzoni RW, Nguyen NH, Khaw HL, Ninh NH. Accounting for genotype by environment interaction in economic appraisal of genetic improvement programs in common carp *Cyprinus carpio*. Aquaculture. 2008;285:47–55.
9. Zuniga-Jara S, Marin-Riffo MC. A bioeconomic model of a genetic improvement program of abalone. Aquacult Int. 2014;22:1533–62.

10. Henryon M, Purvis IW, Berg P. Definition of a breeding objective for commercial production of the freshwater crayfish, marron (*Cherax tenuimanus*). Aquaculture. 1999;173:179–95.
11. Besson M, Komen H, Aubin J, De Boer IJM, Poelman M, Quillet E, et al. Economic values of growth and feed efficiency for fish farming in recirculating aquaculture system with density and nitrogen output limitations: a case study with African catfish (*Clarias gariepinus*). J Anim Sci. 2014;92:5394–405.
12. Besson M, Vandeputte M, van Arendonk JAM, Aubin J, de Boer IJM, Quillet E, et al. Influence of water temperature on the economic value of growth rate in fish farming: the case of sea bass (*Dicentrarchus labrax*) cage farming in the Mediterranean. Aquaculture. 2016. doi:10.1016/j.aquaculture.2016.04.030.
13. Knap PW. Breeding robust pigs. Austr J Exp Agric. 2005;45:763–73.
14. Hietala P, Wolfova M, Wolf J, Kantanen J, Juga J. Economic values of production and functional traits, including residual feed intake, in Finnish milk production. J Dairy Sci. 2014;97:1092–106.
15. van Middelaar CE, Berentsen PBM, Dijkstra J, van Arendonk JAM, de Boer IJM. Methods to determine the relative value of genetic traits in dairy cows to reduce greenhouse gas emissions along the chain. J Dairy Sci. 2014;97:5191–205.
16. Byrne TJ, Amer PR, Fennessy PF, Cromie AR, Keady TWJ, Hanrahan JP, et al. Breeding objectives for sheep in Ireland: a bio-economic approach. Livest Sci. 2010;132:135–44.
17. Barazi-Yeroulanos L. Synthesis of Mediterranean marine finfish aquaculture—a marketing and promotion strategy. In: Studies and reviews. General Fisheries Commission for the Mediterranean. no. 88. Rome: FAO; 2010. p. 1–198.
18. Emmerson DA. Commercial approaches to genetic selection for growth and feed conversion in domestic poultry. Poult Sci. 1997;76:1121–5.
19. Veerkamp RF, Pryce JE, Spurlock D, Berry D, Coffey M, Løvendahl P, et al. Selection on feed intake or feed efficiency: a position paper from gDMI breeding goal discussions. Interbull Bull. 2013;47:15–22.

20. Wolfová M, Wolf J. Strategies for defining traits when calculating economic values for livestock breeding: a review. Animal. 2013;7:1401–13.
21. Iwama GK, Tautz AF. A simple growth model for Salmonids in hatcheries. Can J Fish Aquat Sci. 1981;38:649–56.
22. Jobling M. The thermal growth coefficient (TGC) model of fish growth: a cautionary note. Aquacult Res. 2003;34:581–4.
23. Cho CY. Feeding systems for rainbow trout and other salmonids with reference to current estimates of energy and protein requirements. Aquaculture. 1992;100:107–23.
24. Sae-Lim P, Kause A, Mulder HA, Martin KE, Barfoot AJ, Parsons JE, et al. Genotype-by-environment interaction of growth traits in rainbow trout (*Oncorhynchus mykiss*): a continental scale study. J Anim Sci. 2013;91:5572–81.
25. Trong TQ, Mulder HA, van Arendonk JAM, Komen H. Heritability and genotype by environment interaction estimates for harvest weight, Growth rate, And shape of Nile tilapia (*Oreochromis niloticus*) grown in river cage and VAC in Vietnam. Aquaculture. 2013;384–387:119–27.
26. Lupatsch I, Kissil GW, Sklan D. Comparison of energy and protein efficiency among three fish species gilthead sea bream (*Sparus aurata*), European sea bass (*Dicentrarchus labrax*) and white grouper (*Epinephelus aeneus*): energy expenditure for protein and lipid deposition. Aquaculture. 2003;225:175–89.
27. R Core Team. R: A Language and environment for statistical computing. 2015. http://www.R-project.org/.
28. Houle D. Comparing evolvability and variability of quantitative traits. Genetics. 1992;130:195–204.PubMedPubMedCentral
29. Groen AF. Economic values in cattle-breeding. 2. Influences of production circumstances in situations with output limitations. Livest Prod Sci. 1989;22:17–30.
30. Cacho OJ. Protein and fat dynamics in fish: a bioenergetic model applied to aquaculture. Ecol Model. 1990;50:33–56.
31. Mayer P, Estruch V, Blasco J, Jover M. Predicting the growth of gilthead sea bream (*Sparus aurata* L.) farmed in marine cages under

real production conditions using temperature- and time-dependent models. Aquacult Res. 2008;39:1046–52.

32. Mayer P, Estruch VD, Jover M. A two-stage growth model for gilthead sea bream (*Sparus aurata*) based on the thermal growth coefficient. Aquaculture. 2012;358–359:6–13.
33. Navarro A, Zamorano MJ, Hildebrandt S, Ginés R, Aguilera C, Afonso JM. Estimates of heritabilities and genetic correlations for growth and carcass traits in gilthead seabream (*Sparus auratus* L.), under industrial conditions. Aquaculture. 2009;289:225–30.
34. Quinton CD, Kause A, Koskela J, Ritola O. Breeding salmonids for feed efficiency in current fishmeal and future plant-based diet environments. Genet Sel Evol. 2007;39:431.PubMedCentral
35. Kause A, Tobin D, Dobly A, Houlihan D, Martin S, Mantysaari EA, et al. Recording strategies and selection potential of feed intake measured using the X-ray method in rainbow trout. Genet Sel Evol. 2006;38:389–409.PubMedCentral
36. Falconer DS, Mackay TFC. Introduction to quantitative genetics. 3rd ed. Harlow: Longman; 1989.
37. Vehvilainen H, Kause A, Quinton C, Koskinen H, Paananen T. Survival of the currently fittest: genetics of rainbow trout survival across time and space. Genetics. 2008;180:507–16.PubMedCentral
38. Mulder HA, Bijma P, Hill WG. Prediction of breeding values and selection responses with genetic heterogeneity of environmental variance. Genetics. 2007;175:1895–910.PubMedCentral
39. Hazel LN. The genetic basis for constructing selection indexes. Genetics. 1943;28:476–90.PubMedPubMedCentral
40. Sae-Lim P, Kause A, Janhunen M, Vehvilainen H, Koskinen H, Gjerde B, et al. Genetic (co)variance of rainbow trout (*Oncorhynchus mykiss*) body weight and its uniformity across production environments. Genet Sel Evol. 2015;47:46.PubMedCentral
41. Sonesson AK, Odegard J, Ronnegard L. Genetic heterogeneity of within-family variance of body weight in Atlantic salmon (*Salmo salar*). Genet Sel Evol. 2013;45:41.PubMedCentral
42. Hill WG, Mulder HA. Genetic analysis of environmental variation. Genet Res (Camb). 2010;92:381–95.

43. Sell-Kubiak E, Bijma P, Knol EF, Mulder HA. Comparison of methods to study uniformity of traits: application to birth weight in pigs. J Anim Sci. 2015;93:900–11.

44. Sola L, Moretti A, Crosetti D, Karaiskou N, Magoulas A, Rossi AR, et al. Gilthead seabream—*Sparus aurata*. In: D Crossetti, S Lapègue, I Olesen, T Svaasand, editors. Genetic effects of domestication, culture and breeding of fish and shellfish, and their impacts on wild populations. Genimpact Final Scientific Report. Viterbo; 2007. p. 47–54.

45. EAS-EATiP. Performance of the sea bass and sea bream sector in the Mediterranean. In Minutes of a Workshop held within Aquaculture Europe: 16 October 2014; San Sebastian; 2014.

46. University of Stirling. Study of the market for aquaculture produced seabass and seabream species; 2004. http://ec.europa.eu/fisheries/documentation/studies/aquaculture_market_230404_en.pdf. Accessed 12 April 2016.

47. Wilton JW, Goddard ME. Selection for carcass and feedlot traits considering alternative slaughter end points and optimized management. J Anim Sci. 1996;74:37–45.

48. Goddard ME. Consensus and debate in the definition of breeding objectives. J Dairy Sci. 1998;81:6–18.

49. Besson M, Aubin J, Komen H, Poelman M, Quillet E, Vandeputte M, et al. Environmental impacts of genetic improvement of growth rate and feed conversion ratio in fish farming under rearing density and nitrogen output limitations. J Clean Prod. 2016;116:100–9.

50. Gunsett FC. Linear index selection to improve traits defined as ratios. J Anim Sci. 1984;59:1185–93.

51. Goddard ME. Selection indexes for non-linear profit-functions. Theor Appl Genet. 1983;64:339–44.

52. Dekkers JCM, Birke PV, Gibson JP. Optimum linear selection indexes for multiple generation objectives with nonlinear profit-functions. Anim Sci. 1995;61:165–75.

53. Mulder HA, Bijma P, Hill WG. Selection for uniformity in livestock by exploiting genetic heterogeneity of residual variance. Genet Sel Evol. 2008;40:37.PubMedPubMedCentral

54. Khaw HL, Ponzoni RW, Yee HY, Aziz MA, Mulder HA, Marjanovic J, et al. Genetic variance for uniformity of harvest weight in Nile tilapia (*Oreochromis niloticus*). Aquaculture. 2016;451:113–20.

55. Marjanovic J, Mulder HA, Khaw HL, Bijma P. Genetic parameters for uniformity of harvest weight in the GIFT strain of Nile tilapia estimated using double hierarchical generalized linear models. Genet Sel Evol. 2016;48:41.PubMedCentral

56. Jobling M. Simple indices for the assessment of the influences of social environment on growth performance, exemplified by studies on Arctic charr. Aquacult Int. 1995;3:60–5.

57. Gilmour KM, DiBattista JD, Thomas JB. Physiological causes and consequences of social status in salmonid fish. Integr Comp Biol. 2005;45:263–73.

58. Dumas A, France J, Bureau DP. Evidence of three growth stanzas in rainbow trout (*Oncorhynchus mykiss*) across life stages and adaptation of the thermal-unit growth coefficient. Aquaculture. 2007;267:139–46.

59. Rutten MJM, Komen H, Bovenhuis H. Longitudinal genetic analysis of Nile tilapia (*Oreochromis niloticus* L.) body weight using a random regression model. Aquaculture. 2005;246:101–13.

CHAPTER

15

GENOMICS IN AQUACULTURE TO BETTER UNDERSTAND SPECIES BIOLOGY AND ACCELERATE GENETIC PROGRESS

José M. Yáñez[1,2], Scott Newman[3] and Ross D. Houston[4]

[1] Faculty of Veterinary and Animal Sciences, University of Chile, Santiago, Chile,

[2] Aquainnovo, Puerto Montt, Chile,

[3] Genus plc, Hendersonville, TN, USA,

[4] The Roslin Institute and Royal (Dick) School of Veterinary Studies, University of Edinburgh, Midlothian, UK

Keywords: Aquaculture, genome, breeding programs, QTL, single nucleotide polymorphisms, next-generation sequencing

INTRODUCTION

The production of fish and shellfish through aquaculture is an increasingly important source of high-quality animal protein, with a worldwide production of 66.6 million tons in 2012 (FAO, 2014). Considering the continuously growing global human population and increasing demand for fish products, improvements in the scale, efficiency, and sustainability of aquaculture are essential. To achieve this, several challenges facing the culture of fish and shellfish species need to be overcome. These relate to the diverse biology of the cultured species and their interaction with environmental factors. Examples include outbreaks of infectious diseases, control of sexual maturation, sustainable feed for carnivorous species, and tolerance of diverse and changing environments. This "Frontiers in Livestock Genomics" Research Topic highlights the opportunities offered by recent developments in the field of genomics, and in particular high-throughput sequencing, to contribute to addressing these challenges, with a focus on selective breeding programmes.

The use of selective breeding as a tool to improve the biological efficiency of production in aquaculture generally lags behind plant and farm animal industries, and less than 10% of aquaculture production is based on genetically-improved stocks (Gjedrem et al., 2012). Encouragingly, annual genetic gains reported for aquatic species are in general substantially higher than that of terrestrial farm animals (Gjedrem et al., 2012) and there is considerable scope for achieving significant positive economic impact via improved breeding schemes. However, the status of breeding programs and the level of technology used for aquatic species production are wide-ranging, from use of wild seed stocks through to family-based selection incorporating genomic tools. Family selection and genomic tools can be applied to improve traits that are expensive or difficult to measure on the selection candidates themselves including disease resistance (Yáñez et al., 2014; Ødegård et al., 2014), flesh color (Colihueque and Araneda, 2014; Ødegård et al., 2014) and other appearance traits such as body shape and skin pigmentation (Colihueque and Araneda, 2014) in finfish species. In contrast, despite the global importance of mollusc species for aquaculture, few selective breeding programmes exist and the state of genomic tools and knowledge for these species is typically lacking (Astorga, 2014).

Genomics resources such as whole genome reference sequences, high-density SNP genotyping arrays and genotyping-by-sequencing are in development for several aquaculture species. Fuller characterisation of these resources is underway and is resulting in improved fundamental knowledge of the genome structure and biology, highlighted in this issue by the analysis of repeat elements in the Asian sea bass genome (Kuznetsova et al., 2014). These resources will provide powerful tools for the research community and will aid in the determination of the genetic factors involved in the regulation of complex traits. For example, high-throughput RNA sequencing can give a holistic view of the host response to infectious diseases, and help identify the important genes and pathways defining genetic resistance, as demonstrated in this issue for rainbow trout (Ali et al., 2014; Marancik et al., 2014) and panaeid shrimp (Santos et al., 2014). Sequencing technology has also facilitated the development of abundant genetic markers that have multi-faceted applications for selective breeding of aquatic species, including parentage assignment in mixed-family environments, providing greater control over family representation and inbreeding (Vandeputte and Haffray, 2014). Medium or high-density SNP arrays can be used to predict genomic breeding values for economically-important traits in well-developed breeding programmes, such as Atlantic salmon (Ødegård et al., 2014). For instance, based on simulations of a Pacific white shrimp breeding program, genetic progress of disease resistance traits is faster with genomic-enabled selection compared to conventional phenotype-based selection due to higher accuracy (Castillo-Juárez et al., 2015). Incorporation of genetic marker information can also be a useful asset to optimize genetic diversity and future genetic gain when establishing base populations for breeding programmes (Fernández et al., 2014). Furthermore, these genomic tools can be applied to investigate putative genomic signatures of selection during the domestication process of farmed fish species, thus potentially identifying genomic regions underlying variation in relevant phenotypes in wild and domestic fish populations (López et al., 2015).

Aquaculture species typically have several common features, for example high fecundity and external fertilization, plus a short evolutionary distance from their wild ancestors. The reproductive features enable flexible mating structures to be used for breeding programmes, and can provide a powerful resource for genetic studies of complex traits, such as disease resistance (Yáñez et al., 2014). However, the diversity between these species is enormous and often necessitates the establishment of species-specific reproduction and breeding programmes. For example, there is a remarkable variety of sex-determination systems within aquatic farmed species, and the study of Martínez et al. (2014) highlights various methods of controlling sex ratio with aquaculture breeding programmes. This species diversity also presents an issue for choosing suitable model organisms to inform on the biology of the farmed species of interest. Model finfish species, such as zebrafish, have been well-characterized and Ulloa et al. (2014) highlight their utility for the evaluation of the response to alternative diets. However, due to the vast evolutionary distance between certain farmed aquatic and model species, it is clear that direct research on the species of interest can often be the most feasible and informative.

The aquaculture industry has often been innovative and visionary in their application of new technologies to improve production. Genomics present another major opportunity, and the research published in this special issue provides several excellent examples of their potential or realized application. Using genomic tools to more effectively utilize genetic variation in economically-important traits via sustainable breeding programmes is paramount to the continued successful growth and stability of aquaculture production.

CONFLICT OF INTEREST STATEMENT

The authors declare that the research was conducted in the absence of any commercial or financial relationships that could be construed as a potential conflict of interest.

ACKNOWLEDGMENTS

The authors would like to acknowledge funding from Genus plc, CORFO (11IEI-12843 and 12PIE-17669), Government of Chile, Programa U-Inicia, Vicerrectoría de Investigación y Desarrollo, Universidad de Chile, the UK Biotechnology and Biological Sciences Research Council (BBSRC) (BB/H022007/1) and from the Roslin Institute's BBSRC Institute Strategic Funding Grant.

REFERENCES

1. Ali, A., Rexroad, C. E., Thorgaard, G. H., Yao, J., and Salem, M. (2014). Characterization of the rainbow trout spleen transcriptome and identification of immune-related genes. *Front. Genet.* 5:348. doi: 10.3389/fgene.2014.00348
2. Astorga, M. P. (2014). Genetic considerations for mollusk production in aquaculture: current state of knowledge. *Front. Genet.* 5:435. doi: 10.3389/fgene.2014.00435

3. Castillo-Juárez, H., Campos-Montes, G. R., Caballero-Zamora, A., and Montaldo, H. H. (2015). Genetic improvement of Pacific white shrimp (*Penaeus (Litopenaeus)* vannamei): perspectives for genomic selection. *Front. Genet.* 6:93. doi: 10.3389/fgene.2015.00093

4. Colihueque, N., and Araneda, C. (2014). Appearance traits in fish farming: progress from classical genetics to genomics, providing insight into current and potential genetic improvement. *Front. Genet.* 5:251. doi: 10.3389/fgene.2014.00251

5. FAO. (2014). *The State of World Fisheries and Aquaculture Opportunities and challenges*. Rome: FAO, 243.

6. Fernández, J., Toro, M. Á., Sonesson, A. K., and Villanueva, B. (2014). Optimizing the creation of base populations for aquaculture breeding programs using phenotypic and genomic data and its consequences on genetic progress. *Front. Genet.* 5:414. doi: 10.3389/fgene.2014.00414

7. Gjedrem, T., Robinson, N., and Rye, M. (2012). The importance of selective breeding in aquaculture to meet future demands for animal protein: a review. *Aquaculture* 350, 117–129. doi: 10.1016/j.aquaculture.2012.04.008

8. Kuznetsova, I. S., Thevasagayam, N. M., Sridatta, P. S. R., Komissarov, A. S., Saju, J. M., Ngoh, S. Y., et al. (2014). Primary analysis of repeat elements of the Asian seabass (*Lates calcarifer*) transcriptome and genome. *Front. Genet.* 5:223. doi: 10.3389/fgene.2014.00223

9. López, M. E., Neira, R., and Yáñez, J. M. (2015). Applications in the search for genomic selection signatures in fish. *Front. Genet.* 5:458. doi: 10.3389/fgene.2014.00458

10. Marancik, D., Gao, G., Paneru, B., Ma, H., Hernandez, A. G., Salem, M., et al. (2014). Whole-body transcriptome of selectively bred, resistant-, control-, and susceptible-line rainbow trout following experimental challenge with Flavobacterium psychrophilum. *Front. Genet.* 5:453. doi: 10.3389/fgene.2014.00453

11. Martínez, P., Viñas, A. M., Sánchez, L., Díaz, N., Ribas, L., and Piferrer, F. (2014). Genetic architecture of sex determination in fish: applications to sex ratio control in aquaculture. *Front. Genet.* 5:340. doi: 10.3389/fgene.2014.00340

12. Ødegård, J., Moen, T., Santi, N., Korsvoll, S. A., Kjøglum, S., and Meuwissen, T. H. E. (2014). Genomic prediction in an admixed population of Atlantic salmon (Salmo salar). *Front. Genet.* 5:402. doi: 10.3389/fgene.2014.00402
13. Santos, C. A., Blanck, D. V., and de Freitas, P. D. (2014). RNA-seq as a powerful tool for penaeid shrimp genetic progress. *Front. Genet.* 5:298. doi: 10.3389/fgene.2014.00298
14. Ulloa, P. E., Medrano, J. F., and Feijoo, C. G. (2014). Zebrafish as animal model for aquaculture nutrition research. *Front. Genet.* 5:313. doi: 10.3389/fgene.2014.00313
15. Vandeputte, M., and Haffray, P. (2014). Parentage assignment with genomic markers: a major advance for understanding and exploiting genetic variation of quantitative traits in farmed aquatic animals. *Front. Genet.* 5:432. doi: 10.3389/fgene.2014.00432
16. Yáñez, J. M., Houston, R. D., and Newman, S. (2014). Genetics and genomics of disease resistance in salmonid species. *Front. Genet.* 5:415. doi: 10.3389/fgene.2014.00415

CHAPTER

16

TESTING STRATEGIES FOR GENOMIC SELECTION IN AQUACULTURE BREEDING PROGRAMS

Anna K Sonesson[1] and Theo HE Meuwissen[2]

[1]Nofima Marine AS, 1432 Ås, Norway and

[2]University of Life Sciences, 1432 Ås, Norway

ABSTRACT

Background

Genomic selection is a selection method where effects of dense genetic markers are first estimated in a test population and later used to predict breeding values of selection candidates. The aim of this paper was to investigate genetic gains, inbreeding and the accuracy of selection in a general genomic selection scheme for aquaculture, where the test population consists of sibs of the candidates.

Methods

The selection scheme started after simulating 4000 generations in a Fisher-Wright population with a size of 1000 to create a founder population. The basic scheme had 3000 selection candidates, 3000 tested sibs of the candidates, 100 full-sib families, a trait heritability of 0.4 and a marker density of $0.5N_e/M$. Variants of this scheme were also analyzed.

Results

The accuracy of selection in generation 5 was 0.823 for the basic scheme when the sib-testing was performed every generation. The accuracy was hardly reduced by selection, probably because the increased frequency of favourable alleles compensated for the Bulmer effect. When sib-testing was performed only in the first generation, in order to reduce costs, accuracy of selection in generation 5 dropped to 0.304, the main reduction occurring in the first generation. The genetic level in generation 5 was $6.35\sigma_a$ when sib-testing was performed every generation, which was 72%, 12% and 9% higher than when sib-testing was performed only in the first generation, only in the first three generations or every second generation, respectively. A marker density above $0.5N_e/M$ hardly increased accuracy of selection further. For the basic scheme, rates of inbreeding were reduced by 81% in these schemes compared to traditional selection schemes, due to within-family selection. Increasing the number of sibs to 6000 hardly affected the accuracy of selection, and increasing the number of candidates to 6000 increased genetic gain by 10%, mainly because of increased selection intensity.

Conclusion

Various strategies were evaluated to reduce the amount of sib-testing and genotyping, but all resulted in loss of selection accuracy and thus of genetic gain. Rates of inbreeding were reduced by 81% in genomic selection schemes compared to traditional selection schemes for the parameters of the basic scheme, due to within-family selection.

BACKGROUND

In current family-based aquaculture breeding schemes, many important traits are tested on the sibs of the candidates. The test information is used to calculate breeding values for the selection of parents [1]. Only 50% of the genetic variation *i.e.* the between family variation, is used in these schemes, which are employed for traits that cannot be measured on the selection candidates such as disease challenge testing and slaughter quality traits.

Dense marker maps and high-throughput genotyping have become increasingly available in aquaculture species. Genomic selection is a selection method where the effects of dense genetic markers are first estimated in a test population and later used to predict breeding values of selection candidates [2]. The sib-test design can be used for marker-assisted selection [3] or genomic selection, where the association between markers and phenotypes is estimated in the sibs of the candidates, and the candidates are selected on breeding values that result from summing the estimates of the effects of their marker alleles. For aquaculture breeding schemes, genomic selection may prove very useful, because the breeding goals include many traits that are based on information from the sibs and not from the candidates, and therefore genomic selection can result in increased accuracy of selection for those traits by using both between- and within-family genetic variances.

The aim of this paper was to investigate genetic gains, inbreeding and the accuracy of selection in a general genomic selection scheme for aquaculture, where selection is based on information from the sibs of the candidates. Schemes with different numbers of candidates and test animals (sibs of the candidates) and with different heritabilities of the trait under selection were compared by computer simulation. In addition, the importance of performing the sib-test every generation, which is costly for the breeding program, was assessed. Finally, the effect of selection on the accuracy of selection was evaluated.

METHODS

Population

A population with an effective population size (N_e) of 1000 was simulated for 4000 generations according to the Fisher-Wright population model [4, 5]. Five hundred males and 500 females were randomly selected and mated using sampling with replacement.

Among the individuals of the last of these 4000 generations, 100 males and 100 females were randomly selected to create 100 full-sib families, which each produced 30 or 60 progeny to form generation Gen0. These progeny were selection candidates and were not performance tested.

However, in addition to these selection candidates, every family also produced 30 or 60 full-sibs, which entered into a sib-test where they were performance tested. One hundred sires and 100 dams were selected from the candidates to produce generation Gen1 by either (a) random selection, whereby a sire and a dam were randomly sampled with replacement (RAND) or (b) directional selection, whereby sires and dams with the highest genome-wide breeding values (see Calculation of phenotypic values and true and estimated breeding values) were selected without any restriction on the number of parents selected from each family. Again each of the 100 sires was mated to one of the 100 dams, using sampling without replacement, to produce 30 or 60 full-sib selection candidates and 30 or 60 sib-test progeny in generation Gen1. This scheme was repeated until generation Gen10. Hence, with the number of families, Nfamilies, being 100, the total number of candidates, Ncand, was 3000 or 6000 and the total number of sib-test progeny, Ntested, was 3000 or 6000. In one scheme, Nfamilies = 200, Ncand = 3000 and Ntested = 3000.

Reduction of the Number of Sib-tests

The idea here was to reduce the number of sib-tests by not performing a sib-test every generation. Four different strategies of sib-testing were compared:

- EVERY GENERATION (EVERY-GEN): sib-testing was performed in every generation Gen0-Gen10 as described above in 2.1.
- FIRST GENERATION (FIRST-GEN): sib-testing was performed in generation Gen0.
- EVERY SECOND GENERATION (EVERY-2GEN): sib-testing was performed in the odd generation numbers Gen1, Gen3, .., Gen9.
- FIRST 3 GENERATIONS (FIRST-3GEN): sib-testing was performed only during the first three generations Gen0-Gen2.

Genome

Individuals had a diploid genome with ten 100 cM chromosomes. Recombinations were sampled at random positions on the chromosome assuming the Haldane mapping function. All polymorphisms were generated during the 4000 generations of the Fisher-Wright population model, where a mutation rate of 10^{-9} per nucleotide was assumed and the number of nucleotides per cM was 1000000. This effectively resulted in the infinite sites mutation model [6], *i.e.* every mutation occurred at a unique position and created a bi-allelic SNP. This mutation process generated numerous SNP, among which 100 per chromosome were sampled randomly as a QTL (sampling without replacement from the SNP with minor allele frequency (MAF) >0.05), and among the remaining SNP, the 1000 with the highest MAF were chosen as genetic markers. The latter resulted in a total of markers of Nmarkers = 10000 spread over 1000 cM. Reduced numbers of markers were obtained by taking every 10th marker and every 2nd marker, resulting in a total of Nmarkers = 1000 and 5000 markers, respectively.

The allelic effects of the QTL alleles were sampled from the gamma distribution with a shape parameter of 0.4 and a scale parameter of 1.66 [7]. The QTL effects were assumed to be either positive or negative with a probability of 0.5, because the gamma distribution only gives positive values. After sampling, these QTL allelic effects were standardised so that the total genetic variance was 1.0 in Gen0.

The expected average linkage disequilibrium (R^2) between two adjacent markers can be approximated by Hill [8]:

$$R^2 = \frac{5+2N_e c}{11+26N_e c+8N_e^2 c^2}$$

where c is the distance between adjacent loci, here on average 0.001 Morgan, which resulted in $R^2 = 0.342$. In generation Gen0, the realized average R^2 between adjacent markers was 0.374, which is slightly higher than Hill's approximation, probably due to the selection of the markers on their MAF.

Calculation of Phenotypic Values and True and Estimated Breeding Values

The true breeding value of an individual was calculated as:

$$TBV_i = \sum_{j=1}^{1000} x_{ij1} g_{j1} + x_{ij2} g_{j2}$$

where x_{ijk} is the number of copies that individual i has at the jth QTL position and k^{th} QTL allele, and g_{jk} is the effect of the kth QTL allele at the jth position. The phenotypic values of the individuals in the sib-test were simulated by adding an error term sampled from a normal distribution to the true breeding value (TBV_i):

$$P_i = TBV_i + \varepsilon_i$$

where ε_i is an error term for animal i, which was normally distributed (0, σ^2_e) and σ^2_e was adjusted so the heritability was 0.1 or 0.4.

Marker effects were predicted using the BLUP method described in [2]. The statistical model used to estimate the marker effects was:

$$y_i = \mu + \sum_j^n X_{ij} a_j + e_i$$

where y_i is the record of test individual i; μ is the overall mean; X_{ij} denotes the marker genotype: 0 denotes that the individual is homozygous for the first allele; $1/\sqrt{H_j}$ denotes that it is heterozygous; and $2/\sqrt{H_j}$ denotes that it is homozygous for the second allele, where H_j is the marker heterozygosity and thus dividing by $\sqrt{H_j}$ standardises the variance of the X_{ij} to 1; a_j is the random effect of the jth marker and Var(a_j) is assumed 1/ Nmarkers (total genetic variance was standardised to 1.0); e_i is a random residual.

Genome-wide breeding values were estimated by summing the effects of the markers:

$$EBV_i = \sum_j^n X_{ij} a_j.$$

The accuracy of selection (acc) was calculated as the correlation between true and estimated breeding values. The acc was calculated for all schemes, also for the RAND scheme, although the EBV_i were not used for selection in RAND.

Statistics

Selection schemes were run for ten generations (Gen1-Gen10) and summary statistics for each of the schemes are based on 50 replicated simulations. The breeding schemes were compared for the genetic level (G, expressed in genetic standard deviation units of generation Gen0 (σ_a)), genetic gain (ΔG), accuracy of selection (acc), genetic variance, and level and rate of inbreeding (ΔF). Inbreeding was calculated based on pedigree, assuming that the Gen0 individuals are unrelated base parents. The values of these variables were either shown in figures over generations Gen1-Gen10 or in tables with the values in generation 5, when the Bulmer effect had stabilised, and inbreeding had started to build up in the population.

RESULTS

Basic Scheme

Accuracy of selection was the highest for EVERY-GEN and increased from 0.647 to approximately 0.820 over generations (Figure 1a), due to the increased amount of information on marker effects that becomes available. When phenotypic and genotypic testing was only in the first generation, as for FIRST-GEN, accuracy of selection decreased rapidly over generations and was only 0.304 in generation Gen5. This reduction in accuracy of selection is mainly because of changes in the linkage disequilibrium between marker and QTL. Especially, spurious LD [9], which is not due to linkage, changes quickly over generations. When the sib-test was performed every second generation, as for EVERY-2GEN, the accuracy of selection fluctuated, being almost as high as for EVERY-GEN in the generations with sib-testing and lower in the other generations. However, fluctuations decreased over generations, probably because the selection pressure increased favourable marker alleles, and then, their effects became increasingly accurately estimated. For FIRST-3GEN, the accuracy of selection was as high as for EVERY-GEN until generation 3, as expected, and thereafter, the accuracy of selection decreased. However, the reduction during the first generation after selection had stopped was not as large as for EVERY-2GEN, probably because more generations of information on marker effects had built up. In general, the reduction in accuracy of selection after sib-testing had stopped was larger when there were fewer generations of information *i.e.*the reduction in accuracy of selection was the largest for FIRST-GEN, thereafter EVERY-2GEN, and FIRST-3GEN.

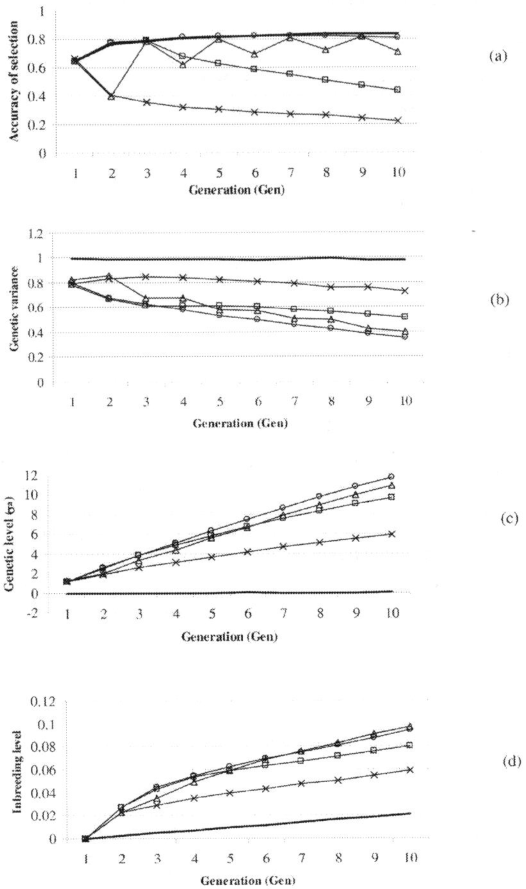

Figure 1: Accuracy of selection, genetic variance, genetic level and inbreeding level for the basic scheme. Accuracy of selection (a), genetic variance (b), genetic level (c) and inbreeding level (d) for schemes when sib-testing was every year (EVERY-GEN; circles), every second year (EVERY-2GEN; triangles), only the first year (FIRST-GEN; cross) or the first three years (FIRST-3GEN; squares); Random selection (solid line); Ncand = 3000, h^2 = 0.4, Nfamilies = 100, Ntested = 3000.

The RAND scheme had an accuracy of selection very similar to EVERY-GEN.

The genetic variance was reduced over generations of selection, except for the RAND schemes, as expected (Figure 1b). The genetic variance decreased most for the EVERY-GEN scheme, with the highest accuracy of selection, and least for the FIRST-GEN scheme. For FIRST-GEN, the largest reduction was between generations Gen1 and Gen2.

The accuracy of selection and genetic variance resulted in the highest genetic level for EVERY-GEN and the lowest for FIRST-GEN, as expected (Figure 1c). In generation Gen5, the genetic level was 6.35σ_a for EVERY-GEN and 3.69 σ_a for FIRST-GEN. For FIRST-3GEN, the genetic level was 5.82 σ_a. However, the genetic level of FIRST-3GEN became over time increasingly lower than that of EVERY-GEN. The genetic level of EVERY-2GEN in generation Gen5 was 5.66 σ_a and the difference in genetic level between EVERY-GEN and EVERY-2GEN increased over generations.

Inbreeding levels were also the highest for EVERY-GEN, probably because it had the highest accuracy of selection (Figure 1d). For FIRST-3GEN, the inbreeding level was as high as for EVERY-GEN until generation Gen4, and decreased thereafter, also because of lower accuracy of selection in the generations after the sib-testing had stopped.

Effect of Marker Density

The effect of different marker densities on the accuracy of selection was large *i.e.* when increasing Nmarkers from 1000 to 5000, the accuracy of selection increased from 0.661 to 0.823 using EVERY-GEN (Figure 2). However, when increasing Nmarkers from 5000 to 10000, the accuracy of selection hardly increased and was 0.842 in Gen5 with Nmarkers = 10000. For FIRST-GEN, the accuracy of selection decreased relatively much faster for the lowest number of markers *i.e.* 1000 (-51%) than for 5000 (-39%) or 10000 (-35%) markers. This suggests that if the markers are sufficiently close to the QTL, the linkage disequilibrium changes less and thus the estimates of the marker effects remain more accurate.

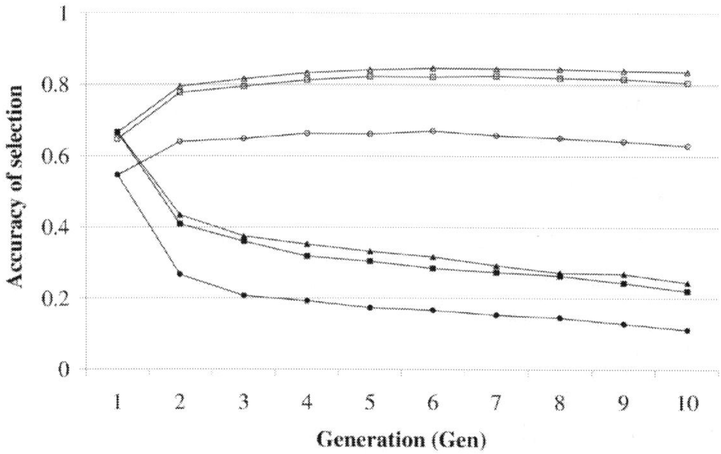

Figure 2: Accuracy of selection with different numbers of markers and sib-testing strategies. Accuracy of selection for schemes with 1000 (circles), 5000 (squares) or 10000 (triangles) markers when sib-testing was every year (EVERY-GEN; open) or only the first year (FIRST-GEN; filled); Ncand = 3000, $h^2 = 0.4$, Nfamilies = 100, Ntested = 3000.

Effect of Heritability and Numbers of Candidates, Tested Sibs and Families

Table 1 shows the sensitivity of the results to lower heritability of the trait, higher number of candidates and tested sibs, and higher number of families. Decreasing the heritability from $h^2 = 0.4$ to 0.1, decreased ΔG_5 by 7%. Similarly, acc_5 was 10% lower for the scheme with low heritability, but ΔF_5 was 50% higher for the scheme with low heritability, because of the increased between-family selection. The vg_5 was 5% higher for the scheme with a low heritability, because of lower selection accuracy.

Table 1: Effect of heritability and numbers of candidates, tested sibs and families

	ΔG_5 (se)	acc_5 (se)	ΔF_5 (se)	vg_5 (se)
Basic scheme[1]	1.22 (0.009)	0.823 (0.003)	0.008 (0.002)	0.532 (0.006)
Lower heritability, $h^2 = 0.1$		0.742 (0.004)	0.012 (0.002)	0.561 (0.009)

Higher number of candidates, Ncand = 6000	1.34 (0.019)	1.13 (0.012)	0.009 (0.002)	0.468 (0.010)
Higher number of tested sibs, Ntested = 6000	1.24 (0.009)	0.845 (0.002)	0.007 (0.002)	0.540 (0.007)
Higher number of families, Nfamilies = 200	0.99 (0.010)	0.749 (0.004)	0.004 (0.002)	0.642 (0.007)

Genetic gain (ΔG_5), accuracy (acc_5), rate of inbreeding (ΔF_5) and genetic variance (vg_5) in generation 5 for schemes with different numbers of candidates (Ncand = 3000 or 6000), tested sibs (Ntested = 3000 or 6000) and families (Nfamilies = 100 or 200) and heritability levels (0.1 or 0.4). EVERY-GEN testing strategy was used.

[1]$h^2 = 0.4$, Ncand = 3000, Ntested = 3000, Nfamilies = 100

Increasing the number of selection candidates from Ncand = 3000 to 6000 increased ΔG_5 by 10%. The increased genetic gain was mainly the result of higher selection intensity. However, this also resulted in larger increases of allele frequencies of favorable alleles, which in turn resulted in more accurate estimates of these alleles. Thus acc_5 was somewhat higher *i.e.* 0.837 for the scheme with 6000 selection candidates compared to 0.823 for the scheme with 3000 selection candidates. Increasing the number of tested sibs from Ntested = 3000 to 6000 increased acc_5 by only 2.7%, which hardly affected neither ΔG_5, vg_5 nor ΔF_5. This suggests that Ntested = 3000 is sufficient when $h^2 = 0.4$.

Increasing the number of families from Nfamilies = 100 to 200, reduced ΔF_5 from 0.008 to 0.004, because of the increased numbers of sires and dams selected. The latter resulted in higher genetic variance, which was 21% higher for the scheme with Nfamilies = 200 than with Nfamilies = 100. The acc_5 was 9% and ΔG_5 19% lower for the scheme with Nfamilies = 200 than with Nfamilies = 100, because selection intensity was reduced.

DISCUSSION

This simulation study examines the accuracies of selection and genetic gain that can be attained with genomic selection sib-testing schemes in aquaculture. Various strategies were evaluated to reduce the amount of sib-testing, but all resulted in a loss of accuracy of selection and thus of genetic gain. Whether these reductions in genetic gains are acceptable will depend on the relative sizes of the reduced benefits and the savings due to less sib-testing. In schemes where selection is for an index combining growth and sib-testing traits, the reduction in accuracy of selection and thus genetic gain will be less than that which was found here due to the reduced importance of the sib-testing traits. How much the accuracy of such an index will be reduced can be assessed using the results in Table 1 and selection index calculations. The general picture that emerges from Figure 1a is, however, that continuous phenotypic and genotypic testing is important to maintain the accuracy of the genome-wide breeding values over generations. The results showed that the number of previous generations of sib-testing affected the level of accuracy proportionally, such that the reduction in accuracy of selection over generations was the largest for FIRST-GEN, thereafter EVERY-2GEN, FIRST-3GEN and finally EVERY-GEN. It may be noted that the reduction in accuracy of selection was substantially larger for the first generation after sib-testing had stopped compared to later generations. This may be because within a generation, markers merely explaining family effects can be used for the prediction of breeding values, whereas across generations, the family effects decay. These results agreed with those of [10]. (Solberg, T.R., Sonesson, A.K., J. A. Woolliams, Ødegård, J., Meuwissen, T.H.E: Persistence of estimates of genome-wide markers over generations when including a polygenic effect, submitted) found a much smaller reduction in the accuracy over generations than we saw here. However, their study did not include the effect of selection. Also, the BayesB method of [2] was used instead of BLUP, which may have increased the weight given to markers that are in close LD with the QTL. Habier *et al.* [11] found that the accuracy of selection did not decrease as much with BayesB as with BLUP, because BayesB gives more weight to the LD when estimating genome-wide breeding values. Hence, the reduction observed in our study may be smaller if the BayesB method was used.

During later generations (say generations Gen6-10), the accuracy of selection was based on LD and its reduction was much smaller, indicating that the LD decayed slowly. Still, the low value of the selection accuracy suggests that in genomic selection breeding schemes also, the prediction of family effects is important and that breeding designs where family effects can be accurately predicted are important. This implies that the relationship between the individuals in the test group and the selection candidates should be as high as possible.

When decreasing heritability from 0.4 to 0.1, acc_5 decreased from 0.823 to 0.742 for the basic scheme with 30 sibs per family, *i.e.* a reduction of 0.08. The expected accuracy of selection for a traditional BLUP scheme is 0.66 for a heritability of 0.4 and 0.55 for a heritability of 0.1, *i.e.* a reduction of 0.11[12]. Hence, the effect of heritability on the accuracy of selection is somewhat larger for traditional BLUP schemes than for a scheme with genome-wide breeding values, which is in agreement with the literature on genomic selection [13] and marker assisted selection ([14] and others). The results also showed that the accuracy of selection increased by 25 and 35% when using genomic selection compared to traditional BLUP for the scheme with a heritability of 0.4 and 0.1, respectively. See [13] for a more detailed comparison of traditional and genomic selection schemes.

Truncation selection for traditional BLUP breeding values [15] resulted in a ΔF of 0.043 for the parameters of the basic scheme. Thus, ΔF was dramatically reduced by 81% for the genomic selection schemes compared to the traditional BLUP breeding schemes, although in practical breeding schemes, a way of reducing rates of inbreeding would probably be used, *e.g.* reducing the use of number of parents from single families or using optimum contribution selection. The reduced ΔF for genome-wide breeding values was probably due to the increased possibilities for within-family selection in genomic selection schemes, whilst this is not possible in traditional sib-testing schemes. It was also due to a stronger Bulmer effect since a more accurate selection increases the Bulmer effect, which leads to less between-family variance and thus more within-family selection [16].

In this study, we used 1000, 5000 and 10000 markers on a genome size of 1000 cM and an effective population size (N_e) of 1000. Hence, the marker density was 0.1, 0.5 and 1N_e/M, which is not very high, but probably in accordance with what a first-generation SNP chip would include for most aquaculture species. For example, with a total genome size of 30M for salmon [17] and an assumed N_e of 1000 (see discussion [18]), 1N_e/M implies a SNP chip of 30000 markers. Furthermore,

Figure 2 shows that increasing the marker density from 0.5 to $1N_e/M$ produces little extra gain. The low number of records relative to the N_e explains this plateau and such a plateau has also been reported by [19] with a SNP marker density of around $4N_e/M$.

The effect of selection on its accuracy was rather small here (EVERY-GEN versus RAND). This result is contrary to that of Muir [10], but he did not update the marker effects every generation, which made the accuracy decay rapidly. In the EVERY-GEN scheme, marker effects were updated every generation, but still one might expect that the Bulmer effect, which reduces the genetic variance (Figure 1b), would also reduce the accuracy of selection, because the signal to noise ratio is reduced. However, a selective genotyping effect occurs, due to selection, resulting in increased frequency and higher accuracy of marker alleles with positive effects. The Bulmer and selective genotyping effects seem to balance each other out approximately, resulting in an accuracy that is hardly reduced by selection.

For practical aquaculture breeding schemes, the results of this study have the following implications, which could lead to a new design of the breeding programs. (i) Genotyping with genomic selection generates extra costs, which can be partly recovered by higher ΔG due to the increased accuracy by genomic selection compared with conventional BLUP breeding values. Here, we found an accuracy of around 0.82 for genome-wide breeding values compared with 0.66 for conventional BLUP breeding values, for a sib-test with 30 sibs tested per family and a heritability of 0.4. (ii) The breeding companies could reduce phenotyping costs if they cancelled all sib-tests of the candidates in one generation, like in EVERY-2GEN. In many salmon breeding schemes, most traits are actually measured on the sibs of the candidates, and phenotypic testing for all these disease and slaughter tests constitute a large part of the cost of the breeding program. However, the reduction of the sib-testing reduces ΔG_5 by 16% from 3.80 σ_a for EVERY-GEN to 3.19 σ_a for EVERY-2GEN. FIRST-GEN reduces ΔG_5 even more. An alternative is to use field data, *e.g.* slaughter house data or practical disease outbreaks, to estimate the association between markers and phenotypes. The data must come from the population under selection and represent animals closely related to the selection candidates, because Figure 1 shows that since the relationship between test individuals and individuals whose EBV are estimated decreases, the accuracy of selection decreases. (iii) Genotyping all selection candidates, *e.g.* 100 families times 30 candidates per family = 3000 individuals, is probably not feasible in practice. Instead, pre-selection on *e.g.* growth and maybe other traits could be a way to reduce the number of candidates to be genotyped. However, this pre-selection step needs to be optimised, in order to get the desired weight on the traits in the pre-selection step relative to the other traits. Also the sib-tested individuals need to be genotyped in the breeding scheme presented. Here the number of genotypes could probably be reduced by a factor of 5 without a substantial loss of accuracy of selection using a selective genotyping strategy [20].

The results show that the accuracy of selection is mainly affected by the following parameters of the breeding scheme:

- In general, accuracies are highly affected by the number of generations with information on marker effects. For the basic scheme, with 3000 selection candidates, 3000 tested sibs of the candidates, 100 full-sib families and a trait heritability of 0.4, accuracy of selection increased from 0.647 to approximately 0.823 over generations for EVERY-GEN. When sib-testing was only in generation 1, as in the FIRST-GEN in order to reduce costs, accuracy of selection dropped rapidly and was only 0.304 in generation 5 for the basic scheme.
- Genomic selection, using EVERY-GEN, showed a higher accuracy of selection than the theoretical maximum of 0.71 of a conventional sib-testing scheme.
- After generation 1, the Bulmer effect will reduce the genetic variance, which indirectly reduces the accuracy of selection. The accuracy of selection increases when more information on the marker effects becomes available over generations, and this effect is larger in the selection scheme than in the random selection scheme, probably because the favourable alleles become more abundant and are thus more accurately estimated. Therefore, the accuracy of selection in generation 5 for genomic selection (0.823) is similar to the accuracy of selection for random selection (0.811) in generation 5 for the basic scheme.

- Increasing Ntested from 3000 to 6000 increased accuracy of selection in generation 5 from 0.823 to 0.845. With a lower heritability of 0.1, the effect of increasing Ntested from 3000 to 6000 (not shown previously), increased the accuracy of selection in generation 5 from 0.742 to 0.763. Hence, the increase in accuracy of selection was similar.
- The reduction in accuracy of selection with FIRST-GEN was somewhat larger for the lowest marker density than for the two other marker densities.

CONCLUSION

The results using the current sib-breeding design of family-based aquaculture breeding schemes, which was not optimised in any way for genomic selection, show that genomic selection yields high genetic gain, accuracy of selection and very low rates of inbreeding, which makes it a promising selection scheme. Various strategies were evaluated to reduce the amount of sib-testing and genotyping, to reduce costs, but all resulted in loss of the accuracy of selection and thus of genetic gain. Genotyping costs may also remain high in a near future, and further research on strategies to reduce the number of fish to genotype is highly needed.

NOTES

Acknowledgements

This study was supported by grant 159831/S40 from the Research Council of Norway.

Calculations were done on the TITAN computer cluster at University of Oslo, Norway.

Authors' Contributions

AKS wrote the main computer program, ran computer programs and drafted the manuscript. THEM wrote computer modules for genome-wide breeding value estimation and for Fisher-Wright populations and helped to draft the manuscript. Both authors have approved the final manuscript.

REFERENCES

1. Gjedrem T: Improvement of productivity through breeding schemes. GeoJournal. 1985, 10: 233-241. 10.1007/BF00462124.
2. Meuwissen THE, Hayes BJ, Goddard ME: Prediction of total genetic value using genome-wide dense marker maps. Genetics. 2001, 157: 1819-1829.
3. Sonesson AK: Within-family marker-assisted selection for aquaculture species. Genet Sel Evol. 2007, 39: 301-317.
4. Fisher RA: The genetical theory of natural selection. 1930, Oxford: Clarendon Press
5. Wright S: Evolution in Mendelian populations. Genetics. 1931, 16: 97-159.
6. Kimura M: Number of Heterozygous Nucleotide Sites Maintained in A Finite Population Due to Steady Flux of Mutations. Genetics. 1969, 61: 893-903.
7. Hayes B, Goddard ME: The distribution of the effects of genes affecting quantitative traits in livestock. Genet Sel Evol. 2001, 33: 209-229.
8. Hill WG: Linkage Disequilibrium Among Multiple Neutral Alleles Produced by Mutation in Finite Population. Theor Popul Biol. 1975, 8: 117-126. 10.1016/0040-5809(75)90028-3.
9. Goddard ME, Meuwissen THE: The use of linkage disequilibrium to map quantitative trait loci. Aust J Exp Agric. 2005, 45: 837-845. 10.1071/EA05066.
10. Muir WM: Comparison of genomic and traditional BLUP-estimated breeding value accuracy and selection response under alternative trait and genomic parameters. J Anim Breed Genet. 2007, 124: 342-355.
11. Habier D, Fernando RL, Dekkers JCM: The impact of genetic relationship information on genome-assisted breeding values. Genetics. 2007, 177: 2389-2397.
12. Cameron ND: Selection indices and prediction of genetic merit in animal breeding. 2007, Wallingford: CAB
13. Nielsen HM, Sonesson AK, Yazdi H, Meuwissen THE: Comparison of accuracy of genome-wide and BLUP breeding value estimates in

sib based aquaculture breeding schemes. Aquaculture. 2009, 289: 259-264. 10.1016/j.aquaculture.2009.01.027.

14. Lande R, Thompson R: Efficiency of Marker-Assisted Selection in the Improvement of Quantitative Traits. Genetics. 1990, 124: 743-756.
15. Henderson C: Applications of Linear Models in Animal Breeding. 1984, Guelph, Canada: Guelph University
16. Daetwyler HD, Villanueva B, Bijma P, Woolliams JA: Inbreeding in genome-wide selection. J Anim Breed Genet. 2007, 124: 369-376.
17. Ng SHS, Artieri CG, Bosdet IE, Chiu R, Danzmann RG, Davidson WS, Ferguson MM, Fjell CD, Hoyheim B, Jones SJM, de Jong PJ, Koop BF, Krzywinski MI, Lubieniecki K, Marra MA, Mitchell LA, Mathewson C, Osoegawa K, Parisotto SE, Phillips RB, Rise ML, von Schalburg KR, Schein JE, Shin H, Siddiqui A, Thorsen J, Wye N, Yang G, Zhu B: A physical map of the genome of Atlantic salmon, Salmo salar. Genomics. 2005, 86: 396-404. 10.1016/j.ygeno.2005.06.001.
18. Hayes BJ, Gjuvsland A, Omholt S: Power of QTL mapping experiments in commercial Atlantic salmon populations, exploiting linkage and linkage disequilibrium and effect of limited recombination in males. Heredity. 2006, 97: 19-26. 10.1038/sj.hdy.6800827.
19. Solberg TR, Sonesson AK, Woolliams JA, Meuwissen THE: Genomic selection using different marker types and densities. J Anim Sci. 2008, 86: 2447-2454. 10.2527/jas.2007-0010.
20. Darvasi A, Soller M: Selective Genotyping for Determination of Linkage Between A Marker Locus and A Quantitative Trait Locus. Theor Appl Genet. 1992, 85: 353-359. 10.1007/BF00222881.

CHAPTER

17

CAPTIVE BREEDING PROGRAMS BASED ON FAMILY GROUPS IN POLYPLOID STURGEONS

Elisa Boscari[1], Jose Martin Pujolar[2], Isabelle Dupanloup[3,4], Riccardo Corradin[5], Leonardo Congiu[1]

[1] Department of Biology, University of Padova, Padova, Italy,

[2] Department of Bioscience, Aarhus University, Aarhus C, Denmark,

[3] Institute of Ecology and Evolution, University of Berne, Berne, Switzerland,

[4] Swiss Institute of Bioinformatics, Lausanne, Switzerland,

[5]Department of Statistics, University of Padova, Padova, Italy

ABSTRACT

In species with long life cycles and discontinuous availability of individuals to reproduction, implementing a long-term captive breeding program can be difficult or impossible. In such cases, managing diversity among familiar groups instead of individuals could become a suitable approach to avoid inbreeding and increase the possibility to accomplish a breeding scheme. This is the case of several sturgeon species including the Adriatic sturgeon, whose recovery depends on the management of a few captive stocks directly descended from the same group of wild parents. In the present study, relatedness among 445 potential breeders was inferred with a novel software for pedigree reconstruction in tetraploids ("*BreedingSturgeons*"). This information was used to plan a breeding scheme considering familiar groups as breeding units and identifying mating priorities. A two-step strategy is proposed: a short-term breeding program, relying on the 13 remaining F0 individuals of certain wild origin; and a long-term plan based on F1 families. Simulations to evaluate the loss of alleles in the F2 generation under different pairing strategies and assess the number of individuals to breed, costs and logistical aquaculture constraints were performed. The strategy proposed is transferable to the several other tetraploid sturgeon species on the brink of extinction.

INTRODUCTION

Conservation management aimed at preserving the genetic variability of wild populations is essential in preventing the extinction of endangered species [1], [2]. Heritable genetic variation is a prerequisite for the long-term survival of species since the rate of evolutionary change able to occur in a group of organisms depends on the amount of genetic variation present in the gene pool; hence, organisms exhibiting low levels of genetic variability will show an impaired evolutionary potential to face future environmental changes [3]. Given the imminent risk of extinction of most critically endangered species and the low number of reproductive populations available in nature, *in situ* conservation efforts (e.g. translocation, breeding in a protected area of wild habitat, supplementary feeding) are often not sufficient to reduce or prevent the effects of anthropogenic and stochastic threats [1]. Therefore, conservation activities must frequently rely on *ex situ* conservation through breeding of captive individuals [1], [3], [4], [5], [6].

Captive breeding has long played an important role in the conservation of threatened and endangered species and generally involves two basic principles. The first one is the complete genetic characterization of all breeders available in captivity in order to quantify the amount of genetic variation and to assess the purity of the stocks. Typically, captive stocks contain only a fraction of the original genetic variability present in the larger wild population. A second goal aims at retaining and transmitting most of the residual genetic diversity to the future generations of captive breeders, thus securing the long-term future of the species [7], [8].

In this regard, pedigree-based management practices are used to identify candidate breeders, *i.e.* the "*breeders unit*", for the production of the future generations [9], [10]. Evidences from empirical studies suggest that the best strategy to retain gene diversity is to minimize average kinship when pairing individuals [11]. Therefore, an accurate pedigree of the captive breeders is critical, together with information on the relatedness among the original founder population, although in most captive programs founders are assumed to be equally unrelated (e.g. "founder assumption") [5]. Due to their high variability, microsatellites are the most adequate marker of choice for parentage assignment in animal conservation [2], [5], [12].

Among fishes, the need of establishing long-term captive programs is especially urgent for sturgeons, which according to the recent report (March 2010) of the *International Union for Conservation of Nature* (IUCN) are the most threatened group of species in the world, with 85% of sturgeons being at risk of extinction (http://www.iucnredlist.org). The decline of sturgeon populations is mostly attributed to human activities, including overexploitation for the harvesting of caviar, habitat degradation, damming of rivers and poaching [13], [14]. So far, recovery actions have been based on supplementation of natural populations with juveniles produced in captivity from wild breeders. However, for several species/populations the low availability of wild animals makes the retention of part of the fingerlings produced necessary to be used as future breeders [15]. Some particular life-characteristics of sturgeons (*i.e.* slow growth, late age at maturity, 2-years or more resting stage between reproductions) complicate breeding plans as selected individuals might not be available when needed for reproduction. Moreover, sturgeons show high juveniles mortality, which might translate into different number of retained fingerlings from each reproduction that reaches maturity. This can result in broodstocks composed by different familiar groups with unbalanced sizes. Consequently, in the absence of pedigree information, the risk of mating close relatives can be high, thus increasing relatedness and distorting founder diversity [8].

With the aim of proposing adequate approaches for the management of broodstocks composed by familiar groups that differ in size and genetic composition, we focused on the Adriatic sturgeon (*Acipenser naccarii*), an endemism of the Adriatic region. Given its tetraploid status [16], we also aim at developing adequate tools for the management of tetraploid genetic data in pedigree-based breeding programs.

The Adriatic sturgeon has been listed by IUCN since May 2010 as "critically endangered" and possibly extinct in the wild (IUCN 2012). A breeding program was initiated in 1977 with the transfer of ca. 50 immature wild individuals (F0) to the private aquaculture plant Azienda Agricola V.I.P. (Orzinuovi, Brescia, Italy). Since the late 1980s, following the first successful reproduction in captivity [17], several F1 broodstocks have been established from randomly paired F0 breeders. Over 250,000 juveniles have been reintroduced in the last 25 years, and although recaptures have been occasional, up to now there is no evidence of natural reproduction. Therefore, the F0 stock at the V.I.P. aquaculture plant, currently reduced to 13 individuals, represents the only living animals of unequivocal wild origin. All Adriatic sturgeons reared in Europe directly descend from this limited stock. The recovery of the Adriatic sturgeon depends mostly on the management of F1 captive stocks that have now reached sexual maturity.

In the present study, we used a pedigree-based approach to conduct the characterization of the stocks available at the Azienda Agricola V.I.P. Plant, including (1) all remaining individuals of the parental F0 stock, and (2) a large number of F1 individuals retained from many breedings performed in the facilities during the last 30 years. After pedigree reconstruction using a new allocation program specific for tetraploid species ("*BreedingSturgeons*"), candidate breeders were identified. A breeding plan was then designed based on family groups as breeding units. The rationale of using families rather than individuals is based on the fact that selected individuals might not be available for reproduction when needed due to the particular complex life history traits of sturgeons (i.e. they reproduce every other year), but related individuals might be available instead. Our breeding program was also supported by simulations in order to estimate the loss of genetic diversity under different scenarios as well as to evaluate the sustainability of the plan from an economic and logistic point of view. The present approach can also be applied to many other species of sturgeons whose recovery depends on the establishment of *ex situ* strategies based on captive broodstock.

MATERIALS AND METHODS

Sample Collection and DNA Extraction

The University of Padova ethic board C.E.A.S.A. (Comitato Etico di Ateneo per la Sperimentazione Animale) exempted this study from review because it was an extra moenia activity. Nevertheless, we tried to minimize the impact of sample collection on the animals. All samples were collected in aquaculture facilities by the owners. Fish were kept into the pond water to minimize stress and were released immediately after sample collection, which consisted of painless clippings from the caudal fin. No mortality, pain or stress was observed.

We analyzed a total of 42 individuals from the parental stock F0, already genotyped at 24 microsatellite loci [10]. All analyses were conducted considering the full data set (referred as Wild; N=42) and considering only those individuals alive at present (referred as Wild-Alive, N=13).

We also analyzed 445 F1 individuals. In July 2003, about half of the F1 individuals died due to a poisoning incident and only 233 animals survived. All analyses were conducted considering the full F1 data set (F1, N=445) and considering those F1 alive at present (F1-Alive, N=233).

Our analysis also includes 133 individuals from the Ticino River Park in Italy for comparison. This individuals are part of a stock established from a very limited number of breeders [18].

Genomic DNA was extracted from fin-clip (10–100 mg) using the DNA Easy Tissue Extraction Mini Kit (Euroclone) and stored at −4°C.

Mitochondrial Analysis

Prior to the analysis, species status of all individuals was assessed using mitochondrial DNA haplotypes plus species-specific SNPs at the S7 and Vimentin genes [19]. Amplification of the mitochondrial control region (d-Loop) was performed for all individuals using the primer pair PRO1F-PHE1R, following the standard experimental conditions [10]. All PCR products were purified by enzymatic reaction with ExoSAP-it (Usb) and sequenced at the external service BMR Genomics using an ABI Prism3730XL automatic sequencer. A multi-alignment was created using ClustalW in Mega5 [20] and a haplotype network was constructed using TCS v. 1.13 [21] following the statistical parsimony approach [22]. Genetic diversity was measured using haplotype diversity (H) and nucleotide diversity (Pi) estimated from number of segregating sites and from mean number of pairwise differences using DnaSP software v. 5 [23].

Finally, in order to detect the presence of other species or interspecific hybrids, the analyses of the mitochondrial control region was supplemented with the scoring of species-specific SNPs in two genes, the ribosomal Protein S7 and Vimentin, following the approach developed in [19].

Microsatellite Analysis

All F1 individuals were genotyped at 7 microsatellite loci: LS-39 [24], AnacE4, AnacC11, AnacA6 [25], An20 [26], AoxD234 and AoxD64 [27], selected among the 24 loci that were originally used to genotype the F0 stock [10]. In case of multi-allocations (when more than one parent pair resulted compatible), 3 additional microsatellite loci were genotyped, An16 [26], AoxD241 and AoxD161 [27]. Microsatellite loci were amplified following the conditions reported in the original references. Microsatellite scoring was performed using GeneMapper software version 1.95 (Soft Genetics LLS^R^). The interpretation of microsatellite patterns in tetraploids is not straightforward due to the presence of multiple alleles at each locus that can be present in more than one copy, thus the true genotype cannot be accurately resolved. As an alternative, we used the classic band-sharing approach as proposed in [10] which considers presence/absence of bands, disregarding the number of alleles present in each individual. Data from all microsatellite loci were combined by creating individual profiles where bands were coded as a string of 0 s (absent) and 1 s (present). Band sharing (BS) between individuals was used as a measure of similarity. Pairwise genetic distances were calculated as 1-BS. Number of alleles at the different loci was used to compare F1 and F0 stocks and to evaluate the efficiency of allele transmission.

Pedigree Reconstruction

A new software ("*BreedingSturgeons*") was implemented to estimate an allocation compatibility index for each F1 individual to all possible parent pairs in the F0 stock, assuming a tetrasomic mode of inheritance with no dominance [28].

For all possible pairs of putative males and females in the F0 stock and the F1 individuals, we inferred all possible tetraploid genotypes compatible with the observed microsatellite profiles at all loci. We defined a single locus compatibility index, *ci*(x), of an individual to a putative parent pair as: $ci(x)=\sum[A(p1)+A(p2)]/4$, where *A(pi)* is the number of alleles that could have been inherited from parent-*i*. *A(pi)* can vary between 0 (if no allele present in parent i is seen in the possible tetraploid genotypes inferred for the F1 individual) and 2 (if 2 alleles could have been transmitted from that parent to the F1 individual). And ci(x) can vary between 0 and 1. Multi locus estimates of compatibility indexes were obtained as average across all microsatellites loci. In this procedure the following assumptions must be satisfied: a) all the offspring alleles should be shared with the parental pair, b) two of the offspring alleles must be shared with the mother and two with the father, c) a complete heterozygote can not transmit two copies of the same allele. The above conditions should be satisfied by at least one of the inferred genotypes in the two parents and in the offspring.

Each F1 individual was allocated to the parent pairs associated with a compatibility index above a threshold (0,857) chosen to account for possible genotyping errors at one locus.

Individuals that could not be assigned to any parent pairs (not-allocated) probably represent the progeny of wild F0 individuals died before the characterization of the F0 stock. All individuals allocated with a compatibility index higher than the threshold value were checked for mitochondrial concordance with the putative mother.

We also used the program to test the power of the set of 7 microsatellite loci used for parental allocation. To do so, we generated 1,000 virtual F1 profiles from the F0 individuals. We identified the allocation power of our microsatellite panel as the fraction of virtual F1 profiles unequivocally allocated back to the correct parental pair.

In order to identify closely related individuals (full-sibs or half-sibs) among not-allocated animals, distributions of pairwise genetic distances between full- and half-sibs were created and threshold values under which 99% of comparisons between animals with a given degree of relatedness were identified [10]. The threshold values estimated for our panel of loci (0.58 for sibs and 0.66 for half-sibs) were applied to the distance matrix between F1 individuals to identify putative groups of related individuals.

Breeding Strategy

A "whole-family" approach (rather than a single-breeder approach) was used. The parent pair of each family was considered to be representative of the genetic diversity of the corresponding progeny. This allowed using a higher amount of genetic information since all the F0 individuals are genotyped at 24 loci [10]. Accordingly, the genetic distance between two families is represented by the genetic distance between the two cumulative parental profiles.

In the case of families with unknown pedigree, estimation of genetic distances from the cumulative profile of parents was not possible. In these cases, parental profiles were inferred cumulating the phenotypes of their offspring using the 7 loci genotyped for parental allocation analysis. Genetic distances estimated on these partial profiles were included in the matrix after observing a significant positive correlation ($r=0,770$; $p<0.001$) between distances estimated at 24 and 7 loci by using the Mantel test implemented in PopTools [29].

The strategy proposed for the breeding plan is based on the following three criteria: (a) prioritize families with higher genetic value, (b) exclude crosses between families with a shared parent and (c) avoid already-performed crosses. Priority of the different families (named "*priority families*") for reproduction was estimated based on how often the parents of each family were already represented in the F1 "*breeders unit*". This value was estimated by averaging the number of progeny of the two parents across all the families in the "*breeders unit*". In this way, the smallest families with under-represented parents were the first to be selected for reproduction. The above values were then converted into an ordinal scale ("priority"). Following the established order, each priority family was then paired with the more distant one (named "*mating family*") selected from among all of the families available in the "breeders unit".

With the aim of excluding crosses between half-sibs families, starting from the F0 profiles at 24 loci, all the possible combinations of virtual families (cumulative profiles) sharing one parent were created and the pairwise genetic distances estimated. The highest distance observed (0.35) was used as threshold value to exclude all crosses involving a shared parent. In the selection of the mating families we imposed the constraint that each *mating family* could not be crossed with more than 3 *priority families*. Consequently, the maximum number of crosses in which a given family can be involved is 4 (3 as "*mating family*" and 1 as "*priority family*"). This choice was necessary to avoid that some families with a high genetic distance from many others were used in too many crosses. This process was implemented in the R software "*BreedingPlanSturgeons*" also used to compare the different pairing strategies described in the following paragraph.

For the families whose parents were unknown, the information at 7 loci was used, which yielded a threshold value of 0.45. These families were excluded from the automated analyses and for each of them the more distant "*mating family*" was manually identified and a single cross suggested.

Simulations

A first R-script, "*CostsBreedingSturgeons*" (script available as additional material), was developed to infer the number of crosses that can be simultaneously performed in a given hatchery. It includes the estimate of costs and yields of a given breeding program based on the number of individuals generated and reared until release. It also considers that at the beginning of the third year (after tagging of all individuals), a fraction of the individuals are retained in the facilities in order to preserve the captive stock, while the rest of individuals can be released into the wild.

We took into consideration the following constraints: number of individuals that can be induced every year, availability of ponds for rearing families separately until tagging and maximum number of animals that can be hosted. We also considered biological features (*e.g.* mortalities and growth rates at different life stages) and costs (*e.g.* reproductions, food, tagging, manipulation, overheads) in order to estimate the total financial cost of the project. Logistic factors can be modified according to the capacities of the different hatcheries and the biological features of the species under consideration. In our example, the different variables were based on the "Storioni Ticino hatchery" located in Cassolnovo (Pavia, Italy), where the F1 V.I.P. Stock will be maintained in the future. The run was performed assuming six tanks available for the first two years of rearing and a common pond of 1,200 square meters for the following years. Biological parameters such as average female fecundity (30,000 eggs), hatching rate (50%) and mortalities rates in the first (87%), second (66%) and following years (10% every year) were based on personal communication by the V.I.P. owners, who have more than 20 years of experience in the reproduction of Adriatic sturgeon in aquaculture.

A second R-script, "*BreedingPlanSturgeons*" (also available as additional material), was developed to estimate the expected fraction of alleles of the F1 stock that are successfully transmitted to the progeny under different breeding strategies. To this purpose, the program "*BreedingSturgeons*" (previously described and used for parental allocation) was applied to generate a virtual F2 progeny from the observed F1 microsatellite phenotypes. The minimum number of individuals to be crossed in order to ensure the transmission of all F1 alleles was estimated. Based on the order of the "*priority families*", four different scenarios were simulated in which the "*mating families*" were chosen by "Maximum Distance", selecting 1, 2, 3 or 4 individuals per family. For each scenario, 100 replicates were performed and the mean cumulative number of alleles successfully transmitted was counted.

Once the optimal number of individuals per family was fixed, we also tested three alternative strategies to select and pair families: random selection of both paired families (Random – Random) vs. priority order of the first family with "*mating family*" selected at random (Priority – Random) or with "*mating family*" selected for having the higher genetic distance (Priority – Maximum distance). The process of random selection of the "*mating family*" was repeated 100 times for each "*priority family*" and for each resulting combination breeders were randomly extracted from the two families. Also the extraction of the breeders was repeated 100 times for each family to minimize possible effects of differences due to intra-family diversity. Finally, from each virtual cross, 20 fingerlings were generated and the total number of alleles counted. The total number of alleles successfully transmitted during the process was averaged among replicates and used to compare the different pairing strategies. The choice of generating 20 virtual individuals is consistent with the number of animals that will be retained in captivity as future breeders from each reproduction.

In all the above simulations, the sex of animals of the "*breeders unit*" was not considered since this information was not available. Moreover, all families for which less than 24 loci were genotyped were excluded from the simulations. For the above simulations RStudio was used.

RESULTS

Genetic Diversity

Prior to the analysis, species status of all individuals was assessed using mitochondrial DNA haplotypes (Table 1) plus species-specific SNPs at the S7 and Vimentin genes. In total, two hybrids were identified (one hybrid white sturgeon female × Adriatic sturgeon male and one hybrid Adriatic sturgeon female × white sturgeon male) and were consequently removed from the analysis.

Table 1: Mitochondrial diversity indices

Sample	N	S	S_i	H	H_d	θ_w	θ_π
Wild	43*	13	3	7	0.783	0.004	0.006
Wild-Alive	13	9	0	4	0.744	0.004	0.006
F1	443	10	0	5	0.665	0.002	0.002
F1-Alive	230	10	0	4	0.681	0.002	0.002

Diversity indices for all samples including number of individuals (N), number of segregating sites (S), number of singletons (S_i), number of haplotypes (H), haplotype diversity (H_d), nucleotide diversity estimated from number of segregating sites (θ_w) and nucleotide diversity estimated from mean number of pairwise differences (θ_π). Samples include the original parental F0 stock (Wild), the current F0 stock (Wild-Alive), the full F1 data set excluding the two detected hybrids (F1) and those F1 alive at present (F1-Alive). *Includes one heteroplasmic female that was considered as two females with different haplotypes.

The number of individuals showing each haplotype in the two stocks (F1/F1-alive) were: 0/0; 118/53, 220/106, 0/0, 75/45, 4/0, 37/23 for haplotype 1 to 7 respectively. A total of 11 individuals (one female F0 and 10 of its progeny) were heteroplasmic, presenting a double peak in the chromatogram sequence, and were counted as two different haplotypes for the analysis. When comparing genetic diversity across generations (Table 1), the F1 sample showed a decrease in haplotype diversity (F1: H=0.67; F1-Alive: H=0.68) compared to the parental stock of wild origin (Wild: H=0.78; Wild-Alive: H=0.74). Two out of the 7 original F0 haplotypes [19] were lost (hapl-1 and hapl-4) due to the random selection of mating pairs, while a third haplotype was lost after the 2013 poisoning event. Similarly, nucleotide diversity declined significantly when comparing the F0 and F1 samples, with dropping from 0.006 to 0.002.

Measures of genetic diversity at microsatellite loci are detailed in Table 2. Number of alleles ranged from 8 to 14 across loci in the wild population. A drop in the average number of alleles was observed when comparing the original Wild and the Wild-Alive (F0) stock, from 10.29 to 8.29. However, the F1 stock showed similar allele numbers in comparison with the original F0 stock (10 before poisoning; 9.86 after poisoning). The lowest average number of alleles (8.57) was observed in the F1-Ticino stock reared at the Ticino River Park. It is worth noting that some of the alleles observed in the F1 stocks were not observed in the F0 generation.

Table 2: Number of alleles per locus in the different stocks and average band sharing (BS) values at 7 microsatellite loci

Locus	Wild (N = 42)	Wild-Alive (N = 13)	F1 (N = 443)	F1-Alive (N = 231)	F1-Ticino (N = 133)
LS-39	8	8	8/1	8/0	7/0
AoxD64	9	8	9/1	9/1	9/0
AnacE4	10	7	10/3	10/3	8/0
AnacC11	11	7	9/1	9/1	8/0
AoxD234	14	12	14/0	14/0	13/0
An20	11	9	11/1	11/0	8/2
AnacA6	9	7	9/0	8/0	7/0
Mean	10.29	8.29	10/1	9.86/0.71	8.57/0.29
BS	0.40	0.39	0.45	0.45	0.50

Samples include the original parental F0 stock (Wild), the current F0 stock (Wild-Alive), the full F1 V.I.P. data set (F1) and those F1 alive at present (F1-Alive). For the F1 stocks, the values presented correspond to alleles shared with the F0 generation and newly observed alleles, respectively. Detected hybrids were excluded from the analysis.

Parental Allocation

Prior to the parental allocation analysis, the power of the markers was assessed using 1,000 virtual F1 individuals. Using the 7 microsatellite set, a correct assignment of 92% was observed. Hence, we can consider the set of microsatellites to be informative enough for parental allocation procedures.

Parental allocations at 7 microsatellite loci for all the F1_V.I.P. individuals (N=445), including mitochondrial haplotypes used to solve ambiguities in case of multiple parental allocations. In total, 382 out of 445 individuals were successfully assigned to a single parent pair (30 families). No multi allocations were observed, which confirms the high resolution power of the markers. No discrepancies were observed between mitochondrial haplotypes and parental allocations. The remaining 63 individuals were not-allocated to any of the possible F0 parent pairs. However, four groups of full-sibs with different mitochondrial haplotypes were detected: Fam-unknownA (haplotype 3, 23 individuals), Fam-unknownB (haplotype 5, 4 individuals), Fam-unknownC (haplotype 2, 11 individuals) and Fam-unknownD (haplotype 6, 4 individuals). In total, the stock is composed of 30 families with known pedigree and at least 4 not-allocated families. Comparison of the F1 and F1-Alive stocks showed the loss of few families, so that the F1-Alive stock is composed of 27 families of known pedigree and 3 not-allocated families (A and C were reduced to 11 and 5 individuals, respectively; B was unchanged; D was lost).

Short-term Breeding Scheme (F0×F0)

A short-term breeding plan including the 13 F0 individuals still alive at present (7 males, 6 females) was designed after evaluation of the genetic relatedness of the individuals (Fig. 1). We aimed at having the contribution of each remaining F0 breeder in at least three families within the F1 generation. Mating pairs were selected among the more distant males and females based on the pairwise genetic distances at 24 loci (Fig. 1), trying to maximize the number of parents represented, with the minimum number of crosses. Following this strategy, 18 crosses were selected as priority (Fig. 1), among which only one was already present within the F1 stock with a limited number of individuals.

Score A	Score B	Score A+B	Alive F0	Sex	Nac4	Nac7	Nac14	Nac20	Nac25	Nac28	Nac5	Nac6	Nac13	Nac17	Nac23	Nac30	Nac31
0	3	3	Nac 4	F	0	0	0	0	0	0	0	0	0	0	0	0	0
1	2	3	Nac 7	F	0,43802	0	0	0	0	0	0	0	0	0	0	0	0
0	4	4	Nac 14	F	0,59259	0,4955	0	0	0	0	0	0	0	0	0	0	0
0	3	3	Nac 220	F	0,53968	0,53488	0,58621	0	0	0	0	0	0	0	0	0	0
0	3	3	Nac 25	F	0,57627	0,57025	0,5	0,50749	0	0	0	0	0	0	0	0	0
1	3	4	Nac 28	F	0,47619	0,48837	0,62069	0,50746	0,57143	0	0	0	0	0	0	0	0
0	3	3	Nac 5	M	0,48718	0,58333	0,60748	0,488	0,60684	0,472	0	0	0	0	0	0	0
1	2	3	Nac 6	M	0,54955	0,5614	0,68317	0,54622	0,65766	0,29412	0,52727	0	0	0	0	0	0
1	2	3	Nac 13	M	0,55	0,5122	0,6	0,46875	0,48333	0,45313	0,54622	0,48673	0	0	0	0	0
2	2	4	Nac 17	M	0,54955	0,54386	0,44554	0,63025	0,54955	0,52941	0,54545	0,59615	0,61062	0	0	0	0
1	3	4	Nac 23	M	0,56667	0,46341	0,49091	0,54688	0,46667	0,5	0,47899	0,61062	0,45902	0,50442	0	0	0
0	3	3	Nac 30	M	0,52542	0,50413	0,44444	0,50794	0,49153	0,49246	0,4359	0,58559	0,55	0,47748	0,43333	0	0
2	3	5	Nac 31	M	0,51724	0,59664	0,60377	0,5	0,53448	0,56452	0,25217	0,59633	0,61017	0,54128	0,52542	0,53448	0

Figure 1: **F0 breeding plan.**

Distance matrix at 24 microsatellite loci between the current F0 stock (Wild-Present) individuals. The already-existing families generated by each F0 parent and represented by at least 10 animals were counted (ScoreA). ScoreB represents the number of crosses per F0 that should be planned in order to get at least three families per breeder in the F1 generation (ScoreA+B). Only crosses between males and females are selected. Cells highlighted in grey indicate crosses selected for the short-term breeding plan on the basis of ScoreA and ScoreB.

Long-term Breeding Scheme (F1×F1) and Selection of Candidate Breeders

The long-term plan was designed based on the F1 progeny, using a family-based approach. A total of 32 crosses was planned, one for each of the 32 families that composed the "*breeders unit*": 30 reared at V.I.P. (27 allocated and 3 not-allocated) plus two families from the Ticino River Park (1 allocated and 1 not-allocated). Pairwise genetic distances among families are shown in Fig. 2, together with details on recommended crosses and those crosses to be avoided (distance values <0.35 for 24 loci and <0.45 for 7 loci).

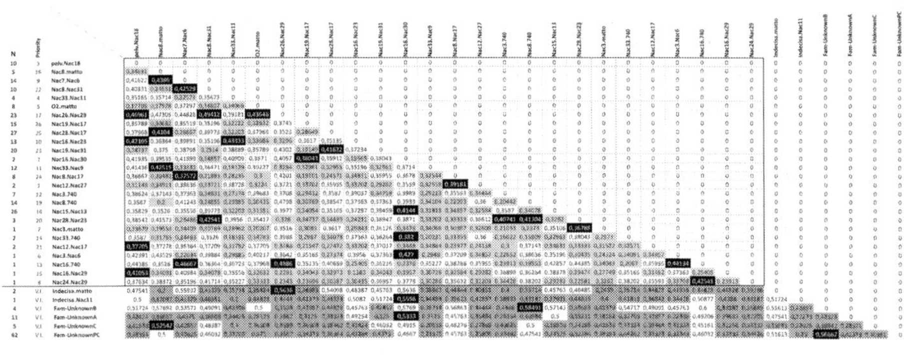

Figure 2: F1 breeding plan.

Pairwise distances among the families of the "*Breeders Unit*" estimated by comparing the cumulative profiles of the parent pairs. The black frame includes distances estimated at 24 loci and excludes the ones based on 7 loci only. Light grey and dark grey cells represent crosses to be avoided based on threshold values of 0,35 (at 24 loci) and 0,45 (at 7 loci), respectively. Black cells represent selected crosses identified by the software "*BreedingPlanSturgeons*" (within the black frame) or by visual inspection (for the families marked with V.I.). N=number of individuals per family.

In the case of 4 not-allocated and 2 allocated families, generated by a single F0 female (reported in Fig. 2 as "Indecisa") whose sample was accidentally loss, a reliable estimation of the genetic distances with the rest of families could not be achieved since they were only genotyped at 7 loci. Thus, these families were excluded from next simulations and only one cross was considered in order to guarantee their contribution to the next generation.

Breeding Plan Simulations

Results obtained from the simulations using the script "*CostsBreedingSturgeons*", including inference of how many animals can be hosted in the facilities and released in the wild, as well as total costs of the project, are shown in Fig. 3.

a) Matrix of crosses

Years		N. crosses per year	Cohorts [1]	[2]	[3]	[4]	[5]	[6]	[7]	[8]	[9]	[10]
	[1]	3	44610	0	0	0	0	0	0	0	0	0
	[2]	3	44652	5979	0	0	0	0	0	0	0	0
	[3]	3	44874	5846	141	0	0	0	0	0	0	0
	[4]	3	44682	5993	141	123	0	0	0	0	0	0
	[5]	3	45528	5966	141	124	107	0	0	0	0	0
	[6]	3	44712	5981	141	122	113	100	0	0	0	0
	[7]	3	45453	5980	141	126	112	102	92	0	0	0
	[8]	3	44838	6061	141	121	114	102	94	84	0	0
	[9]	3	44754	6043	141	132	112	101	93	82	75	0
	[10]	3	44487	5859	141	123	120	102	95	85	77	69

b) Matrix of releases

Years		N. of individuals three years old to release per year [1]	[2]	[3]	[4]	[5]	[6]	[7]	[8]	[9]	[10]
	[1]	0	0	0	0	0	0	0	0	0	0
	[2]	0	0	0	0	0	0	0	0	0	0
	[3]	0	0	1840	0	0	0	0	0	0	0
	[4]	0	0	0	1844		0	0	0	0	0
	[5]	0	0	0	0	1860	0	0	0	0	0
	[6]	0	0	0	0	0	1874	0	0	0	0
	[7]	0	0	0	0	0	0	1870	0	0	0
	[8]	0	0	0	0	0	0	0	1880	0	0
	[9]	0	0	0	0	0	0	0	0	1825	0
	[10]	0	0	0	0	0	0	0	0	0	1896

c) Matrix of costs

Years		Costs: Induction	1st year	2nd year	Chip	Shared tank	Financial support
	[1]	5000	18000	0	0	0	23000
	[2]	5000	18000	12000	0	0	35000
	[3]	5000	18000	12000	9905	0	44905
	[4]	5000	18000	12000	9925	0	44925
	[5]	5000	18000	12000	10005	0	45005
	[6]	5000	18000	12000	10075	0	45075
	[7]	5000	18000	12000	10055	0	45055
	[8]	5000	18000	12000	10105	0	45105
	[9]	5000	18000	12000	9830	0	44830
	[10]	5000	18000	12000	10185	0	45185

Figure 3: Example of output of the R-script "*CostsBreedingSturgeons*".

The run was performed using six tanks available for the first two years of rearing and one common pond of 1200 mq for the following years. (a) Number of individuals for each cohort (columns) maintained in captivity in the different years (lines). (b) Individuals that can be released per year. (c) Annual costs.

Considering 6 ponds available at the facilities of Cassolnovo, where the progenies will be produced and reared, three crosses per year are feasible, which corresponds to about 45.000 larvae per year (15.000 per female). Induction of fish is not straightforward but assuming to induce 5 males and 5 females per cross, at least one successful reproduction is expected. Number of individuals at the 2nd year dropped to ca. 6.000 individuals (Fig. 3a) after assuming a 86.7% mortality.

In order to save pedigree information, fingerlings produced by different family combinations should be kept separated for the first two years (until tagging). At the start of the 3rd year, after tagging, the progeny is split into a fraction that is retained in a common tank with all other families to preserve the captive broodstock and a fraction that can be released into the wild. On the basis of a 10% yearly mortality, *ca.* 140 individuals are retained (Fig. 3a) in order to have an average of 20 individuals per family at the 10th year of age. These individuals will constitute the "captive population" to be kept in the facilities as an "insurance policy" for the species. The remaining individuals (ca. 1.850) can be released into the wild (Fig. 3b). Finally, estimated costs are detailed in Fig. 3c, including costs generated by induction, maintenance and tagging of the individuals as well as total annual amount required for the breeding plan.

Results from simulations using the script "*BreedingPlanSturgeons*" to estimate the optimal number of breeders are presented in Fig. 4a, in which the cumulative number of alleles transmitted to the progeny is reported for different number of breeders. The fraction of alleles from the F0 generation that are successfully transmitted to the F2 is 100% when crossing either 2, 3 or 4 individuals per family. Only when crossing 1 single individual the transmission of alleles is not complete. The best strategy appears to be the use of 3 individuals, since the use of 2 individuals performed worst in the short-them, when the number of crosses was low.

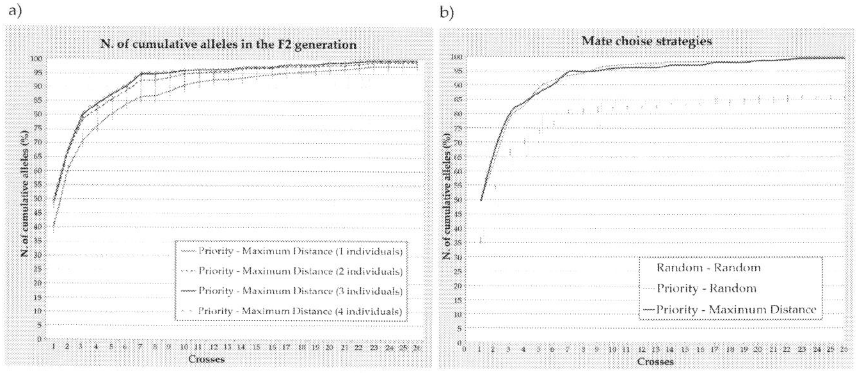

Figure 4: Simulations of allele inheritance under different strategies.

Results of the simulations performed using the R-script "*BreedingPlanSturgeon*" to test: (a) the optimal number of breeders per family and (b) alternative strategies for the mating choice. The cumulative percentages of transmitted alleles are averaged on 100 replicates and reported with the corresponding standard deviation. In figure b, 3 individuals per family were crossed and standard deviation bars are shifted downwards for better visibility.

Finally, results from simulations conducted to compare three alternative mating strategies are shown in Fig. 4b. While the rate of allele transmission was similar across mating strategies, both methods in which at least one family could be selected by chance ("Priority-Random" and "Random-Random") showed larger standard deviations than the "Priority-Maximum Distance" method, particularly when the number of crosses was low, suggesting than the latter is the most conservative mating strategy.

DISCUSSION

We present the first breeding plan for the Adriatic sturgeon, a species which is believed to be possibly extinct in nature and whose future entirely relies on the implementation of an *ex situ*captive breeding plan based on genetic and demographic information upon which to establish management decisions. The program here proposed meets the two principals generally required in captive breeding: (i) the complete genetic characterization of all the individuals and (ii) the planning of a breeding scheme that minimizes genetic relatedness in order to maximize genetic variability. One particular characteristic of our breeding plan is that is "family-based" in contrast with most breeding plans that are "individual-based". One of the difficulties when working with sturgeons is that adults are only reproductively active every other year or more rarely. This represents a serious problem since a selected individual might not be available for reproduction when needed. However, this problem can be avoided by using a family rather than a single individual approach, so that related individuals might be available instead.

Characterization of the F1 V.I.P. Stock

Pedigree reconstruction was applied to characterize the F1 stock retained at the V.I.P. plant and produced by crossing F0 individuals in the last 20 years. The approach implemented by the new software ("*BreedingSturgeons*") and based on the inference of all possible genotypes compatible with observed band profiles, showed a higher allocation efficiency compared with the simple band sharing approach [10]. Accordingly, all ambiguities were completely solved with no F1 individual being multi-allocated. Comparison of genetic variability between the original F0 (Wild) and the F1 stock showed a relatively important loss of genetic diversity at the mitochondrial level, with 2 of the 7 original haplotypes being absent in the F1 stock before poisoning and a third one lost in the F1-alive group. However, at microsatellites, genetic diversity appears to be almost intact as no significant differences in number of alleles were observed between F0 and F1, with both samples presenting an average number of alleles per locus of about 10. By contrast, a lower value of 8.57 was reported for a different F1 stock of Adriatic sturgeon retained as possible future broodstock at the Ticino River Park [18]. The Ticino F1 stock was established without any genetic input and the animals were collected from a limited number of reproductions, causing a significant erosion of genetic diversity in a single generation, as exemplified by a 16.7% loss in average allele number in comparison with the F0 stock.

In synthesis, the V.I.P. Stock resulted to be the major source of variability for this species. Several families were identified and a great part of the F0 parents are represented within the stock. In regards to the relationships among animals, the probability to select for reproduction two related individuals by chance is very low (5,6%), compared to the 35% estimated for the stock reared by the Ticino River Park [18]. However, even if the use of Ticino F1 stock as *ex situ* broodstock is not advisable, this stock includes some families that are absent or under-represented in the V.I.P. plant, so exchange of individuals between plants should be encouraged.

Finally, our results also show how important is to test for the presence of hybrids or alien species. In many aquaculture facilities, different species or interspecific hybrids are reared, so that mixing of individuals can accidentally happen, and those can go unnoticed unless properly tested. In our case, two reciprocal hybrids between Adriatic and White sturgeon were found within the F1 progeny. As expected, hybrids showed microsatellite alleles not present in the *A. naccarii* F0 gene pool but these are not the only new alleles detected. Additional alleles not present in the parental generation were also found among not-allocated F1 individuals. The presence of new alleles in the F1 might be explained by inheritance from F0 breeders that died several years ago and are not present in our sample. Alternatively, F1 individuals carrying new alleles might be hybrids produced in the hatchery and accidentally mixed into the stock. Accordingly, accepting the risk of losing rare variants we chose to exclude from the broodstock all the animals carrying extra alleles.

Is a Short-term Breeding Plan still Possible?

In the case of the Adriatic sturgeon, recruitment of new breeders from the wild is not possible since the species is possibly extinct in nature, so a breeding plan using wild breeders can only be achieved using the original F0 brought to the V.I.P. facilities in 1977. However, the F0 stock at the V.I.P. plant has been greatly reduced in size in recent years, probably due to senescence. While only 13 individuals remain at present, we propose a priority short-term breeding program to be attempted with these remaining F0 individuals. The plan includes 6 females and 7 males for a total of 18 selected crosses on the basis of pairwise genetic distance (Fig. 1). It should be noted that most of the remaining individuals had never been mated before, so that their successful reproduction would represent a relevant contribution to the future diversity of the captive stock of this species. Nevertheless, the fact that some animals have never been crossed in over 20 years of reproduction events might indicate a low reproductive potential. Hence, there is a chance that the proposed crosses between the surviving F0 adults might not be successful, and although we believe that a short-term plan using F0 should be attempted, most efforts should be conducted to the establishment of the long-term plan involving the F1 generation.

A Long-term Breeding Plan to Safeguard the Adriatic Sturgeon

We present a complete long-term breeding plan based on familiar groups that aims at generating enough animals to ultimately allow for the re-introduction of the species back into its natural environment. Considering the low chances of successful reproduction of these animals and their limited number, most efforts were devoted to developing a long-term breeding strategy involving families of F1 generation, assuming that at least some of the individuals are reproducible every year.

The program starts with the exclusion of combinations between families sharing one parent, together with families that do not share any parent but show low genetic distances, which might be indicative of relatives. This allows minimizing possible errors due to the "*founder assumption*", in which wild founders of a captive population are assumed to be equally unrelated [30], although it might not always be the case.

Our choice of keeping separated for two years the animals produced by different family combinations, until the moment of tagging, limits the number of crosses that can be simultaneously performed with a finite number of ponds. However, it allows tracking the pedigree of every single animal providing useful information for monitoring the release program or for managing the following generation of captive broodstock. Alternatively, instead of crossing one individual per family, it could be considered to cross more individuals of the same family that can be pooled and do not require additional ponds. This would also increase the efficiency of allele transmission and guarantee that all F0 alleles of each family are inherited by the F2 generation, as shown by simulations that suggest the use of at least 3 breeders from the same family when possible. However, these numbers might be not easy to reach given the small size of some families and the discontinuous availability of breeders for reproduction (mostly true for females). Moreover, if breeders of the "*priority family*" are not available, one possibility is to postpone the cross until the following reproductive season and to slightly modify the priority order. If the unavailable individuals belong to the "*mating family*", one possibility is to save the priority order and to select an alternative "*mating family*", choosing the second more distant family. As shown by simulations, selection of family combinations based on priority and genetic distance were more efficient in transmitting the higher possible number of alleles to the next generation relative to the random strategies ("Priority – Random" and "Random – Random"). This was particularly true when number of crosses was low, which suggests that random approaches are not recommended for short-term programs. In fact, the higher standard deviation of the two strategies in which at least one family is randomly selected may result in a significant loss of allele diversity after few generation of captive breeding. Accordingly, the breeding plan here proposed was based on the Priority – Maximum distance approach. On one side, the establishment of a priority score allows to first cross those families whose genetic traits have the higher risk to be lost, since they are shared by a lower number of individuals in the population. On the other side, the use of maximum distance minimizes inbreeding and tends to increase the number of crosses involving families that are genetically peculiar, and consequently, are more often selected as "mating families".

Management Implications

The present work represents the first breeding plan for the Adriatic sturgeon based on its detailed genetic characterization and can be considered as a reference guideline for all conservation actions based on controlled reproductions of this species. As in the case of Adriatic sturgeon, many other sturgeon species are critically endangered with greatly reduced wild populations and a consequent difficulty in collecting breeders from the wild. For such species the establishment of *ex situ* conservation plans through the retention of part of the fingerlings as future breeders is the only way forward. In this context, the approach here developed based on the management of familiar groups instead of individuals can become of major interest. Moreover, for those species that share with the Adriatic sturgeon the tetraploid condition, such as the Russian sturgeon (*A. gueldenstaedtii*), the Persian sturgeon (*A. persicus*), the Chinese sturgeon (*A. sinensis*) and many others, the methods here optimized are directly transferable. This would allow a careful identification of familiar groups thus providing the basis for the establishment of breeding programs that takes into account the genetic composition of the available broodstock together with logistic and economic constraints.

Our study also showcases the importance of genetic approaches to *ex situ* management for meeting conservation goals. The low cost of microsatellite markers facilitates the collection of genetic data that can inform breeding programs about the best pairing individuals so as to minimize mean relatedness. In our study, as few as 7 microsatellite markers were sufficient to unambiguously allocate all F1 individuals. In the event of higher number of putative parents, which might be the case in other sturgeon species, the number of microsatellites should be increased to ensure complete allocation.

Since genetic diversity is an essential prerequisite for the long-term adaptive potential, the availability of *ex situ* genetically heterogeneous broodstocks is the starting point for the rehabilitation of self-sustaining natural populations. However, in the short run, the survival rate of the released fingerlings depends on their capability do adapt to the target environment. In this sense, the establishment of adequate rearing protocols aimed to produce individuals with a high probability of survival is strongly encouraged. As clearly reported in FAO guidelines for sturgeon management and release [31], hatcheries involved in juvenile production for restocking purpose should raise fry and juveniles through training and adaptation to natural conditions, in order to maximize their fitness for survival after release.

As important as it is to use genetic studies into conservation practices, those need to be coordinated and not be carried out separately and restrictively. In the case of the Adriatic sturgeon, at present several local administrations are active in the attempt of recovering the species in its historical natural range. Unfortunately, these actions are often represented by short-term programs and are completely isolated and uncoordinated. Therefore, providing a common program like the one presented in our study that can connect and integrate the different actions would represent the first step towards a responsible concerted management of the Adriatic sturgeon. Finally, in order to implement successful conservation strategies, important actions against habitat degradation are urgently needed including the restoration of natural spawning sites and the re-establishment of river connectivity (i.e. fish passages at dams). A relevant challenge for conservation genetics is to go beyond the scientific since reintroduction is not merely a biological issue but a social, cultural and political one as well. An important part of the framework required for the successful incorporation of genetic data into conservation decision-making of culturally-important species is involvement where possible of local people in the program, together with public education about the project.

ACKNOWLEDGMENTS

We thank Giacinto Giovannini and sons, owners of the "Azienda Agricola V.I.P." and of the "Storioni Ticino hatchery". We are grateful to Giorgio Bertorelle, Andrea Benazzo, Paolo Bronzi and Lara Bellini for their help and suggestions.

AUTHOR CONTRIBUTIONS

Conceived and designed the experiments: EB LC. Performed the experiments: EB. Analyzed the data: EB LC ID RC. Contributed reagents/materials/analysis tools: EB LC ID RC. Contributed to the writing of the manuscript: EB LC JMP.

REFERENCES

1. Leus K (2011) Captive breeding and conservation. Zool Middle East 54: 151–158.
2. Witzenberger KA, Hochkirch A (2011) Ex situ conservation genetics: a review of molecular studies on the genetic consequences of captive breeding programs for endangered animal species. Biodivers Conserv 20: 1843–1861.
3. Frankham R, Ballou JD, Briscoe DA (2002) Introduction to conservation genetics. Cambridge University Press, Cambridge, UK.
4. Philippart JC (1995) Is captive breeding an effective solution for the preservation of endemic species? Biol Conserv 72: 281–295.
5. Russello MA, Amato G (2004) Ex situ population management in the absence of pedigree information. Mol ecol 13: 2829–2840.
6. Araki H, Cooper B, Blouin MS (2007) Genetic effects of captive

breeding cause a rapid, cumulative fitness decline in the wild. Science 318: 100–103.

7. Fraser DJ (2008) How well can captive breeding programs conserve biodiversity? A review of salmonids. Evol App 1: 535–586.
8. Kozfkay CC, Campbell MR, Heindel JA, Baker DJ, Kline P, et al. (2008) A genetic evaluation of relatedness for broodstock management of captive, endangered Snake River sockeye salmon, *Oncorhynchus nerka*. Conserv Gens 9: 1421–1430.
9. Goldstein PZ, DeSalle R, Amato G, Vogler AP (2000) Conservation genetics at the species boundary. Conserv Biol 14(1): 120–131.
10. Congiu L, Pujolar JM, Forlani A, Cenadelli S, Dupanloup I, et al. (2011) Managing polyploidy in ex situ conservation genetics: the case of the critically endangered Adriatic sturgeon (*Acipenser naccarii*). PLoS One 6: e18249.
11. Jamie AI, Lacy RC (2012) A comparison of strategies for selecting breeding pairs to maximize genetic diversity retention in managed populations. J Hered 103: 186–196.
12. Hayes B, He J, Moen T, Bennewitz J (2006) Use of molecular markers to maximize diversity of founder populations for aquaculture breeding programs. Aquaculture 255: 573–578.
13. Ludwig A (2006) A sturgeon view on conservation genetics. Eur J Wildlife Res 52: 3–8.
14. Anders PJ, Drauch-Schreier A, Rodzen J, Powell MS, Narum S, et al. (2011) A review of genetic evaluation tools for conservation and management of North American sturgeons: roles, benefits, and limitations. J App Ichthyol 27(Suppl (2)) 3–11.
15. Wuertz S, Gaillard S, Barbisan F, Carle S, Congiu L, et al. (2006) Extensive screening of sturgeon genomes by techniques revealed no sex-specific random screening marker. Aquaculture 258: 685–688.
16. Fontana F, Lanfredi M, Chicca M, Congiu L, Tagliavini J, et al. (1999) Fluorescent *in situ* hybridization with rDNA probes on chromosomes of *Acipenser ruthenus* and *Acipenser naccarii* (Osteichthyes Acipenseriformes). Genome 42: 1008–1012.
17. Bronzi P, Rosenthal H, Arlati G, Williot P (1999) A brief overview on the status and prospects of sturgeon farming in Western and

Central Europe. J App Ichthyol 15: 224–227.

18. Boscari E, Congiu L (2014) The need for genetic support in restocking activities and *ex situ* conservation programmes: the case of the Adriatic sturgeon (*Acipenser naccarii*Bonaparte, 1836) in the Ticino River Park. J App Ichthyol DOI:10.1111/jai.12545.
19. Boscari E, Barmintseva A, Pujolar JM, Doukakis P, Mugue N, et al.. (2014) Species and hybrid identification of sturgeon caviar: a new molecular approach to detect illegal trade. Mol Ecol Res doi:10.1111/1755-0998.12203.
20. Tamura K, Peterson D, Peterson N, Stecher G, Nei M, et al. (2011) MEGA5: Molecular Evolutionary Genetics Analysis using Maximum Likelihood, Evolutionary Distance, and Maximum Parsimony Methods. Mol Biol Evol 28: 2731–2739.
21. Clement M, Posada D, Candrall KA (2000) TCS: a computer program to estimate gene genealogies. Mol Ecol 9: 1657–1659.
22. Templeton AR, Crandall KA, Sing CF (1992) A cladistics analysis of phenotypic associations with haplotypes inferred from restriction endonuclease mapping and DNA sequence data. III. Cladogram estimation. Genetics 132: 619–633.
23. Librado NP, Rozas J (2009) DnaSP v5: a software for comprehensive analysis of DNA polymorphism data. Bioinformatics 25: 1451–1452.
24. May B, Krueger CC, Kincaid HL (1997) Genetic variation at microsatellite loci in sturgeon: primer sequence homology in *Acipenser* and *Scaphirynchus*. Canadian Journal of Fisheries and Aquatic Science 5: 1542–1547.
25. Forlani A, Fontana F, Congiu L (2007) Isolation of microsatellite loci from the endemic and endangered sturgeon *Acipenser naccarii*. Conser Gen 9: 461–463.
26. Zane L, Patarnello T, Ludwig A, Fontana F, Congiu L (2002) Isolation and characterization of microsatellite in the Adriatic sturgeon (*Acipenser naccarii*). Mol Ecol Notes 2: 586–589.
27. Henderson-Arzapalo A, King TL (2002) Novel microsatellite markers for Atlantic sturgeon (*Acipenser oxyrinchus*) population delineation and broodstock management. Mol Ecol Notes 2: 437–439.

28. Boscari E, Barbisan F, Congiu L (2011) Inheritance pattern of microsatellite loci in the polyploidy Adriatic sturgeon (*Acipenser naccarii*). Aquaculture 321: 223–229.

29. Hood GM (2010) PopTools version 3.2.5. Available: http://www.poptools.org. Accessed 2014 October 10.

30. Van Dyke F (2008) Conservation biology: foundations, concepts, applications. Wheaton College, Illinois, USA. Second Editions, Springer Science and Business Media.

31. Chebanov M, Rosenthal H, Gessner J, Van Anrooy R, Doukakis P, et al. (2011) Sturgeon hatchery practices and management for release-Guidelines FAO Fisheries and Aquaculture Technical Paper No 570. Ankara, FAO. 2011: 110.

CHAPTER

18

RESPONSE TO SELECTION FOR BODY WEIGHT IN THE THIRD GENERATION OF MASS SELECTION OF THE AFRICAN CATFISH (CLARIAS GARIEPINUS) AT RESEARCH INSTITUTE FOR FISH BREEDING SUKAMANDI

Bambang Iswanto, Imron, Huria Marnis, and Rommy Suprapto

Research Institute for Fish Breeding. Jl. Raya 2 Sukamandi-Patokbeusi, Subang, West Java, Indonesia

ABSTRACT

Genetic improvement of the African catfish (Clarias gariepinus) in Indonesia for increasing growth performance has been conducted by Research Institute for Fish Breeding at Sukamandi through mass selection. Collection and characterizations of the founder populations, building the synthetic base population, first generation and second generation through mass selection were conducted during 2010-2013. Later, in 2014 it was followed by building the third generation. The present study aimed to find out the genetic gain in the third generation in term of response to selection for body weight. Fifty pairs of the selected (fast growing) individuals from the second generation were mated to produce the third generation. As a comparison, five pairs of average-sized individuals were mated to produce the control population, as a second generation representative.

Larval rearing, nursery and grow-out phases were respectively held for 25 days in the aquaria, 30 days in the concrete tanks and 60 days in the concrete ponds. At the end of each phase, individual samplings of body weight were undertaken. The results showed that mean body weight of the third generation was higher than that of control population at the end of larval rearing phase (0.21 ± 0.26 g versus 0.20 ± 0.15 g), nursery phase (6.12 ± 2.93 g versus 5.80 ± 3.50 g) and grow-out phase (198.67 ± 82.82 g versus 165.22 ± 71.09 g). Those results revealed that response to selection for body weight of the third generation was positive, i.e. about 20.24% (33.45 g).

Keywords: African catfish (Clarias gariepinus); body weight; mass selection; response to selection; third generation

INTRODUCTION

Introduced African catfish (Clarias gariepinus Burchell 1822) recently becomes a great economically important species in Indonesian aquaculture. It has been cultured vastly throughout Indonesia, mainly in Java, Sumatra, and Kalimantan. African catfish has been several times introduced into Indonesia for aquaculture, either directly from Africa or via other countries. All of those introduced strains exhibited superior growth performance at the beginning of their introductions. Unfortunately, their growth performances were then decreasing. Therefore, genetic improvement program for increasing the growth performance of the African catfish in Indonesia should be carried out in order to increase its productivity in aquaculture industry.

Genetic improvement program in establishing a new fast growing strain of the African catfish has been conducted by Research Institute for Fish Breeding at Sukamandi since 2010. The genetic improvement program was conducted using mass selection for the growth rate, in term of body weight. Collections and characterizations of the founder populations were conducted in 2010 (Imron et al., 2010; 2011), for which four strains have been collected, namely Paiton, Sangkuriang, Dumbo, and the Egyptian strains. Then, in 2011 it was followed by building the synthetic (composite) base population through diallele crossing of four strains of the founder populations (Iswanto et al., 2012). In 2012, the first generation was established with response to selection of 20.59% for body weight (Iswanto et al., 2014a). Later, building the second generation was conducted in 2013 resulted in response to selection of 11.80% for body weight (Iswanto et al., 2014b). Finally, in 2014 it was followed by building the third generation to test the growth performance compared to the second generation. The present study aimed to find out the genetic gain resulted in the third generation in term of response to selection (selection response) for body weight.

MATERIALS AND METHOD

The test of response to selection for body weight in the third generation of African catfish compared to the control population was held in May to September 2014 at Research Institute for Fish Breeding (RIFB) Sukamandi. Fifty pairs of the selected (fast growing, in term of body weight) individuals (breeders) of the second generation were mated to produce the third generation. As a comparison, five pairs of average-sized individuals were mated to produce the control population, as a second generation representative. These mating processes were done artificially using ovaprim (Syndel Laboratories Ltd., Canada) to induce final maturation (ovulation and spermiation). The eggs were obtained through stripping, while the sperm were collected through partial gonadectomy techniques. Approximately 100 g eggs obtained from each female breeder were fertilized by pooled sperm from all male breeders, then were incubated in separated 1 m x 1 m tray made from happa placed within indoor concrete tanks with circulated water.

A thousand of two days old larvae obtained from each fertilization process were reared for 25 days within 55 indoor aquaria (60 cm x 40 cm x 20 cm), i.e. 50 aquaria for the third generation and five aquaria for control population. Brine shrimp (Artemia sp.) nauplii was given as first feed during two days of rearing period, then, in the third day it was partially replaced with chopped silkworms (Tubifex sp.). Within following until fifteenth days of larval rearing period, life silkworms was used as natural feed, which was gradually replaced with commercial fish larval feed contained 40% crude protein (HI-PRO-VITE PS-P, PT Central Proteinaprima, Mojokerto). This commercial fish larval feed was given until 25th day, partially combined with commercial shrimp larval feed contained 40% crude protein (BINTANG 581, PT Central Proteinaprima). Those feed were given ad libitum (until apparent satiation) five times daily (at 06:00, 10:00, 14:00, 18:00, and 22:00). At the end of rearing period, individual samplings of body weight of 2% (from initial stock) juveniles from each aquarium were undertaken. Then, juveniles from all aquaria for each third generation and control population were pooled and were graded into < 2 cm, 2-3 cm, 3-4 cm, and > 4 cm size class groups.

Juveniles of each size class group were then stocked into five outdoor concrete tanks (3 m x 5 m x 0.5 m) filled with irrigation water for 30 days (one month) of nursery phase with stocking density of 580 fishes/m2 (equal to 8,700 juveniles/tank). Separation of each size class group juveniles stocked within a same tank (either for the third generation or control populations) was facilitated by net screen. Small crumble sized (BINTANG 581) and crude crumble sized (BINTANG 582) commercial shrimp larval feed with 40% crude protein content were respectively given for small (< 2 cm and 2-3 cm size class groups) and larger (3-4 cm and > 4 cm size class groups) juveniles. These feeds were given ad libitum four times daily (at 08:00, 12:00, 16:00, and 21:00) during first ten days or nursery period. Those feed were then gradually replaced with commercial small sized (0.5- 1.2 mm of diameter) floating fish feed contained 40% crude protein (PRIMAFEED PF 800 and PF 1000, PT Matahari Sakti, Surabaya) given ad libitum thrice daily (at 08:00, 12:00 and 16:00) during 10th-30th days of rearing period. At 15th day, all of juveniles from each the third generation and control populations were pooled and regrouped via grading into < 3 cm, 3-4 cm, 4-5 cm, 5-6 cm, and > 6 cm size class groups. These groups were then restocked into their concrete tanks (separation of each size class group juveniles stocked within a same tank was facilitated by net screen). At the end of nursery phase, individual body weight samplings of 2% population (from initial stock of each the third generation and control populations) were undertaken. Then, the juveniles from all tanks of each the third generation and control populations were pooled and graded into < 3 cm, 3- 5 cm, 5-7 cm, 7-9 cm, 9-11 cm, and > 11 cm size class groups.

Juveniles of each size class group were then stocked into nine outdoor concrete ponds (2 m x 10 m x 0.8 m) filled with irrigation water for grow-out phase of 60 days (two months), using stocking density of 200 fishes/m^2 (equal to 4,000 juveniles/pond). Separation of each size class group juveniles stocked within a same pond (either for the third generation or control populations) was facilitated by net screen. Small sized floating fish feed of 30% crude protein content with ± 2 mm (HI-PRO-VITE 781-1, PT Centralpangan Pertiwi, Karawang) and ± 3 mm (HIPRO-VITE 781-2) of diameter were respectively given for small (< 3 cm, 3-5 cm and 5-7 cm size class groups) and larger (7-9 cm, 9-11 cm, and > 9cm size class groups) juveniles. Those feeds were given ad libitum thrice daily (at 08:00, 12:00, and 16:00) during first ten days of grow-out period. Then, within subsequent ten days of grow-out period, the small size class groups of juveniles were fed with ± 3 mm floating fish feed (HI-PRO-VITE 781-2) and the larger ones with ± 4 mm floating fish feed (HI-PRO-VITE 781). Those feeds were given ad libitum twice daily (at 08:00 and 16:00). Within following days of grow-out period, all of the juveniles were fed with ± 4 mm floating fish feed (HI-PRO-VITE 781), given ad libitum twice daily (at 08:00 and 16:00). During sixty days of grow-out period, individual body weight samplings of 2% population (from initial stock of each the third generation and control populations) were conducted every twenty days.

At the end of grow-out period, individual body weight samplings of 20% population (from initial stock) from each the third generation and control populations were undertaken. Thus, realized heritability and response to selection (Gjedrem & Baranski, 2009) of the third generation could be determined. Response to selection (R) of the third generation was determined as R = (mean body weight of the third generation – mean body weight of control population). While, realized heritability (h^2) was determined as h^2 = (response to selection of the third generation : selection differential of the second generation), with the selection differential of the (previous) second generation was about 179.35% (Iswanto et al., 2014b).

During larval rearing, nursery and grow-out phases of the third and control population of the African catfish held at RIFB Sukamandi, water quality parameters, including water temperature, pH, dissolved oxygen, ammonia and nitrite were recorded at the beginning of the each rearing phase followed by subsequent twice a week recordings. Recordings of water temperature, pH and dissolved oxygen using digital water quality checker (TOA DKK WQC-22A, DKK TOA Corporation, Japan) and the concentration of ammonia and nitrite using spectrophotometer (Hitachi

UV/ Vis Spectrophotometer U-1500, Hitachi Instruments Inc., USA) were carried out twice daily, in the early morning (at 06:00) and the afternoon (at 14:00).

RESULTS AND DISCUSSION

Growth performances of the third generation and control population of the African catfish (C. gariepinus) based on individual samplings of body weight at the end of 25 days of larval rearing phase, 30 days of nursery phase and every 20 days during 60 days of grow-out phase were presented in Figure 1. Those results showed that the third generation attained table-size for commonly Indonesian consumers (about 100-150 g) at 95 days old or 40 days of grow-out period, while the control population attained it at less than two months of grow-out period. Compared to grow-out phase commonly practiced by African catfish farmers in Indonesia, which were commonly carried out for 2-3 months, those growth performances were comparable.

At the end of larval rearing phase, nursery phase and grow-out phase, mean body weight of the third generation was higher than that of control population (Figure 1). Those results suggested that genetic improvement program in term of increasing the growth performance of the African catfish conducted at the RIFB Sukamandi during three generations of mass selection has gained positive results. Response to selection and final mean body weight of the third generation compared to the control population at the end of larval rearing, nursery, and grow-out phases were presented in Table 1.

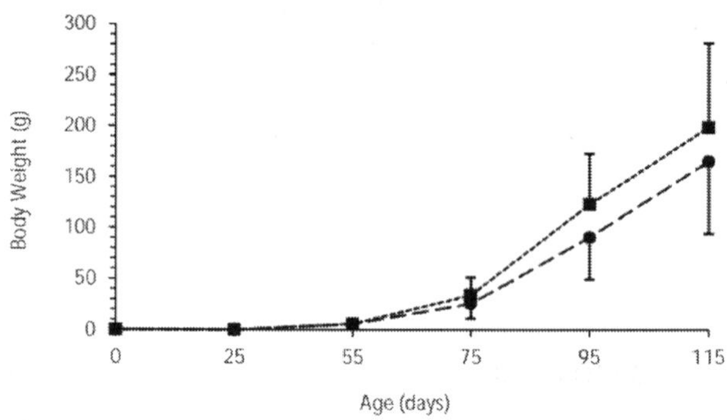

Figure 1: Growth performances based on body weight during 25 days of larval rearing phase, 30 days of nursery phase and 60 days of grow-

out phase (based on samplings of 2% populations) of the third generation (■) and control population (●) of the African catfish (Clarias gariepinus) genetic improvement program held at Research Institute for Fish Breeding, Sukamandi. Vertical lines represent its each standard deviation.

Based on those growth performances, final response to selection for body weight in the third generation of mass selection of the African catfish at RIFB Sukamandi was about 20.24% (33.45 g). This response to selection was similar to that of the first generation, which attained 20.59% (Iswanto et al., 2014a), but was much higher than that of the second generation, which only attained 11.80% (Iswanto et al., 2014b). Thus, cumulative response to selection for body weight during three generations of mass selection of the African catfish at RIFB Sukamandi attained 52.63%

The success of genetic improvement programs for increasing the growth performance using mass selection were also reported in several other cultured fish species. Genetic improvement for growth performance of Salmo trutta fario through four generations of enhanced methods of mass selection in France resulted in maximal response to selection for body weight, which attained 130% (Chevassus et al., 2004). Mass selection for improving the growth performance of Ictalurus punctatus conducted in United States of America resulted in 12% and 18% response to selection for body weight in respectively Kansas and Marion strains (Dunham & Smitherman, 1983). Then, three generations of the mass selection yielded in cumulative response to selection for body weight of 29% and 21% in respectively Kansas and Marion strains (Rezk et al., 2003). In Clarias macrocephalus, improving the growth performance through three generations of mass selection conducted in Thailand resulted in 11.8% response to selection for body weight (Na-Nakorn, 1993). Genetic improvement program for increasing the growth performance of Megalobrama amblycephala conducted in China through five generations of mass selection yielded in 29% response to selection for body weight (Li & Cai, 2003). Three generations of mass selection for improving the growth performance of Barbodes gonionotus in Bangladesh resulted in cumulative response to selection of 21.9% (Hussain et al., 2002).

Genetic improvement program to increase the growth performance using mass selections in the past years, however, were reported to be unsuccessful in Cyprinus carpio (Moav & Wohlfarth, 1976) and

Oreochromis niloticus (Hulata et al., 1986; Huang & Liao, 1990). Those failed genetic improvement programs by applying the mass selection were, however, caused mainly by low genetic variation in the base populations due to limited number of broodstocks used to produce the offsprings, increasing the inbreeding level and the influence of maternal effects since the use of mass spawnings (Huang & Liao, 1990; Lutz, 2001; Chevassus et al., 2004; Gjedrem & Baranski, 2009).

Table 1: Response to selection and final mean body weight of the third generation compared to the control population of the African catfish Clarias gariepinus at the end of larval rearing, nursery and grow-out phases

Phases	Periods (days)	Final mean body weight (g)		Response to selection	
		Third generation	Control	Gram (g)	Percentage (%)
Larval rearing	25	0.19 ± 0.10	0.19 ± 0.07	-	-
Nursery	30	6.12 ± 2.93	5.80 ± 3.50	-	-
Grow-out	60	198.67 ± 82.82	165.22 ± 71.09	33.45	20.24

Based on the response to selection for body weight in the third generation of 20.24% and selection differential for body weight in the establishing the second generation of 179.35% (Iswanto et al., 2014b), realized heritability for body weight at the end of grow-out phase in the third generation of mass selection of African catfish at RIFB Sukamandi was about 0.12. This realized heritability was relatively similar to that of the first generation of about 0.10 (Iswanto et al., 2014a), and somewhat higher than that of the second generation of about 0.08 (Iswanto et al., 2014b). Applying the mass selection for improving performance would be effective and produced positive response to selection when heritability of respective trait was high and/or the population size was large, thus high selection differential could be applied (Kapuscinski & Miller, 2007; Beaumont et al., 2010). Although realized heritability in three generations of mass selection of African catfish at RIFB Sukamandi were low, however, selection differential in establishing the base population (Iswanto et al., 2012), the first (Iswanto et al., 2014a) and second (Iswanto et al., 2014b) generations were high (respectively 180.40%, 179.35%, and 167.50%), therefore still always produced positive response to selection. Mass selection in improving growth performance of O. niloticus at Sukabumi produced 2.3% response to selection for total length, while its realized heritability was 0.12 (Brzeski & Doyle, 1995). Similar results was also reported for mass selection in improving growth performance of O. niloticus in the Philippines, which produced 8% response to selection for total length, while its realized heritability was 0.16 (Basaiao et al., 2005). The successful genetic improvement program for O. niloticus through combined family selection namely GIFT (Genetic Improvement of Farmed Tilapias) project conducted in the Philippines resulted in improved growth performance with cumulative response to selection of 85% for body weight during five generations, while its heritability were respectively 0.18, 0.25, 0.18, 0.07, 0.13, and 0.09 (Eknath & Acosta, 1998).

Table 2: Water quality during 25 days of larval rearing phase, thirty days of nursery phase and sixty days of grow-out phase of the third generation and control population of the African catfish (Clarias gariepinus) genetic improvement program held at Research Institute for Fish Breeding (RIFB), Sukamandi

Parameters	Phases		
	Larval rearing	Nursery	Grow-out
Temperature (°C)	27.4-28.9	25.5-33.9	28.0-31.5
pH	7.61-8.02	6.10-7.54	5.06-6.98
Dissolved oxygen (mg/L)	5.00-5.80	0.40-11.40	0.44-3.69
Ammonia (mg/L)	0.0320-0.0522	0.1055-0.3753	0.1688-3.8953
Nitrite (mg/L)	0.0178-0.0671	0.0103-0.1728	0.0032-0.1321

Survival rates of the third generation and control population of African catfish during 25 days of larval rearing phase and 30 days of nursery phase carried out at RIFB Sukamandi were quiet high, respectively 80.14 ± 10.76% and 70.58 ± 11.37% in the larval rearing phase and respectively 83.16% and 77.39% in the nursery phase. Whereas, survival rates during 60 days of grow-out phase were lower, i.e. 68.19% for the third generation and 66.09% for the control population.

Those slightly low survival rates in the grow-out phase were caused by relatively high ammonia concentration in the water, which almost reached 4 mg/L. Ammonia concentration over 0.34 mg/L was not recommended for grow-out phase of the African catfish, even though the threshold concentration was 2.5 mg/L (Schram et al., 2010). Except for high con centration of ammonia in the grow-out phase, other water quality parameters during 25 days of larval rearing phase, 30 days of nursery phase and 60 days of grow-out phase in both third generation and control population of the African catfish at RIFB Sukamandi were still in the acceptable ranges required for optimum survival and growth of the African catfish (Table 2), as were recommended by several authors. Recommended optimum water temperature for larval rearing and nursery phases of African catfish ranged 27.5°C-32.5°C (Britz & Hecht, 1987), while for growout phase ranged 27°C-30°C (Degani et al., 1988). Well oxygenated water was recommended for larval rearing phase (Britz & Hecht, 1988), while for nursery and grow-out phases it was not necessary, since African catfish is an air breathing fish that used direct aerial respiration (Teugels, 1986). Median lethal pH water for African catfish juveniles was reported at 4.3 and 9.2 (Ndubuisi et al., 2015). For nitrite, median lethal concentration for the African catfish was reported at 0.2 mg/L (Ajani & Adeyemo, 2012).

CONCLUSION

Growth performance of the third generation of mass selection of the African catfish (Clarias gariepinus) during 25 days of larval rearing phase, 30 days of nursery phase and 60 days of grow-out phase conducted at RIFB Sukamandi was higher than that of control population, resulted in a positive response to selection for body weight.

ACKNOWLEDGEMENT

Thanks to all research assistants of the Clarias team of RIFB Sukamandi for their invaluable technical assistances during larval rearing, nursery, and growout phases. This research was funded by Ministry of Marine Affairs and Fisheries.

REFERENCES

1. Ajani, F., & Adeyemo, O.K. (2012). Nitrite intoxication of Clarias gariepinus at different water temperatures. International Journal of Fisheries and Aquaculture, 4(4), 77-80.
2. Basaiao, Z.U., Arago, A.L., & Doyle, R.W. (2005). A farmer-oriented Nile tilapia, Oreochromis niloticus L., breed improvement in the Philippines. Aquaculture Research, 36, 113-119.

3. Beaumont, A.R., Boudry, P., & Hoare, K. (2010). Biotechnology and genetics in fisheries and aquaculture. Second Edition. Blackwell Science. Singapore, 202 pp.
4. Britz, P.J., & Hecht, T. (1987). Temperature preferences and optimum temperature for growth of African sharptooth catfish (Clarias gariepinus) larvae and post-larvae. Aquaculture, 63, 205-214.
5. Britz, P.J., & Hecht, T. (1988). Artificial propagation and fry production. In Hecht, T., Uys, W., & Britz, P.J. (Eds.). The culture of sharptooth catfish, Clarias gariepinus in Southern Africa. South African National Scientific Programmes Report No. 153 (p. 36-46). Pretoria, Republic of South Africa.
6. Brzeski, V.J., & Doyle, R.W. (1995). A test of an onfarm selection procedure for tilapia growth in Indonesia. Aquaculture, 137, 219-230.
7. Chevassus, B., Quillet, E., Krieg, F., Hollebecq, M.G., Mambrini, M., Faure, A., Labbe, L., Hiseux, J.P., & Vandeputte, M. (2004). Enhanced individual selection for selecting fast growing fish: the "PROSPER" method, with application in brown trout (Salmo trutta fario). Genetics of Selection and Evolution, 36, 643-661.
8. Degani, G., Ben-Zvi, Y., & Levanon, D. (1988). The effect of different dietary protein sources and temperatures on growth and feed utilization of African catfish Clarias gariepinus (Burchell). The Israeli Journal of Aquaculture – Bamidgeh, 40(4), 113-117.
9. Dunham, R.A., & Smitherman, R.O. (1983). Response to selection and realized heritability for body weight in three strains of channel catfish, Ictalurus punctatus, grown in earthen ponds. Aquaculture, 33, 89-96.
10. Eknath, A.E., & Acosta, B.O. (1998). Genetic Improvement of Farmed Tilapias (GIFT) Project. Final Report ICLARM. Manila, 173 pp.
11. Gjedrem, T., & Baranski, M. (2009). Selective breeding in aquaculture: an introduction. Springer. Netherland, 221 pp.
12. Huang, C.M., & Liao, I.C. (1990). Response to mass selection for growth rate in Oreochromis niloticus. Aquaculture, 85, 199-205.

13. Hulata, G., Wohlfarth, G.W., & Halevy, A. (1986). Mass selection for growth rate in the Nile tilapia (Oreochromis niloticus). Aquaculture, 57, 177-184.
14. Hussain, M.G., Islam, M.S., Hossain, M.A., Wahid, M.I., Kahinoor, A.H.M., Dey, M.M., & Mazid, M.A. (2002). Stock improvement of silver barb (Barbodes gonionotus Bleeker) through several generations of genetic selection. Aquaculture, 204, 469-480.
15. Imron, Tahapari, E., Sari, R.D., & Hudaidah, S. (2010). Growth and survival of catfish, Clarias gariepinus, having different domestication history in indoor nursery. Proceedings of International Conference of Aquaculture Indonesia, Hang Tuah University, October 2010.
16. Imron, Sunandar, D., & Tahapari, E. (2011). Microsatellite genetic variation in cultured populations of African catfish (Clarias gariepinus) in Indonesia. Indonesian Aquaculture Journal, 6(1), 1-10.
17. Iswanto, B., Imron, Suprapto, R., & Marnis, H. (2012). Establishment of a rapid growth strain of African catfish Clarias gariepinus (Burchell, 1822) by means of individual selection: Building the synthetic base population. In Haryanti, Rachmansyah, Sugama, K., Sudradjat, A., Imron, Sunarto, A., Sumiarsa, G.S., Azwar, Z.I., & Kristanto, A.H. (Eds.). Prosiding Forum Inovasi Teknologi Akuakultur 2012. p. 1197- 1210 (in Indonesian).
18. Iswanto, B., Imron, Suprapto, R., & Marnis, H. (2014a). Establishment of a rapid growth strain of African catfish Clarias gariepinus (Burchell, 1822) by means of individual selection: Production the first generation population. J. Ris. Akuakultur, 9(3), 343- 352. (in Indonesian with English abstract).
19. Iswanto, B., Imron, Suprapto, R., Marnis, H., & Ridzwan, N.S. (2014b). Establishment of a rapid growth strain of African catfish Clarias gariepinus (Burchell, 1822) by means of individual selection: Building the second generation. In Sugama, K., Kusnendar, E., Rachmansyah, Giri, I N.A., Yuhana, M., Kristanto, A.H., Imron, Radiarta, I N., & Dewi, R.R.S.P.S. (Eds.). Prosiding Forum Inovasi Teknologi Akuakultur 2014. p. 837-847 (in Indonesian).
20. Kapuscinski, A.R., & Miller, L.M. (2007). Genetics guidelines for fisheries management. University of Minnesota Sea Grant Program.

Minnesota, 116 pp.

21. Li, S.F., & Cai, W.Q. (2003). Genetic improvement of the herbivorous blunt snout bream (Megalobrama amblycephala). Naga, WorldFish Center Quarterly, 26(1), 20-23.
22. Lutz, C.G. (2001). Practical genetics for aquaculture. Fishing News Book. London, 252 pp.
23. Moav, R., & Wohlfarth, G. (1976). Two-way selection for growth rate in the common carp (Cyprinus carpio L.). Genetics, 82, 83-101.
24. Na-Nakorn, U. (1993). Review of aquaculture genetic researches in Thailand. Biotropia, 6, 45-54.
25. Ndubuisi, U.C., Chimezie, A.J., Chinedu, U.C., Chikwem, I.C., & Alexander, U. (2015). Effect of pH on the growth performance and survival rate of Clarias gariepinus fry. International Journal of Research in Biosciences, 4(3), 14-20.
26. Rezk, M.A., Smitherman, R.O., Williams, J.C., Nichols, A., Kucuktas, H., & Dunham, R.A. (2003). Response to three generations of selection for increased body weight in channel catfish, Ictalurus punctatus, grown in earthen ponds. Aquaculture, 228, 69-79.
27. Schram, E., Roques, J.A.C., Abbink, W., Spanings, T., de Vries, P., Bierman, S., van de Vis, H., & Flik, G. (2010). The impact of elevated water ammonia concentration on physiology, growth and feed intake of African catfish (Clarias gariepinus). Aquaculture, 306, 108-115.
28. Teugels, G.G. (1986). A systematic revision of the African species of the genus Clarias (Pisces: Clariidae). Annales Musee Royal de l'Afrique Centrale, 247, 1-199.

CHAPTER

19

EFFECT OF INBREEDING ON MORPHOLOGICAL TRAITS OF POPULATIONS OF GIANT FRESH WATER PRAWN, MACROBRACHIUM ROSENBERGII: IMPLICATIONS FOR SELECTIVE BREEDING

Imron[1], Harry Wuwungan[2], Dinar Tri Soelistyawati[2]

[1]Research Institute for Fish Breeding

[2] Department of Aquaculture, Faculty of Fisheries and Marine Sciences, Bogar Agricultural University

ABSTRACT

Genetic improvement through selective breeding relies on the genetically-controlled phenotypic variability in the character of interest. Therefore, the extent of phenotypic variability in the population to be selected is an important parameter, as it potentially influences the population's response to selection. This study was aimed to investigate the effect of inbreeding on the survivability, growth, and phenotypic variability of giant freshwater prawn (GFP), Macrobrachium rosenbergii. Two GFP populations of different inbreeding levels, namely 25% and 0%, respectively, were formed by mating of individual broodstock with known pedigree. Study was conducted for two months starting from newly hatched larvae up to nursery stage (30 day-old post larva). Phenotypic variability profile, expressed in the morphometric mean and coefficient of variation (CV) of twenty one morphometric characters were measured and evaluated. Results showed that in general, the inbred populations had lower values in the mean of all characters (100%), indicating that they suffered from inbreeding depression. Similarly, a lower CV values were observed in sixteen (75%) of the morphometric characters measured, indicating a potential reduced of genetic gain when they are used in selective breeding program. These results suggest the importance of controlling inbreeding level in breeding population that adverse effects resulted resulted from inbreeding can be minimized.

Keywords: Inbreeding; phenotypic variability; coefficient of variation

INTRODUCTION

The raw material for genetic improvement through selective breeding is the existence of heritable-genetic variation within the character of interest. In practice, this variation is manifested in the form of phenotypic variation, either in morphometric or meristic traits. Population's phenotypic variability, most commonly expressed as size variation, may result from both exogenous and endogenous sources. The exogenous factors include such circumstances as feed competition, social stress, increased motor activity, and dominance cost (Goldan *et al.,* 1997). The endogenous factors such as differences in strain, populations or family (De March, 1997), and differences in genetic variation may also influence the population's size variation.

Inbreeding, which is mating of closely related individuals, reduces genetic variation of populations. Studies using molecular tools in several fish taxa have documented and confirmed this trend. For the purpose of breeding program, particularly those carried out through conventional approach however, information of those effect on quantitative traits is more preferable. There are at least two main reasons for this concern. Firstly, most conventional breeding programs work with those traits. Most traits of commercial interest such as growth, disease resistance, and feed consumption efficiency are quantitative traits. Secondly, due to the nature of relatively small and closed population, breeding population could not avoid from the occurrence of inbreeding.

In preliminary study, lmron (2009) reported the effect of inbreeding on the performance of larval stage of GFP including the rate of larval development, survival rate, and growth of the newly hatched larvae up to post larvae (PL) stage. He found that the performance of the inbred group was lower that of the non-inbred counterparts. However, no information was obtained as to the phenotypic variability within that life stage and beyond. This study was aimed to investigate the effect of inbreeding on GFP's phenotypic variability at the later stage of the life.

MATERIALS AND METHODS

Preparation of Test Populations

Test animal consisted of two populations of GFP that were set to have different levels of inbreeding coefficient. The development of the groups was allowed due to the availability of live biological collections of different GFP germ plasm in the Research Institute of Fish Breeding (RIFB). The first group, called inbred population, was population of progenies produced by mating of individual broodstocks derived from the same parental pair. Given this relationship, inbreeding level in the first group would be at least 2S% (Nakadate *eta/.*, 2003). The actual inbreeding level may exceed that figure depending on the previous parental (grandparent) relatedness. However, due to lack of that information, it was assumed that individuals within grandparent generation were not related. The second group, called outbred population, was progenies produced by mating individual broodstocks derived from the same population but of different parental pair (Figure 1). Under these circumstances, the inbreeding level of the second group would be zero.

The two groups of progeny populations differing in inbreeding level, each with four replicates, were reared for two months. Rearing activity was carried out in two stages, namely larval and nursery stages. Larval rearing period started from the newly hatched larvae up to newly metamorphosed-post larvae. Rearing at this stage was conducted in 50 L conical tanks placed inside the hatchery. Subsequent rearing (nursery) was conducted in 1 m x 1 m square hapas installed in outdoor pond. Rearing management including stocking density, feeding practice, and water quality management during larval rearing period across groups was made equal as described by Imron (2009), that comparison among groups was allowable with a minimum bias (Lin & Ritland, 1997).

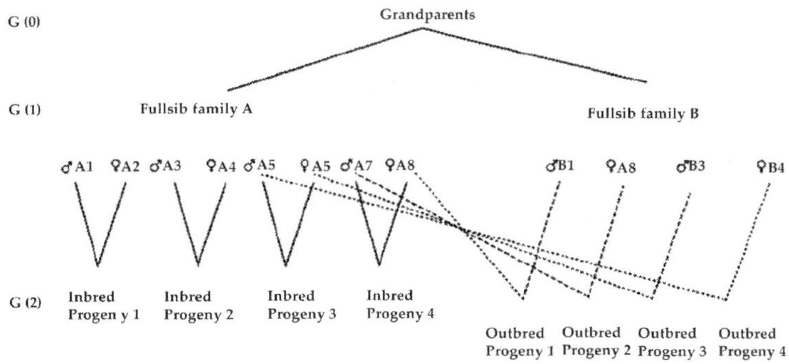

Figure1: Mating design used to produce groups of inbred and outbred progenies. Groups of inbred progenies were produced by brother-sister mating while groups of outbred progenies were produced by mating of individual derived from different families. Population tests were those belong to G (2).

Parameter Examined

The main parameters examined were morphometric traits as presented in Table 1. Measurement was carried out at three day interval using a light microscope equipped with micrometer. Evaluation of phenotypic variability was focused on the measurement of mean and coefficient of variation of twenty two morphometric characters as described in Table 1. Measurement of these characters was conducted at the end of the nursery period by sampling in random thirty individuals per replicate. Growth parameter was assessed using measurement of total length and limited to nursery period. Total 240 individual samples representing the inbred and outbred groups were measured for their phenotypic variability.

Data Analyses

The data of morphometric characters were analyzed in two ways, i.e. descriptive-graphical presentation and statistical significance tests. Inbreeding depression coefficient within any characters was assessed using formula $\partial = 1 - X_1/X_0$ (Roff, 1998; Wright *et al.*, 2008); where ∂ was the difference between the inbred and outbred morphometric means and X_1 and X_0 represent the mean value of a morphometric character within the inbred and the outbred, respectively. Comparisons of mean of morphometric values between the inbred and the outbred populations were carried out using Kruskai-Wallis and t-test, respectively.

Statistical significance tests for coefficient of variation were conducted by applying the guideline explained by Lewontin (1966). In brief, this analysis is similar to the F-test, namely by comparing the counted F-value against the distribution of F-values in the table (counted F against F table values) with corresponding degrees of freedom. With respect to the comparison of CV values, the counted F-values was obtained using Lewontin's (1 966) formula: S^2_{logX}/S^2_{logY}; N_X-1; ; N_Y-1; where S^2_{logX} and S^2_{logY} were logarithmic transformed of morphometric variances of the first and second populations, while the N_x-1 and N_v-1 represented the degrees of freedom of the respective populations.

Table 1: Description of twenty two morphometric characters measured within the inbred and outbred populations of *Macrobrachium rosenbergii*

Traits	Description	Ref
Rostrum length (RST)	Distance from median posterior margin to the rostrum tip	2
Head dorsal length (HDL)	Dorsal-side distance from the orbital tip to the posterior margin of cephalic carapace	*
First segment dorsal length (FDL)	Dorsal-side distance between posterior margin of cephalic carapace and posterior margin of the first abdominal segment	*
Second segment dorsal length (SDL)	Dorsal-side distance from the posterior margin of the first segment to the posterior margin of the second abdominal segment	1
Third segment dorsal length (TDL)	Dorsal-side distance from the posterior margin of the second to the posterior margin of the third abdominal segment	1
Fourth segment dorsal length (ODL)	Dorsal-side distance from the posterior margin of the third segment to the posterior margin of the third abdominal segment	1
Fifth segment dorsal length (FDL)	Dorsal-side distance from posterior of the fourth segment to the posterior margin of the fifth abdominal segment	1
Sixth segment dorsal length (XDL)	Dorsal-side distance from posterior of the fifth segment to the posterior margin of the sixth abdominal segment	1
Telson (TLS)	Distance from the tip to the base of the telson	2
Prosertema (PST)	Distance from the tip to the base of prosertema	2
Antennulles Length (UTL)	Distance from the tip to the base of antennule	

Antenna length (ATL)	Distance from tip to the base of antenna	
Total length (TOL)	Distance from the tip of the rostrum to the tip of the telson	2
Head ventral lenth (HVL)	Ventral side distance from the base of antenna to the posterior margin of the cephalic carapace	*
First segment ventral length (FVL)	Ventral-side distance between posterior margin of cephalic carapace and posterior margin of the first abdominal segment	*
Second segment ventral length (SVL)	Ventral-side distance from the posterior margin of the first segment to the posterior margin of the second abdominal segment	*
Third segment (TVL)	Ventral-side distance from the posterior margin of the second to the posterior margin of the third abdominal segment	*
Fourth segment (FVL)	Ventral-side distance from the posterior margin of the third segment to the posterior margin of the third abdominal segment	*
Fifth segment (IVL)	Ventral-side distance from posterior of the fourth segment to the posterior margin of the fifth abdominal segment	*
Sixth segment (XVL)	Ventral-side distance from posterior of the fifth segment to the posterior margin of the sixth abdominal segment	*
Exopod (EXL)	Distance from the posterior margin of the sixth segment to the tip of the exopod.	*
Abdominal length (ABL)	Distance from the anterior margin of the abdomen to the tip of uropod	2

Notes: Reference: (1) Lester (1983); (2) Dall (195 7); and single asterisk (*) new measurements developed within the current study

RESULT AND DISCUSSION

Inbreeding and the Mean of Morphometric Characters

Mean values of the 22 morphometric characters of both inbred and outbred populations are presented graphically in Figure 2 and statistical significance tests are presented in Table 2. The data presented in both the Figure 2 and Table 2 show two important features, namely a general trend of lower mean of morphometric values within the inbred population (1), and variation in the magnitude of inbreeding effect among characters (2).

Except for XVL character, all means of morphometric characters within the inbred group were lower than those of the outbred counterparts, although not all the differences were statistically significant. This general trend suggests that inbreeding depression appeared to be in place. This result is consistent with those obtained by lmron (2009) in previous study who worked on the same species but in earlier stage. He found that survival rate, the rate of developmental stage and growth of the inbred was lower than those of the outbred group.

Inbreeding depression coefficient between these two groups ranged from 3.7%-23.0%. The least and the most different characters mean values between the two groups were XVL and EXL, respectively. Further analysis by applying statistical significance test (Table 2) suggested that 36% among these mean value differences were statistically significant. The variation in morphometric characters in response to inbreeding depression suggest that the characters had a variable sensitivity in responding to the genetic make-up. Some characters were more sensitive than others. In a study aiming at investigating the effect of inbreeding on the fluctuating asymmetry and developmental stability (Mazzi *et al*., 2002) found that the level of asymmetry in pectoral fin was higher than those in pelvic fin and gill racker.

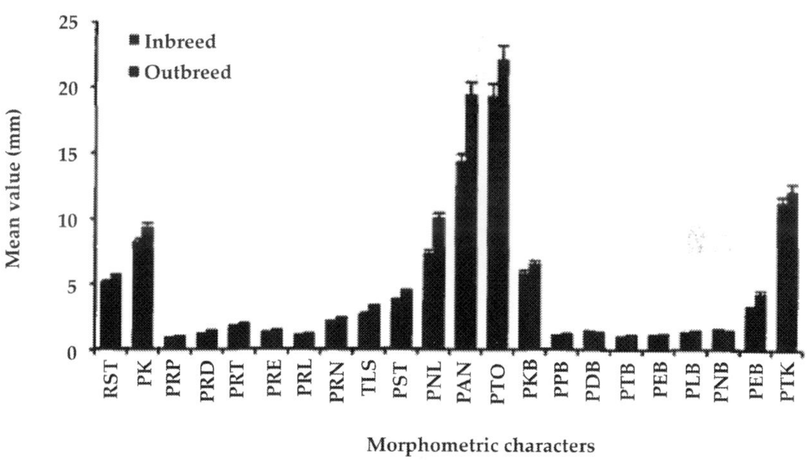

Figure 2: Profile of mean value of 22 morphometric characters of inbred and outbred groups of *Macrobrachium rosenbergii.* Abbreviation of morphometric characters along horizontal axis refer to those explained in Table 1

Table 2: Summary of mean value of twenty two morphometric characters, mean differences, and statistical significance test between populations of inbred and outbred of GFP. Ns, single asterisk and double asterisks represent statistical symbols for no significant difference, significant at 5% and significan at 1%, respectively

Characters	Inbreed		Outbreed		Inbreeding depression (%)	P-value	Statistical significance
	Mean	S.D.	Mean	S.D.			
RST	5.10	1.08	5.624	1.129	9.36	0.195	ns
CAL	8.14	1.52	9.098	1.499	10.57	0.034	*
FDL	0.89	0.23	0.926	0.143	3.69	0.964	ns
SDL	1.25	0.26	1.422	0.286	12.01	0.014	*
TDL	1.81	0.34	1.968	0.365	8.01	0.237	ns
ODL	1.37	0.31	1.475	0.298	7.12	0.167	ns
IDL	1.07	0.18	1.230	0.196	12.70	0.029	*
XDL	2.17	0.30	2.323	0.368	6.42	0.208	ns
TLS	2.72	0.45	3.052	0.487	11.00	0.014	*
PST	3.68	0.82	4.304	0.740	14.55	0.003	ns
UTL	7.28	3.33	9.068	3.362	19.77	0.054	ns
ATL	14.28	5.23	18.534	5.781	22.97	0.053	ns
TOL	19.33	3.22	21.488	2.866	10.04	0.031	*
HVL	5.92	1.16	6.390	1.048	7.27	0.149	ns
FVL	1.14	0.23	1.215	0.253	6.50	0.151	ns
SVL	1.44	0.34	1.474	0.367	2.35	0.405	ns
TVL	1.00	0.25	1.031	0.226	3.05	0.544	ns
OVL	1.11	0.24	1.136	0.242	2.68	0.0339	ns
IVL	1.31	0.25	1.393	0.270	6.04	0.099	*
XVL	1.52	0.29	1.495	0.273	1.42	0.705	ns
EXL	3.25	0.52	4.000	0.927	18.83	0	**
ABL	11.23	1.85	12.812	0.326	12.36	0.045	*

Inbreeding and Phenotypic Variation

In contrast to the relatively significant effect of inbreeding on the mean of morphometric characters, its effect on the phenotypic variation, as expressed as CV values, was generally not quite substantial. Although in majority of the characters inbred shows a slightly higher CV (Figure 3), their magnitude are not significantly different from those of the outbred (Table 3). Despite insignificant, the observed pattern in which inbred's CV were slightly higher than those of the outbred's counterpart, was actually conform with that predicted under homeostasis theory in which developmental instability increase with inbreeding (Deng, 1997).

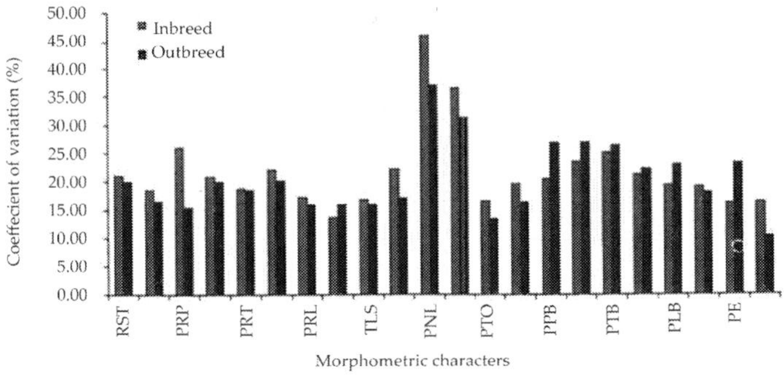

Figure 3: Profile of coeffecient of variation (CV) across 22 morphometri c characters of inbred and outbred populations of *Macrobrachium rosenbergii.* Abbreviation of morphometric characters along horizontal axis refer to those explained in Table 1

Table 3: Summary of CV value of morphometric characters within the inbred and the out bred groups and their statistical significance values

Character	CV of inbred (%)	CV of outbred (%)	S2log(I)/ S2log(O)	Statistical significance
RST	21.125	20.069	0.611	NS
CAL	18.658	16.472	1.032	NS
FDL	26.013	15.424	0.751	NS
SDL	21.097	20.100	1.064	NS
TDL	18.747	18.526	1.071	NS
ODL	22.261	20.224	0.982	NS
IDL	17.218	15.975	1.037	NS
XDL	13.585	15.859	1.221	NS
TLS	16.802	15.974	1.092	NS
PST	22.180	17.186	0.675	NS
UTL	45.827	37.072	0.993	NS
ATL	36.647	31.189	0.943	NS
TOL	16.681	13.340	1.112	NS
HVL	19.622	16.396	3.237	**
FVL	20.494	20.801	1.060	NS
SVL	23.404	24.891	1.085	NS
TVL	25.086	21.884	0.929	NS
OVL	21.300	21.326	1.020	NS
IVL	19.368	19.347	1.047	NS
XVL	19.103	18.264	0.954	NS
EXL	16.120	15.794	1.363	NS
ABL	16.440	10.351	2.172	**

Implications for Selective Breeding Program

The objective of selective breeding pro· gram is to improve performance in the character of interest through genetic gain over generations. With respect to genetic gain, two parameters of quantitative genetic, namely differential selection and heritability, determine population's ability to respond to selection (Falconer & Mackay, 1996). Genetic gain or response to selection is a product of those two parameters, as expressed in the equation R=h2S; where R, h2, and S represent for response to selection, heritability, and differential selection, respectively. The maximum response of a trait to selection will be achieved when a population is characterized by a high differential selection and heritability. While heritability estimate of both populations need to be investigated further, the current data, particularly mean and CV of morphometric characters, may give insight as to the magnitude of differential selection potential.

Differential selection is a difference between the maximum or minimum values and the corresponding population mean value of a given character within a breeding population. Assuming heritability values of a given character in both populations are similar, a higher differential selection within a population means the likely to gain higher selection response is also higher. With a few exceptions, the data obtained in the current study of variety suggested two important features: lower level of mean and a slightly higher level of coefficient of variation in the inbred group than those of the outbred. Under circumstances of equal morphometric means, population with a higher CV will have a wider range of phenotypic values, which means higher potential of differential selection. However, this was not the case with the inbred group. In addition to the lack of significant magnitude of CV differences, the inbred also were characterized by lower values in both mean and ranges of morphometric values. Combination of these properties implies that the inbred groups have a lower potential of differential selection. Consequently, the use of inbred group as brood stock within selective breeding program will prevent the breeding program from obtaining maximum response to selection. Letting the inbreeding to occur generation after generation will result in the accumulation of inbreeding coefficient to a level that may deteriorate performance.

Despite the quite convincing results, it should be noted that the present data is of preliminary in nature due to limited scope of life stage. Whether the pattern observed in the current study will remain consistent at later stage, e.g. at marketable size by which time the selection for trait of interest taken place, need to be explored further. To address this query, the life specimens of both groups are currently being maintained for further observations.

CONCLUSION

Inbreeding significantly reduced survival of *Macrobrachium rosenbergii* during post larval and first nursery stages. The effect on nursery stage was more significant than that on post larval stage. Inbreeding also reduced growth in the form of lower level of morphometric variability means and increased phenotypic variability within the inbred implying the use of it for selective breeding is susceptible of yielding a lower response to selection. Over all, inbreeding coefficient within breeding population need to be controlled that the ad· verse effect of inbreeding can be minimized.

ACKNOWLEDGEMENT

Authors would like to thank all technical staffs and research team of giant freshwater prawn for invaluable assistance during the preparation and execution of the study. The study was part of selective breeding project of GFP conducted at the RIFFBA in which the first author serves as coordinator for the breeding program.

REFERENCES

1. Dall, W. 1957. The revision of Australian species of Penaeinae (crustacea decapoda: Penaeidae). *Aust.). Mar. Freswat. Res.,* 8: 136-231.
2. De March, B.G.E. 1997. Social and genetic determinants of size variation in tanks of Nauyuk, Norwegian, and hybrid Arctic charr, *Sa/velinus a/pinus* (L.). *Aquaculture Research,* 28: 305·315.
3. Deng, H.W. 1997. Increase in developmental instability upon inbreeding in Daphnia. *Heredity,* 78: 1 82·1 89.
4. Falconer, D.S. & Mackay, T.F.C. 1996. Introduction to quantitative genetics. Fourth ed. Longman Group Ltd, Essex, England.
5. Goldan, 0., Popper, D., & Karplus, I. 1997. Management of size variation in juvenile gilthead sea bream *(Sparus aurata).* 1: Particle size and frequency of feeding dry and live food. *Aquaculture,* 152: 181-190.

6. lmron. 2009. Effect of Inbreeding on the early performance of freshwater prawn, *Macrobrachium rosenbergii. Indonesian Aquaculture journal,* 4(1): 2 5-31 .
7. Lester, LJ. 1983. Developing a selective breeding program for penaeid shrimp mariculture. *Aquaculture,* 33: 41-50.
8. Lewontin, R.C. 1966. On the Measurement of Relative Variability. *Systematic Zoology,* 15:141-142.
9. Lin, J.Z. & Ritland, K. 1997. Quantitative trait loci differentiating the outbreeding *Mimulus guttatus* from the inbreeding *M. platycalyx. Genetics,* 146: 1 , 11 5·1 , 1 21 .
10. Mazzi, D., Largiader, C.R., & Bakker, T.C.M. 2002. Inbreeding and developmental stability in three-spined sticklebacks *(Gasterosteus aculeatus* l.). *Heredity,* 89: 293-299.
11. Nakadate, M., Shikano, T., & Taniguchi, N. 2003. Inbreeding depression and heterosis in various quantitative traits of the guppy, *Poeci/ia reticulata. Aquaculture,* 220: 219-226.
12. Roff, D.A. 1998. Effects of inbreeding on morphological and life history traits of the sand cricket, *Cry/Ius firmus. Heredity,* 81: 28-37.
13. Wright, L.l., Tregenza, T., & Hosken, D.j. 2008. Inbreeding, inbreeding depression and extinction. *Conservation Genetics,* 9: 833-843.

CITATION

CHAPTER 1

Bruno Augusto A. Borges, breeding programs on aquaculture, Aquaculture Engineer Florianopolis, Brazil.

CHAPTER 2

Agim Rexhepi, Kurtesh Sherifi, Hysen Bytyqi, Behlul Behluli, Review: importance of establishing and running a breeding program in the developing fish farming industry, University of Prishtina, Faculty of Agriculture and Veterinary, Kosova.

CHAPTER 3

João Costa Filho and Leila de Genova Gaya, Recent approaches of fish breeding, DOI:10.5777/ambiencia.2012.01.03rb.

CHAPTER 4

Zaijie Dong, Nguyen Hong Nguyen and Wenbin Zhu, Genetic evaluation of a selective breeding program for common carp Cyprinus carpio conducted from 2004 to 2014, DOI 10.1186/s12863-015-0256-2.

CHAPTER 5

Hisham Abdelrahman, Mohamed ElHady , Acacia Alcivar-Warren , Standish Allen , Rafet Al-Tobasei , Lisui Bao , Ben Beck , Harvey Blackburn , Brian

Bosworth , John Buchanan , Jesse Chappell , William Daniels , Sheng Dong , Rex Dunham , Evan Durland, Ahmed Elaswad , Marta Gomez-Chiarri, Kamal Gosh , Ximing Guo, Perry Hackett, Terry Hanson , Dennis Hedgecock, Tiffany Howard , Leigh Holland , Molly Jackson, Yulin Jin , Karim Khalil , Thomas Kocher, Tim Leeds, Ning Li , Lauren Lindsey , Shikai Liu , Zhanjiang Liu, Kyle Martin, Romi Novriadi , Ramjie Odin , Yniv Palti, Eric Peatman , Dina Proestou, Guyu Qin , Benjamin Reading, Caird Rexroad, Steven Roberts, Mohamed Salem , Andrew Severin, Huitong Shi , Craig Shoemaker , Sheila Stiles, Suxu Tan , Kathy F. J. Tang, Wilawan Thongda , Terrence Tiersch, Joseph Tomasso , Wendy Tri Prabowo , Roger Vallejo, Hein van der Steen, Khoi Vo , Geoff Waldbieser , Hanping Wang, Xiaozhu Wang , Jianhai Xiang, Yujia Yang , Roger Yant, Zihao Yuan , Qifan Zeng and Tao Zhou, Aquaculture genomics, genetics and breeding in the United States: current status, challenges, and priorities for future research, DOI 10.1186/s12864-017-3557-1.

CHAPTER 6

Fernández J, Toro MÁ, Sonesson AK and Villanueva B (2014) Optimizing the creation of base populations for aquaculture breeding programs using phenotypic and genomic data and its consequences on genetic progress. *Front. Genet*. **5**:414. doi: 10.3389/fgene.2014.00414.

CHAPTER 7

Yaw B. Ansah, Emmanuel A. Frimpong and Eric M. Hallerman, Genetically-Improved Tilapia Strains in Africa: Potential Benefits and Negative Impacts, doi: 10.3390/su6063697.

CHAPTER 8

Kang Xu, Wei Duan, Jun Xiao, Min Tao, Chun Zhang, Yun Liu and ShaoJun Liu, Development and application of biological technologies in fish genetic breeding, doi: 10.1007/s11427-015-4798-3.

CHAPTER 9

JinGou Tong and XiaoWen Sun, Genetic and genomic analyses for economically important traits and their applications in molecular breeding of cultured fish, doi: 10.1007/s11427-015-4804-9.

CHAPTER 10

JianFang Gui and ZuoYan Zhu, Molecular basis and genetic improvement of economically important traits in aquaculture animals, doi: 10.1007/s11434-012-5213-0.

CHAPTER 11

Jovana Marjanovic, Han A. Mulder, Hooi L. Khaw and Piter Bijma, Genetic parameters for uniformity of harvest weight and body size traits in the GIFT strain of Nile tilapia, DOI 10.1186/s12711-016-0218-9.

CHAPTER 12

Cheryl D. Quinton, Antti Kause, Juha Koskela and Ossi Ritola, Breeding salmonids for feed efficiency in current fishmeal and future plant-based diet environments, DOI: 10.1051/gse:2007013.

CHAPTER 13

Cecilia J. Jaspe, Christopher M. A. Caipang, Increasing salinity tolerance in tilapias: selective breeding using locally available strains, http://agris.fao.org/agris-search/search.do?recordID=DJ2012069556.

CHAPTER 14

Kasper Janssen, Paul Berentsen, Mathieu Besson and Hans Komen, Derivation of economic values for production traits in aquaculture species, DOI 10.1186/s12711-016-0278-x.

CHAPTER 15

Yáñez JM, Newman S and Houston RD (2015) Genomics in aquaculture to better understand species biology and accelerate genetic progress. *Front. Genet.* **6**:128. doi: 10.3389/fgene.2015.00128.

CHAPTER 16

Anna K Sonesson and Theo HE Meuwissen, Testing strategies for genomic selection in aquaculture breeding programs, doi:10.1186/1297-9686-41-37.

CHAPTER 17

Boscari E, Pujolar JM, Dupanloup I, Corradin R, Congiu L (2014) Captive Breeding Programs Based on Family Groups in Polyploid Sturgeons. PLoS ONE 9(10): e110951. https://doi.org/10.1371/journal.pone.0110951.

CHAPTER 18

Bambang Iswanto, Imron, Huria Marnis, and Rommy Suprapto, Response To Selection For Body Weight In The Third Generation Of Mass Selection Of The African Catfish (Clarias Gariepinus) At Research Institute For Fish Breeding Sukamandi, DOI: http://dx.doi.org/10.15578/iaj.11.1.2016.15-21.

CHAPTER 19

Imron, Harry Wuwungan, Dinar Tri Soelistyawati, Effect of Inbreeding on Morphological Traits of Populations of Giant Fresh Water Prawn, *Macrobrachium Rosenbergii:* Implications for Selective Breeding, DOI: http://dx.doi.org/10.15578/iaj.8.2.2013.127-134.

INDEX